Dahlem Workshop Reports
Life Sciences Research Report 28
Changing Metal Cycles and Human Health

The goal of this Dahlem Workshop is:
to evaluate the potential health consequences from alterations of biogeochemical cycles of metals by society

Life Sciences Research Reports
Editor: Silke Bernhard

Held and published on behalf of the
Stifterverband für die Deutsche Wissenschaft

Sponsored by:
Deutsche Forschungsgemeinschaft
Senat der Stadt Berlin
Stifterverband für die Deutsche Wissenschaft

Changing Metal Cycles and Human Health

J. O. Nriagu, Editor

Report of the Dahlem Workshop on
Changing Metal Cycles and Human Health
Berlin 1983, March 20 – 25

Rapporteurs:
M. O. Andreae · B. A. Fowler · J. F. Jaworski
T. E. Kjellström

Program Advisory Committee:
D. F. Hornig and J. O. Nriagu, Chairpersons · M. Bernhard
R. A. Duce · C. G. Elinder · E. D. Goldberg · H.-J. Hapke
J. H. R. Kägi · A. L. Page · A. J. M. Schoot Uiterkamp

Springer-Verlag
Berlin Heidelberg New York Tokyo 1984

Copy Editors: M. A. Cervantes-Waldmann, K. Geue
Photographs: E. P. Thonke

With 4 photographs, 56 figures, and 70 tables

ISBN-13: 978-3-642-69316-8 e-ISBN-13: 978-3-642-69314-4
DOI: 10.1007/978-3-642-69314-4

CIP-Kurztitelaufnahme der Deutschen Bibliothek:

Changing metal cycles and human health: report of the Dahlem Workshop on Changing Metal Cycles and Human Health, Berlin 1983, March 20-25 /. J. O. Nriagu, ed. Rapporteurs: M. D. Andreae ... [Dahlem Konferenzen. Held and publ. on behalf of the Stifterverb. für d. Dt. Wiss. Sponsored by: Dt. Forschungsgemeinschaft ...], – Berlin; Heidelberg; New York; Tokyo: Springer, 1984.

(Life sciences research report; 28) (Dahlem Workshop reports)

NE: Nriagu, Jerome D. [Hrsg.]; Andreae, M. O. [Mitverf.]; Workshop on Changing Metal Cycles and Human Health <1983, Berlin, West>; Dahlem Konferenzen; 1. GT

Softcover reprint of the hardcover 1st edition 1984

EDP and Photocomposition: Satz-Rechen-Zentrum, D-1000 Berlin 30

2131/3014 - 5 4 3 2 1 0

TABLE OF CONTENTS

THE DAHLEM KONFERENZEN

Founders

Recognizing the need for more effective communication between scientists, especially in the natural sciences, the Stifterverband für die Deutsche Wissenschaft*, in cooperation with the Deutsche Forschungsgemeinschaft**, founded Dahlem Konferenzen in 1974. The project is financed by the founders and the Senate of the City of Berlin.

Name

Dahlem Konferenzen was named after the district of Berlin called "Dahlem", which has a long-standing tradition and reputation in the arts and sciences.

Aim

The task of Dahlem Konferenzen is to promote international, interdisciplinary exchange of scientific information and ideas, to stimulate international cooperation in research, and to develop and test new models conducive to more effective communication between scientists.

Dahlem Workshop Model

Dahlem Konferenzen organizes four workshops per year, each with a limited number of participants. Since no type of scientific meeting proved effective enough, Dahlem Konferenzen had to create its own concept. This concept has been tested and varied over the years, and has evolved into its present form which is known as the *Dahlem Workshop Model*. This model provides the framework for the utmost possible interdisciplinary communication and cooperation between scientists in a given time period.

The main work of the Dahlem Workshops is done in four interdisciplinary discussion groups. Lectures are not given. Instead, selected participants write background papers providing a review of the field rather than a report on individual work. These are circulated to all participants before the meeting to provide a basis for discussion. During the workshop, the members of the four groups prepare reports reflecting their discussions and providing suggestions for future research needs.

* *The Donors Association for the Promotion of Sciences and Humanities*
** *German Science Foundation*

Topics
The topics are chosen from the fields of the Life Sciences and the Physical, Chemical, and Earth Sciences. They are of contemporary international interest, interdisciplinary in nature, and problem-oriented. Once a year, topic suggestions are submitted to a scientific board for approval.

Participants
For each workshop participants are selected exclusively by special Program Advisory Committees. Selection is based on international scientific reputation alone, although a balance between European and American scientists is attempted. Exception is made for younger German scientists.

Publications
The results of the workshops are the Dahlem Workshop Reports, reviewed by selected participants and carefully edited by the editor of each volume. The reports are multidisciplinary surveys by the most internationally distinguished scientists and are based on discussions of new data, experiments, advanced new concepts, techniques, and models. Each report also reviews areas of priority interests and indicates directions for future research on a given topic.

The Dahlem Workshop Reports are published in two series:
1) Life Sciences Research Reports (LS), and
2) Physical, Chemical, and Earth Sciences Research Reports (PC).

Director
Silke Bernhard, M.D.

Address
Dahlem Konferenzen
Wallotstrasse 19
1000 Berlin 33, F.R. Germany

Changing Metal Cycles and Human Health, ed. J.O. Nriagu, pp. 1–6.
Dahlem Konferenzen 1984. Berlin, Heidelberg, New York, Tokyo: Springer-Verlag.

INTRODUCTION

J.O. Nriagu
National Water Research Institute
Environment Canada
Burlington, Ontario L7R 4A6, Canada

It is becoming increasingly evident that human activities have modified the global cycles of the heavy metals. It has recently been estimated that about 0.5, 20, 240, 250, and 310 million tonnes of cadmium, nickel, lead, zinc, and copper have so far been mined and ultimately wasted in the biosphere (3). Recent estimates ((1), and Buat-Ménard, this volume) put the quantities of anthropogenic As, Cd, Pb, Cu, and Zn that are currently being broadcast annually via the atmosphere to distant ecosystems at about 22,000, 7,300, 400,000, 56,000, and 214,000 tonnes, respectively. In many instances, the emissions from anthropogenic sources exceed the contributions from natural sources by severalfold. Such massive redistribution orchestrated by mankind has apparently overwhelmed the natural reservoirs and flux rates for some of the toxic metals in many ecosystems. In view of the close linkage of the metal cycles to biological processes, there is a growing body of evidence which shows a general elevation of pollutant metal burdens in many marine and land biota. Indeed, recent studies suggest that the human body burdens of lead and cadmium in contemporary urban residents are severalfold higher than those of the prehistoric populations (1, 4). Whether such elevated levels of metals in man and his food chain constitute a threat to health was the main objective of this Workshop which was held in Berlin from March 20 to 25, 1983.

One of the outstanding features of the Dahlem Workshops is their multidisciplinary nature, and the present workshop was no exception. The participating experts were drawn from the fields of geochemistry, soil science, nutrition, epidemiology, microbiology, environmental engineering, toxicology, environmental and occupational health, and biostatistics. The participants shared a common interest in their concern for the contamination of the human environment with toxic metals. The wide ranging views, insights, and perspectives which were brought to bear on the topic clearly resulted in very stimulating discussions. Detailed summaries of the key problems and areas of concern pertaining to basic themes of the workshop are given in the group reports. What follows are my personal perceptions, arguably biased, of the general concerns expressed during the week.

One incontestable issue that emerged during the workshop was that there are critical gaps in the data needed to estimate the extent of the health problems associated with changes in the global metal cycles. If a health effect is defined as any measurable metal-induced change in the biochemistry, physiology, or psychology of an individual, the number of children in the United States suffering from lead poisoning could well number over one million. The number of elderly people in Japan suspected of experiencing cadmium poisoning (itai-itai disease) may conceivably be several thousand. Also, the cases of mercury poisoning (Minamata disease) in Japan presumably number several thousand, and a large number of persons are similarly afflicted in Iraq. When these numbers are converted to a worldwide scale, it is clear that large numbers of persons may be afflicted with the silent epidemic of metal poisoning. Such an epidemic has apparently not been a matter of public concern because the number of people whose health has actually been impaired by metal poisoning is small in terms of the current medical criteria used in defining adverse or critical effect.

Actually, the threat of metal pollution to the public health is an age-old but unappreciated problem. The environmental levels of the metals rarely produce acute injury, and any chronic damage they produce may not become evident for months or years and is hardly manifested by any pathognomonic symptoms. For the high risk groups, such as occupationally-exposed workers, chronic poisoning may result in obvious illness, but when comparable ailments are encountered in the general population, they are more likely to be attributed to other etiological factors. Even today, medical diagnosis is bedeviled by conflicting claims on the significance attached to some of the early biological signs of subclinical poisoning by metals. The early biological indicators of metal interactions with subcellular biochemical systems usually either are metabolites of the system affected (porphyrinurias) or represent some specific function of a cellular system being impaired (proteinurias). One typically finds a continuum of symptoms, from the subtle or so-called "no effect" biochemical and physiological indicators of exposure to severe clinical disease and death. This continuum is the basis of much of the controversy since many health officials follow the traditional practice of applying the "threshold-health-effect" concept in evaluating the problems of environmental exposure to metals.

The past decade or so, however, has seen a vast increase in our understanding of the effects of elevated concentrations of toxic metals in local populations and ecosystems. At the same time, there is a growing awareness that the effects of the metals which occur naturally in the environment must be distinguished from those imposed by the pollutant fraction. This point was amply documented in a recent study of cadmium intake and cadmium in a number of human tissues in Sweden, Japan, and the United States, which showed fairly conclusively that the background exposure in Japan was about threefold higher than in the other two countries (2). One immediate implication is that any health effect studies of cadmium in Japan using control groups within that country are liable to underestimate the difference between the exposed and the control groups simply because of the the high "background" intake.

A related matter of much concern is the change in rate of human exposure engendered by the increasing intensities of metal release to the environment. Historical increase in exposure could (a) negate the usual medical assumption that the current "normal" exposures are synonymous with "no effect" exposures, and (b) affect the interpretation of observed accumulation curves in body tissues with long biological half-lives for the metals. Historical changes in background metal exposure also depend very much on changes in societal diet over time. Since the metal concentrations in different food items vary considerably, the metal intake with food will vary substantially with changing eating habits.

Human activities likely to alter the future patterns of metal emissions and the exposed populations are many. The exhaustion of currently known ore deposits will likely mean the exploitation of either new mineralized districts or lower-grade ore deposits which may increase the metal pollution per kilogram of product. In view of the differences in metal content of fossil fuels, a change in the mix of fuels used for heating, producing electricity, and in transportation will produce changes in metal emissions. Deforestation on a large scale and draining of major marshes will reduce the number of sinks available for the mobilized toxic metals. Salting of waterways and agricultural soils can markedly affect the mobility and availability of metals in soils and water. The addition of large quantities of chemical fertilizers and composts may likewise introduce unwanted metals into the soils or greatly affect the availability of metals in soils. Increased urbanization will put more people at a risk of increased exposure to such urban pollutants as lead. Changing dietary habits (increased consumption of processed foods, nontraditional foods, etc.) can greatly affect the intake of metals. It would be otiose to list other instances discussed at the workshop.

The information about metal exposure in developing countries was generally recognized to be very inadequate, and the environmental levels of toxic metals in such parts of the world could be expected to rise. Sociologically, the populations in these countries may be particularly at risk as a result of: (a) high population density and poor hygienic conditions in the crowded cities, (b) consumption of a high proportion of local foods produced in local metal-contaminated areas, and/or (c) the preponderance of the most susceptible population groups related to age (children and pregnant women) or nutritional or health status (interaction with preexisting endemic diseases). The elucidation of the undue effects of metal exposure in these countries will have to take the "total health" approach since other endemical health problems than those related to the metals in question may be of greater importance for the population affected or may obscure the metal-induced effects.

There was some disagreement as to the pathways of exposure to metals in a multimedia context. Those people concerned with regulations would want the pathway to be defined in such a manner that a specific environmental compartment which is regulatable is identified. On the other hand, scientists more interested in intoxication mechanisms would identify only the processes involved in transfer from source to receptor. The subsequent metabolism and in vivo toxicity of many metals are still improperly understood. Studies on the biochemistry of metals in human systems are stymied by major gaps in the establishment of the true bioavailability and bioactivity of each metal. The bioavailability of a given metal depends critically on its chemical forms during the exposure commitment, and this is an area that has hardly received the proper study that it deserves.

A major concern in the understanding of the environmental cycling of toxic metals and the estimation of their intake by the human population is the quality of the available data. Dramatic improvements certainly have been made with respect to the quality of environmental data both through the development of "clean" techniques and in the reduction of analytical detection limits. Much attention, however, still needs to be paid to contamination problems in the sampling process and during sample storage and analysis. It was emphasized that the protocol for contamination-free sampling needs to be examined for each element individually; a "clean" sample for one element may be "dirty" for another.

Besides the debate on scientific issues, there was also a recognition of the serious difficulties in establishing any set of criteria for legislative action to eliminate or control the metal pollution problem. As noted already, experimental verification of the long-term effects of low-level exposure to toxic metals is a difficult and an expensive proposition. Since human subjects are rarely candidates for study, there are still many unresolved uncertainties in extrapolating the data from animal models to human beings. Epidemiological studies, on the other hand, estimate the probability of health effects based on the interpretation of experimental and statistical data. The researcher's preconceptions and predispositions often influence the interpretations so that while a public health official might attribute a certain chronic illness to metal pollution, an industrial hygienist, using the same information base, would absolve the contaminant of any blame. Moreover, the interpretations are not limited to the same data, for all the interested parties are free to design and conduct their own experiments and surveys, which serves to further exacerbate the difference of opinion. And the debate does not become more rational when the battle lines are drawn around the questions of risks and benefits.

Given these constraints, it is not surprising that there was no consensus on how a prudent and rational society should act when faced with evidence of risk from a pollutant metal but possesses a body of evidence which does not prove a causal influence. In the final analysis, the regulatory action has to be judgemental, and it was generally agreed that any window of vulnerability should not compromise the public's health.

The outcome of this workshop depended in no small part on the efficiency and helpfulness of Silke Bernhard and the entire staff of Dahlem Konferenzen. Their unflinching support was appreciated by all who participated in the meeting.

The meeting benefitted greatly from the chairmanship and general guidance of D. Horning of Harvard University. The structure of the workshop was in the hands of the program advisory committee which met May 9–10, 1982. Any success of this volume belongs to this advisory committee, to the moderators, rapporteurs, and to all the participants. The authors of the background papers share the credit for the quality of the material in the book.

REFERENCES

(1) Drasch, G.A. 1983. An increase of cadmium body burden for this century — an investigation on human tissues. Sci. Total Env. **26**: 111–119.

(2) Kjellström, T. 1979. Exposure and accumulation of cadmium in people from Japan, U.S.A. and Sweden. Env. Health Perspect. **28**: 169–197.

(3) Nriagu, J.O. 1979. Global inventory of natural and anthropogenic emissions of trace metals to the atmosphere. Nature **279**: 409–411.

(4) Settle, D.M., and Patterson, C.C. 1980. Lead in albacore: a guide to lead pollution in America. Science **207**: 1167–1176.

Changing Metal Cycles and Human Health, ed. J.O. Nriagu, pp. 7–25.
Dahlem Konferenzen 1984. Berlin, Heidelberg, New York, Tokyo: Springer-Verlag.

HISTORICAL RECORDS OF METAL POLLUTION IN THE ENVIRONMENT

I. Thornton and P. Abrahams
Applied Geochemistry Research Group
Imperial College, London SW7 2BP, England

Abstract. Metals have been mined and processed since prehistoric times. Early records are primarily of the effects of industrial exposure, though adverse effects on farm livestock in Britain were recorded in the 18th and 19th centuries. There are many recent studies into the quantification and extent of historical pollution; this may still present a hazard to plant, animal, and human health.

INTRODUCTION

Man's first use of metals for the manufacture of simple tools, weapons, and ornaments dates back to the Bronze Age, and during the period of the Roman Empire mining of iron, gold, silver, tin, copper, and lead was active. The expansion of the manufacturing industry during the Industrial Revolution brought about an increasing need for metals, a need which has continued to increase particularly in the 20th century.

The toxicity of metals to plants, animals, and man has been known for centuries, though it is probable that toxic effects due to either industrial or environmental exposure were frequently not recognized nor diagnosed correctly at the time. Relatively recently it has been suggested that the fall of the Roman Empire was in part due to endemic lead poisoning caused by the extensive use of lead lined pots used for cooking and storing wine (21), and it is speculated that Napoleon may have died of arsenic poisoning resulting from the use of copper arsenite as a cheap and stable pigment for coloring wallpaper (28). Lewis Carroll's description of the Mad Hatter in *Alice in Wonderland* corresponded with classic symptoms of chronic mercury poisoning, a problem quite common during the 19th century.

BACKGROUND LEVELS OF METALS IN THE ENVIRONMENT

All the heavy metals of concern in terms of environmental pollution are present in rocks, soils, waters, and air in small amounts from natural geological sources which in turn may influence the chemical composition of potable water supplies, foodstuffs for human consumption, and the nature of airborne particulates and dusts inhaled or ingested. It has been said that the geochemists' primary contribution to the understanding and measurement of pollution is in determining the average background amounts of each element to be expected in the natural uncontaminated geological setting (8). Before attempting to assess the industrial contribution of heavy metal pollutants to the environment, it is in fact essential to establish reliable baseline information on the natural amounts present.

From a practical viewpoint, the high end of the natural range may at times be deleterious; for instance, large amounts of molybdenum in marine black shale formations and associated soils may give rise to excess molybdenum in the diet of grazing cattle and result in molybdenosis or induced copper deficiency (42). Soils developed in mineralized areas frequently contain high concentrations of one or more of the heavy metals copper, lead, zinc, and cadmium, though in Britain much of the metal is probably derived as a result of mining and smelting rather than the natural weathering of underlying material. Typical normal ranges of metals in soils and examples of high concentrations due to natural geochemical sources in Britain are shown in Table 1.

SOURCES OF METAL POLLUTION

In Britain, the oldest and most extensive sources of metal pollution are related to the history of metalliferous mining which commenced in Roman and earlier times, thrived until the end of the 19th century, and then declined rapidly. Peak output was in the mid-19th century when Britain produced 75% of the world's copper, 60% of the tin, and 50% of the lead. It has been estimated that

	Typical normal range in soil (ppm)	Metal-rich soils (ppm)	Sources	Possible effects
As	<5-40	up to 2500	Mineralization	Toxicity in plants and livestock; excess in food crops
		up to 250	Metamorphosed rocks around Dartmoor	
Cd	<1-2	up to 30	Mineralization	Excess in food crops
		up to 20	Carboniferous black shale	
Cu	2-60	up to 2000	Mineralization	Toxicity in cereal crops
Mo	<1-5	10-100	Marine black shales of varying age	Molybdenosis- or molybdenum-induced hypocuprosis in cattle
Ni	2-100	up to 8000	Ultra-basic rocks in Scotland	Toxicity in cereal and other crops
Pb	10-150	1% or more	Mineralization	Toxicity in livestock, excess in foodstuffs
Se	<1-2	up to 7	Marine black shales in England and Wales	No effect
		up to 500	Namurian shales in Ireland	Chronic selenosis in horses and cattle
Zn	25-200	1% or more	Mineralization	Toxicity in cereal crops

TABLE 1 — .Metals in soils derived from normal and geochemically anomalous parent materials in Britain.

over 4000 km^2 of agricultural land in England and Wales is contaminated with one or more metals due to this legacy from our forebears (43).

Other sources of metal pollution are more recent, and their possible adverse effects have been poorly documented unless they gave rise to obvious symptoms of toxicity in agricultural crops, farm animals, or man, until the current wave of awareness commenced in the early 1970s, in response to the then epidemic proportions of mercury poisoning in Japan, the much documented Minamata disease.

We have defined the scope of this paper to follow the course of man's concern with metal pollution and its effects from early times up to the start of this present era and have included reference to the initial "happening" which triggered present-day interests for specific metals. In addition to actual historical records, the subject area has been broadened to include records of historical pollution, based on recent and present-day analysis of soils, dated sediment cores, ice cores, etc.

EARLY MINING AND SMELTING

Probably the oldest known metal mine in the world is that of Bomvu Ridge, Swaziland, where some 50,000 years ago hematite (Fe_2O_3) was taken, initially from surface deposits and then underground, for use in fumerary, as primitive man aiming for immortality was buried in a mass of red ochre powder with lumps of "red stone" scattered around the grave (22). However, the iron ores hematite, limonite, and goethite were used as red, yellow, and brown pigments well before that time, probably as long as 300,000 years ago by Homo erectus (40). The next metal to attract attention was gold, originally mined from visible placer deposits in stream beds, used by the Egyptians in ornaments and jewelry, as early as 6,000 B.C. and found by archeologists in the tombs of the Egyptian pharoahs. The first recorded mining "boom town" was Coptus in the Nile Valley, and the ruins of over 1000 miners' stone huts may still be seen in Foakhir.

Copper was first used around 5500 B.C. for ornamental purposes, and copper ores were smelted in many parts of the Middle East around 4300 B.C., and then after 4000 B.C. in the Great Lakes region of North America by primitive Indians who found native copper locally and also obtained ores through Alaska from Siberia. Archeological evidence indicates that copper ores were also smelted in the Sinai district where large piles of smelter slag (containing around 2.75% Cu) still remain.

The discovery of tin in the form of cassiterite heralded the beginning of the Bronze Age around 3000 B.C., when it was found that by adding a small amount of tin to copper a metallic alloy much harder than copper was produced. This invention of bronze played a major role in the evolution of man over the following 2000 years.

Rich deposits of lead and silver were first worked in Turkey and the lead ore galena (PbS) containing silver was smelted by crude cupellation to obtain the silver which was first used as weights and then as coinage. Lead was used widely by the Egyptians for making human and animal figures, sinkers for fishing nets, rings, dishes and ornaments, and as an ingredient of eye-paints. Lead played a dominant role in the technology of the Roman Empire, being obtained mainly from Gaul, Britain, Germany, and Spain. A large amount of Roman lead was used in the manufacture of aquaducts and water mains, and large water reservoirs for public supply were lead lined. Lead was also used extensively to line cooking pots and wine pots leading to considerable contamination of food and drink (21,26).

Mercury was recovered from cinnebar ore (HgS) by the Greeks from about 3000 B.C. and used as medication and in religious ceremonies (16). The Romans imported large amounts of cinnebar to manufacture the pigment vermillion to decorate their villas and as an ingredient of cosmetics such as rouge (5), and by 77 A.D. with the discovery of amalgam, 10,000 lb. of mercury were imported into Rome each year. This was mined mainly in Spain and led to the poisoning of countless slaves and criminals in the mercury mines.

This brief section is by no means comprehensive, and detailed descriptions of early mining, processing, and utilization of metals are presented in volumes by Aitcheson (2), Forbes (20), Tylecote (47), and Nriagu (35,36).

EARLY RECORDS OF CONTAMINATION IN BRITAIN

Historical metal pollution in Britain was discussed by Borlase (6) who referring to Cornwall states, "the sterility may be great where metals and minerals abound, of which no place perhaps affords more frequent instances than Cornwall; for here the coarsest grounds abound most in metals, and on the other hand there is the greatest plenty of corn, grass, plants and trees, where no metals or minerals have ever appeared." Beach sands were commonly added as soil ameliorants in Cornwall, but Borlase (6) mentions that these could not be used if they contained mine waste as "their acrimonial mineral juices are mortal enemies to vegetation." Farey (17), writing about mining and agriculture in Derbyshire, describes early smelting techniques and mentions noxious fumes emitted from Cupolas or low arched reverbatory furnaces which were used from about 1747. These fumes poisoned herbage for $^1/_4$ mile of more around the Cupolas, and owners were obliged to pay a rent to the farmers for the damage caused to their land. The process of Buddling — separating lead ore from gangue materials using water — used to poison the streams, and Farey writes, "I have myself seen every part of the Derwent River at Matlock Bath, which was before beautifully clear, suddenly made thick yellow as a strong solution of Gumbong and so continues for home from the Buddling operation of the miners." He described how this killed the fish and concluded "surely the antiquated and inapplicable custom or laws as they are called, which permit such doings, ought to be revised and amended."

The second volume by Farey (18) describes poisoning of cattle in Derbyshire, locally known as Belland, caused by grazing too long in the vicinity of the lead smelter furnaces or Cupolas and sometimes from drinking the water that had been used for Buddling. Lead poisoning could also affect sheep, poultry, and small animals such as dogs. Human lead poisoning, then called Belland, Flight, Mill-reak, and Lead-colic, was quite common in people working in the smelters and was also recorded in people living near the Cupolas. However, like many of the early records, there are no statistics attached to any of this information, and no measurements were made of levels of lead or other metals in the environment.

ARSENIC AND TIN

The first commercial use of arsenic as an insecticide was in the late 1860s when Paris green was used effectively to kill the Colorado potato beetle, though by the end of the century this had mostly been replaced by lead arsenate (50). However, even before its commercial use, arsenic's association with tin ores was of prime concern to the mining industry in southwest England, as the arsenic impurity produced a relatively hard, less ductile tin product (4). The arsenic had to be removed by roasting the ore. Once a demand for arsenic arose, its production became an important by-product of many tin mines. In Cornwall, arsenic occurred as the ore arsenopyrite; this was crushed and calcined, and the vapor passed through chambers and flues where arsenious oxide accumulated as a gray powder known as "arsenic soot" or "crude arsenic" (15).

Historically, southwest England was the premier metalliferous mining center of the world. In Cornwall, more than 600 mines have operated in the past, mainly within a belt some 75 miles long and 10 miles wide. Although tin, copper, and arsenic were the main metals extracted, W, Pb, Zn, U, Fe, Sb, Ag, and Mn were also economically recovered, with small amounts of Bi, Ni, and Co. Up to the 18th century, output was relatively small, however, copper production increased markedly from the 1720s and by 1867 tin had exceeded copper, production peaking in the 1870s. Total output has been estimated at two million tons of metallic tin and 920,000 tons of metallic copper (15). From about 1860 to 1900, Cornwall was the world's major producer of arsenic and at the height of the industry also imported arsenical ores for smelting.

The history of mining and smelting in Cornwall has left behind a legacy of dereliction and pollution, with mine tailings and other wastes abundant. Barton (4) records a celebrated legal case reported in the *Royal Cornwall Gazette* in 1851 in which action was taken against the owners of a tin calciner near Restronguet Creek as a result of cattle and other animals being poisoned by eating herbage downwind of the stacks. Barton (4) states that the fumes from arsenic smelters were of an extremely poisonous and voluminous white smoke with an accompanying stench of garlic. Actual amounts lost and the extent and degree of environmental contamination was not recorded at this time, though it has been estimated that the average loss of tin in the course of mining and processing was on the order of 30% or more. Present-day evidence of these losses has been well documented. For example, the Devon Great Consuls mine in the Tamar Valley, separating Devon and Cornwall, was at one time Europe's richest copper mine and from 1845 to 1901 produced some 70,000 tons of copper and 72,000 tons of refined arsenic. Alluvial soils downstream of this mine still contain much greater concentrations of arsenic, copper, lead, and zinc than those upstream (Table 2), due to down-wash of metal-rich mine waste and suspended solids in adit waters, runoff from contaminated upland soils, etc. (43).

	No. of Samples	As	Cu	Pb	Zn
Alluvium contaminated downstream of Devon Great Consuls Mine	12	309 90-900	620 35-2000	435 60-1008	368 114-1020
Alluvium upstream of mine	5	25 20-30	44 36-40	85 72-96	121 67-180

TABLE 2 — Ranges and mean total concentrations of arsenic and selected heavy metals in agricultural soils (0–15 cm) from the Tamar Valley, Cornwall (13).

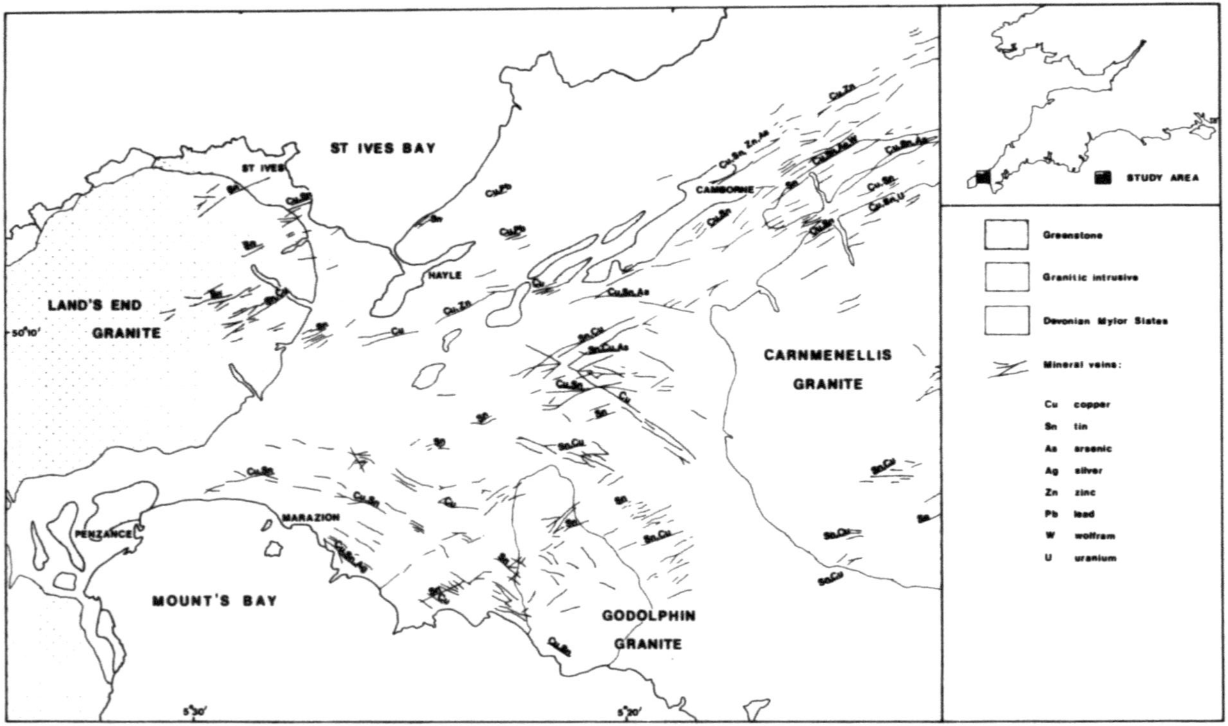

FIG. 1 — Mineral occurrences in Western Cornwall, England.

At one downstream site of agricultural land, the alluvium contained in excess of 300 ppm As, 1000 ppm Cu, 700 ppm Pb, and 700 ppm Zn to a depth of 75 cm (the maximum depth sampled). In the western Cornish peninsula (Fig. 1), present-day anomalies for arsenic, copper, lead, zinc, and cadmium in surface stream sediments extend for some 250 km^2, with peak values of 14000 ppm As and 8000 ppm Cu. Extensive contamination of agricultural soils has been documented (1) and is illustrated by present-day levels of arsenic in surface soils (mostly farmed) around the stack of the arsenic calcining works at Newmill, Cornwall, which ceased production in 1890 (Fig. 2). River waters in the Cornwall mining districts are at some locations contaminated with arsenic, ranging up to 3000 ppb As, and this naturally limits their use for potable supplies. Damage to crops is mainly restricted to small areas adjacent to mine waste, and it has even been suggested that increased intake of arsenic by ruminants in this area is beneficial as it adds a sheen to their coats.

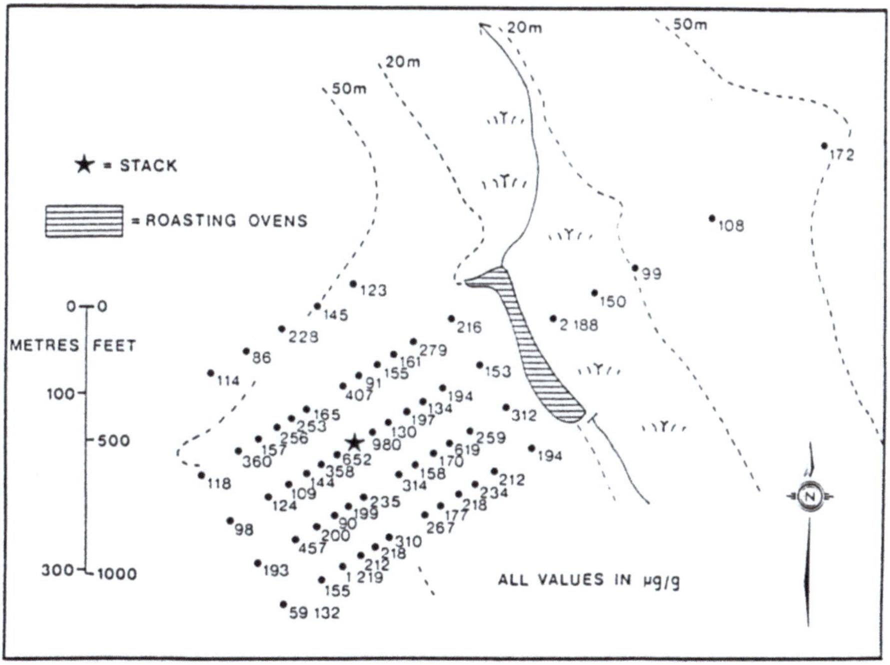

FIG. 2 — Concentrations of arsenic in surface soils (0–15 cm) around an old calcining works at Newmill, Cornwall.

There are no proven relationships recorded between human disease and the high levels of historical pollution in Cornwall, though relationships with various cancers and multiple sclerosis have been proposed but are not supported by acceptable statistical data. An article in the British Medical Journal in 1980 suggested that the high prevalence of skin cancer in Cornwall might well be due to high environmental exposure to arsenic. This may well be relevant in the light of current concern about the possible relationship between selenium deficiency and impairment of the immune response, as arsenic is a well-known selenium antagonist.

COPPER

Nriagu (34), in discussing the global cycling of copper, lists estimated figures for copper production and atmospheric copper emissions; the total (all-time) amount of copper emitted to the atmosphere is quoted as 3.2 million tons, of which about 13% was pre-1900. Hutchinson (27) traces the history of copper mining and smelting and discusses its impact on the environment. Smelting of copper ores mined in Cornwall commenced in the Lower Swansea Valley, Wales, in 1717, which in the 19th century became a world smelting center for copper, zinc, and lead, declining from the 1880s, until copper smelting ceased in 1913. The dereliction of this valley and subsequent reclamation and revegetation has been well documented, and today there is little evidence of the 10 million tons of waste which blighted this landscape only a decade ago.

Hutchinson ascribes the extreme devastation of countryside in North America by early smelting (e.g., commencing in Clifton, Arizona, in 1872; Sudbury, Ontario, 1886) to the "open roasting of ore and uncontrolled or poorly controlled smelting." The damage was documented by Hedgecock (25); little vegetation was left in the worst areas, and species diversity was affected over a considerable area. The area around Ducktown was still barren 50 years after the smelter closed. Large-scale deaths of livestock and crop damage around the world's largest copper smelter at Anaconda, Montana (capacity of 10,000 tons of ore per day), was investigated by Harkins and Swain (23), who analyzed flue gases and estimated emissions of copper and other metals in smoke. Hutchinson (27) draws attention to the fact that this early work set the scene for the now well established principle that aerial discharges from smelters offer the possibility of multi-element toxicities as, of course, more than one metal is emitted. Quoting other authors, Hutchinson (27) lists specific cases of livestock deaths including up to 1000 head of cattle and 800 sheep on one ranch 12 miles from the smelter. Haywood (24) provides some of the first actual analytical data on smelter pollution, including a value of 12.1 ppm insoluble copper and 0.77 ppm soluble copper in the water of the Deer Lodge River, 2.4 km below the smelter.

Tyler (48) has determined contamination leading to copper concentrations of 1.1 to 1.7% in organic layers of a podzol near a brass foundry that operated for some 300 years at Gusum in Sweden. Hutchinson (27) concludes that the presence of forest vegetation on these highly contaminated soils suggests that either the selection of metal-tolerant individuals has occurred over the centuries or that metals are bound up in the organic matter of the soils and are thus in a largely unavailable form for root uptake. We consider this to be an important factor in the present-day vegetation/land utilization of many metal contaminated sites in Britain.

LEAD

Actual accounts of the production, utilization, and health effects of lead in ancient times have been given by Nriagu (35,36), who estimates worldwide production from earliest times (Copper Age) to the end of the Roman Empire to be about 39 million tons. It is also estimated that the production of lead during the period of the Roman Empire (50 B.C.–500 A.D.) was around 540 g per person per year which may be compared with present-day industrial consumption in the United States of 5500 g per person per year. Sources of lead in the environment have been discussed by Walden and Stofen (49) who review some of the studies on historical aspects. Early uses of lead by the Egyptians and Romans no doubt gave rise to local environmental contamination, but it was not until the Industrial Revolution in the 19th century that large-scale losses from lead smelters coupled with the development of furnaces with tall chimneys significantly increased aerosol lead levels leading to pollution on a global scale. Losses from smelters from the 1750s to the 1960s have been estimated by Murozumi et al. (31) who quote as much as 2% Pb converted in the course of smelting to aerosol in the period 1750–1880, falling to 0.06% in the 1960s due to more efficient processing. Studies on Greenland snow profiles clearly show increasing industrial lead pollution with time since 800 B.C. (31) and point to the fact that lead concentrations in rainfall have risen in recent times. Arctic snows contained 10 times the amounts of lead of those in the Antarctic, reflecting greater uses of lead in the northern compared with the southern hemisphere. Controversy surrounding the validity of the conclusions reached in this classical study has been refuted in a recent paper by Ng and Patterson (32) in which the authors verify the interpretation of the earlier findings of a 300-fold increase in lead concentration in the Greenland ice sheet during the past 3000 years. They summarize, "virtually all the present-day

300-fold excess of Pb above natural levels in Greenland ice can be shown to be caused by industrial Pb emissions to the atmosphere on the basis of the following factors: (1) the historic increase of Pb in snow strata coincides with the historic increase of industrial Pb production and atmospheric emissions; (2) mass inventories of industrial emissions can account for the excess Pb in polar snow; (3) new quantitative measurements of Pb emissions from volcanic plumes and from sea spray show that these natural sources cannot account for 99% of the excess Pb above contributions by silicate dusts observed today in the atmosphere; and (4) the historic increase of Pb in snow strata is paralleled by analogous increases of excess Pb shown by isotopic tracers to be industrial in water-laid sediments in a remote continental region."

Using a collection of mosses collected in Sweden over the period 1860–1968, Rutling and Tyler (38) established a lead gradient which they inferred to represent increases in lead pollution. From concentrations of around 10 ppm over the period 1860–1865, concentrations doubled between 1875 and 1900 and increased to an average of 80–90 ppm by 1968. Pollen analysis from soil profiles in the old lead-mining area of Derbyshire, England, have confirmed major accumulations of lead in soils commencing in the late 17th and early 18th century, reaching a maximum in the late 19th century (30). Chow and Patterson (12) showed lead concentrations in open ocean water to decrease from 0.07 ug/kg at the surface to about 0.04 ug/kg at 800 m depth, suggesting that lead pollutants have penetrated to depths of about 500 m; they suggested that preindustrial lead concentrations in surface waters would be about 0.01–0.02 ug/kg. More recent work shows that lead is supplied to the open north Pacific largely from the atmosphere at a present-day rate of around 60 ng/cm^2/year. This is attributed to anthropogenic inputs from smelter emissions, leaded gasoline, etc., which greatly exceed lead inputs in prehistoric times, thought to be largely through rivers. The values recorded lie more than an order of magnitude below previously published values for the open ocean; deep water concentrations as low as 1 ng/kg were recorded, with those at the surface ranging from 5 to 27 ng/kg (39).

Changes in acid soluble lead concentration in dated basin sediments offshore from the Los Angeles complex show the industrial component to increase markedly after 1900, partly attributable to directly deposited sewage particles (33).

Studies on dated sediment cores from a remote pond in the Yosemite National Park, California, showed lead concentrations to decrease 4-fold from the surface to layers 130 years old at depth. There was also a change in the Pb^{206}/Pb^{207} ratio in the sediments from "industrial site" values of 1.18 near the surface to "natural" values of 1.24 at depth (41).

Present-day effects of historical lead pollution have been reported by numerous authors (including Davies (14) and Thornton (43)). Effects of pollution from lead mining on river fauna and flora in the Aberystwyth area of Wales were studied as long ago as 1919 by Carpenter (9–11) who provides detailed descriptions of fauna in several rivers in relation to concentrations of lead both in water and in bottom sediments; in particular fish were affected when lead was present in a "diffusible" form. Cases of lead poisoning in cattle at times leading to death are still occasionally reported around historical mining sites in Derbyshire, when animals stray onto mine spoil. Nowadays the old lead rakes are sometimes intentionally grazed. In this area, blood samples taken from 15 animals on 12 farms comprising background (100–200 mg/kg Pb), moderately contaminated (600–800 mg/kg Pb), and heavily contaminated (1000–2000 mg/kg Pb) locations, selected on the basis of the soil lead content, showed blood lead levels clearly reflecting lead in the soil, with herd means in the early summer ranging from around 10μg Pb/100 ml blood on background farms to over 30μg Pb/100 ml on high-lead farms. Lead levels in pasture sampled monthly over the grazing season showed no significant difference between the three groups of farms, and it has been shown that on these historically contaminated soils, the cattle take in the majority of the lead in the form of accidentally ingested soil (45).

Of course, the present controversy around human lead exposure centers around lead in fuel additives and the contribution of lead from vehicle emissions to lead in air, dusts, etc. While this subject can hardly be called historical, it is nonetheless of interest that as long ago as 1924, street dusts in New York City were examined for lead content at the time tetraethyl gasoline was introduced; further sampling in 1934 indicated a 50% increase in lead content, though too few samples were analyzed to provide a reliable comparison (29).

CADMIUM AND ZINC

All commercial cadmium is a by-product of zinc production, and there are no ores mined and processed exclusively for cadmium. The two metals are geochemically linked and in most minerals and soils occur in ratios of 1:100 to 1:1000.

The well documented Japanese itai-itai disease brought the human effects of cadmium pollution to the notice of the world in the 1950s and 1960s (reviewed by Tsuchiya (46)), but we would draw attention to recent studies in Britain relating to environmental and health effects of cadmium pollution resulting from historical mining. Over the period 1700–1850, zinc was mined as the ore smithsonite ($ZnCO_3$) at Shipham, a village in Somerset, southwest England, and used with copper to make brass. The ore was washed in ponds and calcined within the village. Surface agricultural soils reclaimed from and near old mine workings at present range from 30–80 μg/g Cd together with 2500 μg/g to over 6% Zn, and a survey of household garden soils showed that 60% exceeded 60 μg/g Cd, far in excess of the levels in polluted paddy soils in Japan (44). These soils also contained large amounts of zinc with Zn:Cd ratios of around 90:1. On average, housedust samples contained 26 μg/g Cd and 2300 μg/g Zn. Comprehensive studies on metals in vegetables and diets showed an average human intake of 200 μg Cd per week (compared with the average U.K. intake of 140 μg Cd per week), with individual intakes rarely exceeding the WHO provisional tolerable weekly intake of 400–500 μg Cd. Health inventories and biochemical tests on 548 residents of Shipham and 543 controls from a nearby uncontaminated village showed only slight differences attributable to cadmium (3). We wonder if the large amounts of zinc and calcium present in this environment are affording some degree of protection against possible adverse effects of cadmium, perhaps by saturating exchange sites on plant roots and on human and animal tissues.

MERCURY

Historical aspects of mercury utilization, industrial exposure, and more recent environmental incidents have been reviewed by D'Itri and D'Itri (16). The use of mercuric nitrate for the "carotting" of animal fur to make felt was brought from France to Britain by the Huguenots in the late 17th century (37), and mercury poisoning in hatters was still common in the earlier part of the present century.

Industrial exposure in those operating mercury mines has been recognized since Roman times. It has been suggested that the Roman authorities closed mines at Monte Amiata in northern Italy and imported mercury from Spain (Almaden), perhaps to protect Italian workers from mercury poisoning or to prevent the devastation of the Italian landscape. Spain was the world's major exporter of mercury, and it has been estimated that Almaden has produced over 500 million lb. Hg. In 1921, 2100 miners and 400 furnace men were affected by mercury poisoning due to vapor escaping from the furnaces; environmental consequences are not recorded. However, mining at Idria in what is now Yugoslavia led to mercurialism in all inhabitants of the town in 1893 due to a fire in the mine tunnel; poisoning was also recorded in farm livestock. Before industrial safety procedures were adopted, children were said to experience mercury poisoning if they slept in the same bed as their fathers who worked in the mines. It has been recorded that when roasting furnaces were moved from the valley floor to a nearby hilltop, the land stayed green and local people suffered no effects of mercury translocated into vegetables, though drops of the metal could still occasionally ooze out of the ground and collect on grass or foliage (7).

Historically mercury was also used in South America in the refining of silver, using a system of amalgamation know as the *patio* process, which was introduced in New Spain in 1571 and remained the basis of silver refining in both Mexico and Peru until the late 19th century (19). Considerable amounts of mercury were lost in waste materials.

CONCLUSIONS

Contamination of the environment and adverse effects of exposure resulting from the mining, processing, and utilization of metals have been recognized for centuries, though only during about the last 80 years has the degree of pollution been quantified by chemical analysis. Conversely, recent studies in historically polluted areas have enabled the degree, extent, and effects of metal pollution to be assessed in terms of present-day land and water use and present-day hazards to farm livestock, local communities, etc. Historical mining and smelting in Britain has left a heritage of metal pollution which will be of significance far into the future.

Acknowledgements. Several examples are cited from the continuing program in agricultural geochemistry of the Applied Geochemistry Research Group supported by grants from the Agricultural Research Council.

REFERENCES

(1) Abrahams, P. 1983. Distribution, dispersion and agricultural significance of metals in soils of the mining regions of southwest England. Ph.D. Thesis, University of London.

(2) Aitchison, L. 1960. A History of Metals. London: MacDonald & Evans Ltd.

(3) Barltrop, D., and Strehlow, C.D. 1982. Cadmium and health in Shipham. Lancet **Dec. 18.**

(4) Barton, D.B. 1967. A history of tin mining and smelting in Cornwall. Truro, Cornwall: D. Bradford Barton Ltd.

(5) Bennett, E. 1948. Almaden — world's greatest mercury mine. Trans. Inst. Min. Metall. **29**: 6.

(6) Borlase, W. 1758. The Natural History of Cornwall. Oxford: Oxford University Press.

(7) Byrne, A.E., and Kosta, C. 1971. Studies on the distribution and uptake of mercury in the area of the mercury mine at Idrija, Slovenia (Yugoslavia). Chem. Abstr. **74**.

(8) Cannon, H.C., and Anderson, B.M. 1971. The geochemists involvement with pollution problems. **In** Environmental Geochemistry in Health and Disease. Geological Society of America Memoir 123, eds. H.L. Cannon and H.C. Hopps, p. 155–177. Boulder, CO: Geological Society of America Inc.

(9) Carpenter, K.E. 1924. A study of the fauna of rivers polluted by lead mining in the Aberystwyth district of Cardiganshire. Ann. Appl. Biol. **11**: 1–23.

(10) Carpenter, K.E. 1925. On the biological factors involved in the destruction of river fisheries by pollution due to lead mining. Ann. Appl. Biol. **12**: 1–3.

(11) Carpenter, K.E. 1926. The lead mine as an active agent in river pollution. Ann. Appl. Biol. **13**: 395–401.

(12) Chow, T.J., and Patterson, C.E. 1966. Concentration profiles of barium and lead in Atlantic waters off Bermuda. Earth Planet. Sci. Lett. **1**: 397–400.

(13) Colbourne, P.; Alloway, B.J.; and Thornton, I. 1975. Arsenic and heavy metals in soils associated with regional geochemical anomalies in south-west England. Sci. Total E. **4**: 359–363.

(14) Davies, B.E. 1980. Base metal mining and heavy metal contamination of agricultural land in England and Wales. **In** Inorganic Pollution and Agriculture, Ministry of Agriculture, Fisheries and Food Reference Book 326, pp. 142–156. London: H.M.S.O.

(15) Dines, H.G. 1956. The metalliferous mining region of south-west England, vol. 1. Memoirs of the Geological Survey of Great Britain. London: H.M.S.O.

(16) D'Itri, P.A., and D'Itri, F.M. 1977. Mercury Contamination: A Human Tragedy. New York: John Wiley and Sons.

(17) Farey, J. 1811. A General View of the Agriculture and Minerals of Derbyshire, vol. 1. London: Macmillan.

(18) Farey, J. 1817. A General View of the Agriculture and Minerals of Derbyshire, vol. 2. London: Macmillan.

(19) Fisher, J.R. 1977. Silver Mines and Silver Miners in Colonial Peru, 1776–1824. Liverpool: University of Liverpool Center for Latin-American Studies.

(20) Forbes, R.J. 1971. Studies in Ancient Technology, vols. I–VIII. Leiden: E.J. Brill.

(21) Gilfillan, S.C. 1965. Lead poisoning and the fall of Rome. J. Occup. Med. **7**: 53–60.

(22) Gregory, C.E. 1980. A Concise History of Mining. New York: Pergamon Press.

(23) Harkins, W.D., and Swain, R.E. 1907. Papers on smelter smoke. I. J. Am. Chem. Soc. **29**: 970–998.

(24) Haywood, J.K. 1907. Injury to vegetation and animal life by smelter fumes. J. Am. Chem. Soc. **29**: 998–1009.

(25) Hedgecock, G.G. 1914. Injury by smelter smoke in Southeastern Tennessee. Wash. Acad. Sci. J. **4**: 70–71.

(26) Hofmann, K.B. 1883. Die Getränke der Griechen und Römer von hygienischen Standpunkten. Arch. Geschichte Med. **6**: 26.

(27) Hutchinson, T.C. 1979. Copper contamination of ecosystems caused by smelter activities. **In** Copper in the Environment, ed. J.O. Nriagu, pp. 451–502. New York: John Wiley & Sons.

(28) Jones, D. 1982. The singular case of Napoleon's wallpaper. New Scientist **96**: 101–104.

(29) Kaye, S., and Reznikoff, P. 1947. A comparative study of the lead content of street dusts in New York City in 1924 and 1934. J. Indust. Hygiene **29**: 178–179.

(30) Lee, J.A., and Tallis, J.H. 1973. Regional and historical aspects of lead pollution in Britain. Nature **245**: 216–217.

(31) Murozumi, M.; Chow, T.J.; and Patterson, C. 1969. Chemical concentrations of pollutant lead aerosols, terrestial dusts and sea salts, in Greenland and Antarctic snow strata. Geochim. Cosmochim. Acta **33**: 1247–1294.

(32) Ng, A., and Patterson, C.C. 1981. Natural concentrations of lead in ancient Arctic and Antarctic ice. Geochim. Cosmochim. Acta **45**: 2109–2121.

(33) Ng, A., and Patterson, C.C. 1982. Changes of lead and barium with time in California off-shore barium sediments. Geochim. Cosmochim. Acta **46**: 2307–2321.

(34) Nriagu, J.O. 1979. The global copper cycle. **In** Copper in the Environment, ed. J.O. Nriagu, pp. 1–17. New York: John Wiley & Sons.

(35) Nriagu, J.O. 1983. Lead and Lead Poisoning in Antiquity. New York: John Wiley & Sons.

(36) Nriagu, J.O. 1983. Occupational exposure to lead in ancient times. Sci. Total E., in press.

(37) Porter, C. 1902. Felt hat making: its processes and hygiene. Brit. Med. J. **1**: 377.

(38) Ruhling, A., and Tyler, G. 1968. An ecological approach to the lead problem. Bot. Notiser. **121**: 321–338.

(39) Schaule, B.K., and Patterson, C.C. 1981. Lead concentrations in the north-east Pacific: evidence for global anthropogenic perturbations. Earth Planet. Sci. Lett. **54**: 97–116.

(40) Schmandt-Besserat, D. 1980. Ocher in prehistory: 300,000 years of the use of iron ores as pigments. **In** The Coming of the Age of Iron, eds. T.A. Wertime and J.D. Muhly, pp. 127–150. New Haven and London: Yale University Press.

(41) Shirahata, H.; Elias, R.W.; and Patterson, C.C. 1980. Chronological variations in concentrations and isotopic compositions of

anthropogenic atmospheric lead in sediments of a remote subalpine pond. Geochim. Cosmochim. Acta **44**: 149–162.

(42) Thompson, I.; Thornton, I.; and Webb, J.S. 1972. Molybdenum in black shales and the incidence of bovine hypocuprosis. J. Sci. Fed. Agric. **23**: 878–891.

(43) Thornton, I. 1980. Geochemical aspects of heavy metal pollution and agriculture in England and Wales. **In** Inorganic Pollution and Agriculture. Ministry of Agriculture, Fisheries and Food Reference Book 326, pp. 105–125. London: H.M.S.O.

(44) Thornton, I.; John, S.; Moorcroft, S.; Watt, J.; Strehlow, C.D.; Barltrop, D.; and Wells, J. 1980. Cadmium at Shipham — a unique example of environmental geochemistry and health. **In** Trace Substances in Environmental Health, vol. XIV, ed. D.D. Hemphill, pp. 27–37. Columbia: University of Missouri.

(45) Thornton, I.; Kinniburgh, D.; Clegg, F.G.; Lewis, G.; Barltrop, D.; and Strehlow, C.D. 1983. Lead uptake by cattle in the old lead-zinc mining area of Derbyshire. J. Agric. Sci., in press.

(46) Tsuchiya, K., ed. 1978. Cadmium Studies in Japan. Amsterdam: Elsevier/North-Holland Biomedical Press.

(47) Tylecote, F.F. 1976. A History of Metallurgy. London: The Metals Society.

(48) Tyler, G. 1974. Heavy metal pollution and soil enzyme activity. Plant Soil **41**: 303–311.

(49) Waldron, H.A., and Stofen, D. 1974. Sub-clinical Lead Poisoning. London: Academic Press.

(50) Whorton, J.C. 1971. Insecticide spray residues and public health: 1865–1938. Bull. Hist. Med. **45**: 219.

Changing Metal Cycles and Human Health, ed. J.O. Nriagu, pp. 27–41.
Dahlem Konferenzen 1984. Berlin, Heidelberg, New York, Tokyo: Springer-Verlag.

ANTHROPOGENIC PERTURBATION OF METAL FLUXES INTO THE ATMOSPHERE

W.H. Zoller
Chemistry Dept., University of Maryland
College Park, MD 20742, USA

Abstract. A review of the data on the levels of metals in atmospheric particulate material shows that on a local or regional basis anthropogenic emissions dominate natural sources for the volatile or enriched elements. There is strong evidence of the long-range transport of pollutant aerosols to some areas of the northern hemisphere downwind of large anthropogenic sources. Data from the Greenland ice cap show the enhanced concentrations of several elements commonly associated with pollution (Cu, Zn, Sb, and Pb) during the last few decades. For the southern hemisphere, lead is the only element which shows a potential impact of anthropogenic sources. The source data on both anthropogenic and natural emissions for most elements would predict a very significant impact for many metals from pollution sources. Observations do not show such a strong impact, suggesting that anthropogenic materials either are not transported long distances or are effectively removed from the atmosphere before they reach remote areas.

INTRODUCTION

During the last decade there has been a good deal of interest in the chemical composition of atmospheric particulates in both urban and remote areas around the earth. To accurately assess the impact of anthropogenic sources of heavy metals on the natural cycles of the elements, one needs to look at the general atmospheric cycles of the elements to have an understanding of the system, especially the inputs and removal mechanisms. To evaluate the impact of anthropogenic sources, one must know their magnitude, distribution around the world, and the mechanisms of removal of the aerosols from the atmosphere. A simple model can be set up in which there are essentially four reservoir areas, the continental and marine-free troposphere, and the continental and marine-boundary layer as shown in Fig. 1.

In this system, the continental areas contain a number of major sources such as dust, or soil erosion, biological processes, vegetative burning, and volcanoes, in addition to anthropogenic sources. Of these sources, only volcanoes and vegetative burning routinely transport their emissions directly into the free troposphere without vertical mixing through the boundary layer. Removal processes are most active within the boundary layer, or just above the layer by clouds, so that particles released into the lower atmosphere are more effectively

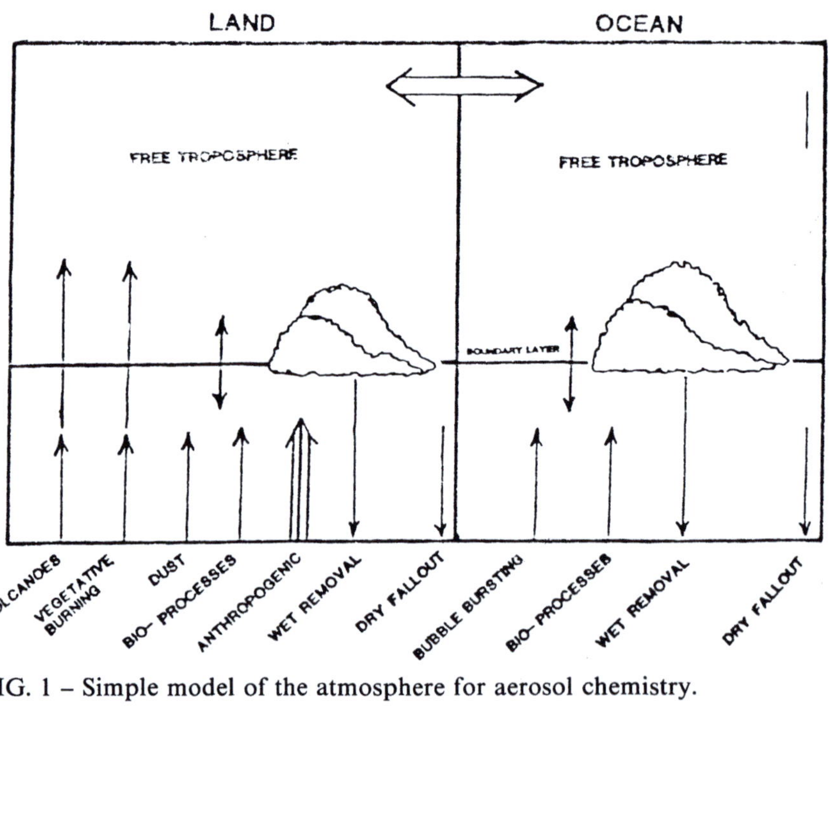

FIG. 1 – Simple model of the atmosphere for aerosol chemistry.

scavenged from the atmosphere. For this discussion, this means that pollutants which are released at low elevations and are, therefore, more prone to scavenging near the source, may not be effectively transported over long distances. When pollutants are mixed to higher altitudes, long-range transport is more effective. For the oceanic system, anthropogenic particulates are brought into the system usually in the free troposphere and then removed through the boundary layer to the ocean. In both cases, natural sources also release particles predominantly in the boundary layer, so that the differences between the chemical and physical properties of the particles are important in their relative removal rates.

TRACE METAL DISTRIBUTIONS

Much of the work in urban regions has been to measure the impact of various anthropogenic emissions on the air quality in the surroundings. For example, the studies of atmospheric particulate material in urban areas of North America and Europe all show the presence of fly ash from power plants and heavy metals from industrial processes to be very common. From these studies it is quite evident that anthropogenic activities do, indeed, have a large impact on the aerosol composition close to the large population centers of the world (14). One way of interpreting data on trace metals in urban or remote areas has been to use crustal enrichment factors (EF_c). The EF_c is defined as:

$$EF_c = \frac{(C_x/C_{Al})\text{ air sample}}{(C_x/C_{Al})\text{ crust}},$$

where C denotes the concentrations of element X and Al in aerosol material and in crustal material of the Earth, respectively. A substantial fraction of the total suspended particulate material (TSP) in continental areas is entrained soil and rock dust, which is expected to have a composition similar to that of average crustal material. Elements contributed mainly by this source have EF value close to unity. Those with significant additional sources, the so-called "anomalously enriched elements," have EF values much larger than unity. High concentrations and enrichments of Pb, Fe, Cd, Zn, Cu, Sb, As, and Se, as well as other elements, are always observed near population or industrial centers. The individual sources of many of the metals observed in urban areas have been identified using receptor models or chemical element balances. For this work, one needs to know the chemical composition of each potentially important source of aerosols in urban areas, and then individual aerosol samples collected in the region can be resolved into components. In a study of the Washington aerosol (5), the elements Zn, Cd, and Sb were primarily associated with refuse incineration; Va and Ni from oil combusion; I, As, and Se from coal combustion; Pb, Br, and Ba from motor vehicles; and the majority of the other elements measured came from soil with small contribution from other sources. From the studies of atmospheric particulate composition in urban areas throughout the world, it is quite evident that anthropogenic emissions are the dominant source of the enriched elements. Even though a substantial fraction of some metals are associated with crustal dust, this dust is mobilized into the atmosphere by anthropogenic activities.

When one leaves the urban scale of pollution and looks at large regions, the impact of anthropogenic emissions can still be easily observed, especially if large industrial or urban centers are not too far away. For example, rural areas in the United States and Europe still show the effects of the industrial centers nearby. This means that attempts to control air quality require a regional approach, and local solution will probably not be entirely effective. An example of this type of problem can be identified with the acid rain in the northeastern United States and Canada, as well as Scandinavia and Europe.

As the size of the region is increased, longer-range transport of anthropogenic particulates becomes necessary in order to have a significant impact on the larger region. For example, there is a great deal of evidence for the long-range transport of pollutant aerosols into the Arctic during the early spring. This has been shown using the ratio of trace metals, such as the Mn/V ratio discussed by Rahn (15). Pollutant transport to the Arctic has become a problem because of the potential of carbon that is a significant portion of the particulates to absorb solar radiation.

Another example of long-range transport of material in the atmosphere is the Saharan dust transport to Europe, which is found in glaciers along with heavy metal pollutants from sources in Europe. The Saharan dust is also transported across the Atlantic Ocean to the Caribbean area (13). This dust was also found to contain pesticides transported from Africa to the western hemisphere, so that it is possible for pollutants to hitch a ride with the dust plumes. Another example of this type of transport has been documented for the Pacific Ocean where dust from the Asian mainland is carried to the central Pacific basin and observed each spring in Hawaii (12). The air mass trajectories point back to the Asian mainland during the spring of each season and are correlated with large storm systems.

Many of the metals associated with crustal dust show strong correlations with this seasonal transport. Some of the anomalously enriched elements correlate, to some extent, with the dust transported across the Pacific Ocean. In Fig. 2, the seasonal patterns of dust transport to the Hawaiian Islands is shown for the years between 1979 and 1982.

The top portion of the plot shows the large increase in dust concentrations during each spring season. In general, the metals also show a general increase during this same time period, although these are different spikes and there is a lot more variation in their concentration patterns. Obviously, the elements Zn, Sb, and As are fairly well correlated with each other, even at times of low

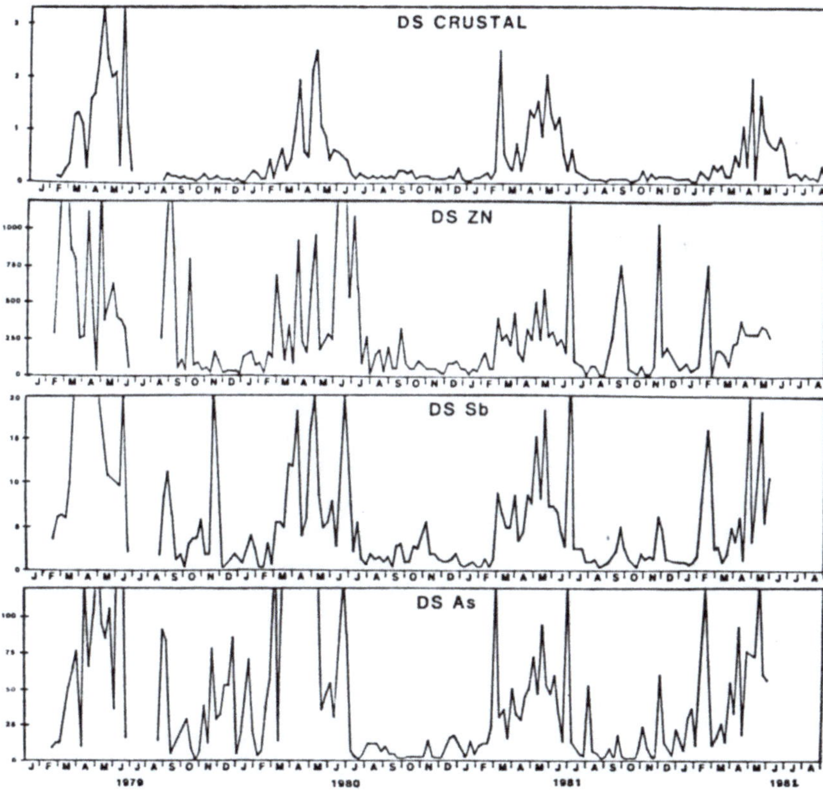

FIG. 2 – The seasonal variation of concentrations of crustal dust (μg/m³), zinc (pg/m³), antimony (pg/m³), and arsenic (pg/m³) for nighttime downslope winds at the Mauna Loa Observatory in Hawaii.

Asian dust. There are several points to be made from these data. First, the fact that there is a general increase in these enriched elements when the dust level is higher may suggest that these elements are pollutants transported along with the dust from China and Japan. Second, the association of these elements with dust may only indicate that desert dust in Asia is highly enriched in these elements, a rather unlikely explanation. When comparing these observations with the simple model of Fig. 1, this transport represents a special case when the dust, and possibly anthropogenic material, is transported to the free troposphere by intense storm activity. Finally, the correlation could be entirely fortuitous and the higher concentrations of Zn, Sb, and As are from an increase in a natural source such as biological activity that increases the output substantially each spring at the same time as long-range transport is maximized.

To get a better understanding of the potential sources of these and other enriched elements in the atmosphere, we can look at data from very clean and remote areas, as well as ice cores from polar areas. The best clean and remote area is the Antarctic where work on aerosol chemistry at the South Pole has been conducted since 1970. Other clean air sampling sites in the central Pacific also have yielded excellent data for comparison. Ice cores from both Antarctica and Greenland can also be examined to look for increases over the past decade which may then correlate with the increase in anthropogenic emissions over the past few decades.

Results of studies at the South Pole in Antarctica show that there are seasonal variations in the concentration of metals in the Antarctic atmosphere which are associated with changes in transport (2). The concentrations of most of the enriched elements do follow the seasonal changes in crustal dust and sea salt, but in some cases they are not correlated. This information may eventually lead to a better understanding of the sources of some of these elements. Others such as Zn, Se, As, and Sb have shown large increases in certain years due to volcanic emissions, such as the large increase in 1975 which is shown in Table 1.

The concentrations shown in Table 1 cover a wide range of locations and long time periods. For the South Pole it appears as if the concentrations of Zn, As, and Sb are higher in the summer than winter season and, therefore, correlate closer with crustal dust transport into the Antarctic. The much higher values for the winter of 1975 are due to a volcanic injection from an eruption in New Zealand. The arsenic levels for all periods in Hawaii are higher than the clean winter levels in Antarctica but are close to the summer values. For As this

	Concentration (pg/m^3)				
	As	Sb	Se	Zn	Pb
South Pole, Antarctica[a]					
summer (70-71)	-	1.7	5.6	30	
(74-75)	7.1	0.84	6.2	33	
(75-76)	8.8	4.4	7.0	38	
(77-78)	9.2	1.1	6.3	40	
(78-79)	2.8	3.5	7.9	-	
(79-80)	2.9	1.5	10.1	437	
(80-81)	3.1	3.2	6.1	-	
winter (75)	28	2.4	5.0	-	
(76)	5.3	0.60	8.9	77	
(78)	1.9	0.65	5.5	-	
(79)	2.1	1.50	4.9	143	
(80)	2.8	0.8	5.7	311	
(81)	1.2	0.6	4.4	284	
Mauna Loa, Hawaii[b]					
upslope dust (80)	138	7.5	25	390	
(81)	31	5.6	24	425	
upslope clean (80)	41	5.1	23	280	
(81)	24	3.9	21	250	
downslope dust (80)	598	9.4	26	340	
(81)	68	8.9	16	270	
downslope clean (80)	13	3.2	12	88	
(81)	11	1.5	7.4	77	
Enewetak (1981)[c]					
wet season (Fall)	-	2.4	110	110	96
dry season (Spring)	-	5.2	150	222	130 (88-260)[d]
Samoa (1980)[e]	20	~0.50	~70	~30	~50

[a] Cunningham and Zoller (2) plus newer data.

[b] Parrington et al. (12) plus newer data.

[c] Duce et al. (3).

[d] Settle and Patterson (16).

[e] Duce (personal communication).

TABLE 1 – Concentration of volatile elements in remote areas.

would indicate a continental source or possibly anthropogenic perturbation due to long-range transport as indicated by the spikes in Fig. 2, as well as substantial contributions to the Pacific area during the spring Asian dust episodes. Antimony is similar to As, but the changes and correlations with season are not as strong. The higher levels during dust transport and correlation with crustal dust would point to a continental and/or anthropogenic source. The selenium concentrations at the South Pole and Hawaii are remarkably similar, as there is no strong seasonal variation. The only significant variation in the concentrations are for low level samples collected in Enewetak (sea level) and for upslope samples from Hawaii. It would appear as if Se is not significantly impacted by anthropogenic sources but has a major source in the marine environment. Selenium is probably released to the atmosphere by biomethylation.

The results of studies on ice cores in Antarctica are inconclusive. The very low levels observed in the ice cores make contamination very easy and the results of studies generally have very large errors. But, in general, the results show that there has not been a measurable increase in the concentration of most elements in the snow and ice cores (8). The results of Ng and Patterson (9) would indicate that there might have been a slight increase in lead levels during the last few decades, so that for the Antarctic, and other remote areas in the southern hemisphere, anthropogenic contributions may not be important for the heavy metals at this time (with the exception of lead).

One of the major problems encountered in the study of ice cores is that they are contaminated by the drilling process and handling. Ng and Patterson (9) have shown that the outer portion of ice cores are highly contaminated, and that the level of contamination decreases toward the center of the core. In some cases, the concentration of a species decreases all the way to the center of the core, so that there may not be any portion of the core that is useful. Unfortunately, most studies of historical records from ice cores never worried about this phenomenon, and the literature is filled with very inconclusive data on ice core chemistry. The case for Greenland ice cores is considerably different since most of the anthropogenic sources are in the northern hemisphere, and the distances from these sources to Greenland are relatively short. There is definite evidence for a large increase in lead (9) and, depending upon the individual results, small possible increases in Cu, Zn, Sb, and S in current snow and ice samples, although the data are far from conclusive (8). The results for the other elements indicate that they probably have strong natural sources that enrich the elements over crustal abundances and overwhelm anthropogenic sources. Although the data on ice core analyses give no clear evidence one way or the other for the importance of anthropogenic sources, except for Pb, the results indicate which elements have the highest potential for anthropogenic impact. There are several problems that must be resolved before the results of ice core analyses can be used to estimate anthropogenic source impact: fractionation during precipitation processes, nonhomogeneity across the ice sheet, and local (nearby) large natural sources (Iceland volcanoes, etc.), in addition to the contamination problem.

SOURCES OF METALS TO THE ATMOSPHERE

Another method to assess the importance of anthropogenic sources to the atmosphere is to make up reasonable anthropogenic source strengths and then compare these with natural sources. During the past few years, there have been several review articles with estimates of anthropogenic sources. Table 2 shows the estimates of global anthropogenic emissions from four of these studies.

The first column in Table 2 is from a report detailing the emissions of United States pollution sources for 1969. To compare these emissions with those of the other studies, the emission factors are multiplied by a factor of ten which is the approximate fraction of production of the United States as compared with the rest of the world. The second column represents recent data from Pacyna (11) who collected the emission data from Eastern and Western Europe for 1979. These data have been multiplied by a factor of five to derive the global data from the European input to the atmosphere. From this table it is evident that the different studies have come up with substantially different estimates of global anthropogenic sources. In many cases the estimated emission rates are within a factor of two which would be quite acceptable, except that many of the estimates are based on the same data bases. But they do give useful

Element	Flux Estimated by Authors (10^9 g/yr)			
	USA x 10 (17)	Pacyna x 5 (11)	Lantzy & MacKenzie (7)	Nriagu (10)
Cu	137	78	263	56
Cd	30	14	5.5	7.3
Zn	1,600	400	840	314
As	89	33	78	
Se	9.9	2.1	14	
Sb		1.9	38	
Hg	7.3	0.24	11	
Pb	180[a]	615	2,030	449

[a]nonautomotive

TABLE 2 – Literature estimates of global anthropogenic emissions.

estimates that can be compared with the natural sources to gauge their potential importance. Two of the reports listed in Table 2 also list the source strengths of natural systems and thus can be examined to check on their consistency. These natural source strengths are listed in Table 3 for comparison.

Several things are very obvious from Table 3. In some cases there is excellent agreement for the estimated fluxes of an element from a given source. For example, the fluxes of Zn from windblown dust are all the same and for most of

	Cd	Zn	Hg	Pb
Windblown Dust				
(L&M)[a]	0.25	25	0.03	5.0
(N)	0.1	25	-	16
(G)[a]	0.25	25	0.005	5.0
Biomass Burning				
(L&M)	1.5	250	0.04	6.8
(N)	0.012	2.1	-	0.5
(G)	0.012	2.1	0.16	0.5
Vegetation				
(N)	0.2	9.4	-	1.6
Volcanic				
(L&M)	0.04	10.8	0.01	0.87
(N)	0.52	7.0	-	6.4
(G)	1.3	24	>0.24	16
Sea Salt*				
(N)	~0.001	0.01	-	0.02
(G)	0.003	0.03	0.001	0.001
Total Natural Emissions				
(L&M)	1.79	286	0.08	12.7
(N)	0.83	43.5	-	24.5
(G)	1.6	51	>0.4	22

[a]Goldberg flux of 5×10^{14} g/yr
[b]Assumes forest fires plus cosmic dust
(L&M) = Lantzy and Mackenzie (7)
(N) = Nriagu (10)
(G) = Gordon et al. (6)
*Large particle material only

TABLE 3 – Comparison of major natural sources to anthropogenic emissions (10^9 g/yr).

the other elements they are fairly close. In general, this is because all of the studies quoted here in one way or another go back to a common piece of data, or estimate. In the case of windblown dust most estimates go back to Goldberg's flux of dust to the atmosphere of $5x10^{14}$ g/yr. The results for forest fires show a great variability between Lantzy and Mackenzie (6), and Gordon et al. and Nriagu (6,10). In general, this is due to the lack of any real data upon which to base these estimates for forest fire emissions. Of the unknown sources, vegetation is a problem, as is biomethylation which is not listed in any of these studies. The fluxes due to sea salt are all very small and consistent, but this is due to the fact that they are all based on similar data on the flux of sea salt to the atmosphere. In general, for natural sources, we really do not know much about the fluxes of materials to the atmosphere. One of the sources that may be very important to the atmosphere for volatile elements is volcanic emanations. From the estimate in Table 3 it is obvious that the flux estimates for volcanic emissions are very different and, in most cases, based on *no* measurements, just wild guesses.

For the case of volcanic emissions, a great deal of data on volcanic emissions have been gathered during the last few years by research groups around the

	Cd	Zn	Se	As	Hg
Individual Volcanic Emission Rates (10^3 g/day)					
Mt. St. Helens (USA)	5.0	1,000	100	800	80
Poas (Costa Rica)	2.4	40	7.6	3.0	76
Arenal (Costa Rica)	50	50	14	7.0	55
Colima (Mexico)	6.7	10	6.2	3.1	
El Chichon (Mexico)		20	4	200	83
Global Volcanic Emission Rates (10^9 g/yr)					
Mt. St. Helens (USA)	0.056	10	0.10	8.0	0.90
Poas (Costa Rica)	0.032	0.54	0.10	0.039	1.00
Arenal (Costa Rica)	2.7	2.4	0.76	0.38	3.00
Colima (Mexico)	0.23	0.34	0.21	0.11	
El Chichon (Mexico)		0.25	0.046	2.5	2.20
Lantzy and MacKenzie (7)	0.040	10.8	0.11	0.31	0.011
Nriagu (10)	0.51	7.0			
Walsh (18)				7.0	
Gordon et al. (6)	1.3	24	17	39	>0.24
Fitzgerald (4)					0.43
Arnold (1)			0.28		

TABLE 4 – Comparison of volcanic emission estimates.

world. In Table 4, an example of the data on the fluxes of some metals from some volcanoes and recent literature data are given.

Data in this table show that the emission rates of individual volcanoes can vary by more than two orders of magnitude. When one uses data of this type to calculate the global flux by assuming a value for the volcanic SO_2 flux (10^7 T/yr) and for other elements using their ratio to sulfur, an interesting data set is produced. For elements such as Cd and Zn the fluxes from volcanoes vary by a factor of one hundred, whereas for an element such as Hg the global flux appears to agree very well. This is probably due to the chemistry of mercury versus that of the other elements. Unfortunately, a comparison with the literature values even for an element such as mercury shows vast disagreements. Therefore, we come to the unfortunate conclusion that our data base is not as good as is really needed to evaluate most of the natural sources, and therefore, we cannot accurately estimate the impact of anthropogenic sources. A direct comparison of the emission rates of anthropogenic sources (Table 2) with natural sources (Table 3) suggests that natural sources should be overwhelmed easily. However, the data from remote areas do not show this impact. Based on the atmospheric and ice core data, several elements (Cu, Zn, As, and Sb) appear to have inputs from anthropogenic sources that may be significant on a regional scale. For most of the other elements, specifically, Cd, Hg, and Se, the impact from anthropogenic sources is probably minimal, and contributions from natural sources must dominate. The only element that is clearly enhanced by anthropogenic emissions on a global scale is lead.

CONCLUSIONS

Based on the data in the literature reported by different research groups working in remote areas of the world, we can at least set out some preliminary conclusions as to the impact of anthropogenic emissions on the atmospheric burdens of many trace elements:

1. Lithophilic elements, usually associated with crustal dust, are not impacted significantly by anthropogenic emissions on a global or regional scale.

2. Lead is the only element which shows good evidence, on a global basis, for an impact by anthropogenic emissions.

3. For elements (except Pb) which are emitted in vast quantities from anthropogenic sources, the evidence for a significant impact is seen only for regional scales.

4. Estimates of anthropogenic and natural emissions indicate that many of the anomalously enriched elements should see vast impacts of man's mobilization of the elements. This is not justified by the data base on atmospheric concentration and records in glaciers.

5. Data from ice caps and remote areas suggest that very few elements (in the northern hemisphere) show evidence of significant anthropogenic impact. The elements that may be impacted are Cu, Zn, and Sb.

6. The atmospheric burdens of the elements Se, As, Hg, and Cd appear to be dominated by natural sources, most of which are currently unknown.

7. On a regional basis, many metals are drastically impacted by anthropogenic emissions. The largest impacts are found downwind of large industrial complexes.

8. Urban regions have massive impacts due to the anthropogenic release of metals, mostly via high temperature processes.

Based on the results of past studies and the most recent reviews of our knowledge on global cycles, the following points must be investigated to obtain a better understanding of what sources are important contributors of metals in the atmosphere:

1. Better data are needed on the natural sources of metals to the atmosphere, especially the biogenic sources.

2. Data on the anthropogenic emissions are needed for developing countries.

3. More accurate measurements are need on ice core samples to determine the historical trends.

Acknowledgement. Supported in part by the U.S. National Oceanographic and Atmospheric Administration grant NA 79 RAC0050 and the U.S. National Aeronautics and Space Administration grant NAG-1-200.

REFERENCES

(1) Arnold,M.; Buat-Menard, P.; Chesselet, R. 1981. An estimate of the input of trace metals to the global atmosphere by volcanic activity. IAMAP, Third Scientific Assembly, Hamburg, F.R. Germany.

(2) Cunningham, W.C., and Zoller, W.H. 1981. The chemical composition of remote area aerosols. J. Aerosol Sci. **12:** 367–384.

(3) Duce, R.A.; Arimoto, R.; Ray, B.J.; Unni, C.K.; and Harder, P.J. 1983. Trace elements in marine aerosols from Enewetak Atoll: I. Concentrations, sources and temporal variability. J. Geophys. Res., in press.

(4) Fitzgerald, W.F. 1981. Volcanic mercury emissions and the global mercury cycle. IAMAP, Third Scientific Assembly, Hamburg, F.R. Germany.

(5) Gordon, G.E. 1980. Receptor models. Env. Sci. Technol. **14:** 792–800.

(6) Gordon, G.E.; Moyers, J.L.; Rahn, K.A.; Gatz, D.F.; Dzubay, T.G.; Zoller, W.H.; and Corrin, M.H. 1983. Atmospheric trace elements: cycles and measurements. Rev. Geophys. and Space Phys., submitted.

(7) Lantzy, R.J., and Mackenzie, F.T. 1979. Atmospheric trace metals: global cycles and assessment of man's impact. Geochim. Cosmochim. Acta **43:** 511–525.

(8) Li, Y.-H. 1981. Geochemical cycles of elements and human perturbation. Geochim. Cosmochim. Acta **45:** 2073–2084.

(9) Ng, A., and Patterson, C. 1981. Natural concentrations of lead in ancient Arctic and Antarctic ice. Geochim. Cosmochim. Acta **45:** 2109–2121.

(10) Nriagu, J.O. 1979. Global inventory of natural and anthropogenic emissions of trace metals to the atmosphere. Nature **279:** 409–411.

(11) Pacyna, J.M. 1983. Trace element emission from anthropogenic sources in Europe. Technical Report 10/82, Norwegian Institute for Air Research.

(12) Parrington, J.R.; Zoller, W.H.; and Aras, N.K. 1983. Asian dust: seasonal transport to the Hawaiian islands. Science **220:** 195–197.

(13) Prospero, J.M., and Nees, R.T. 1977. Dust concentrations in the atmosphere of the equatorial North Atlantic: possible relationship to the Sahelian drought. Science **196:** 1196–1198.

(14) Rahn, K.A. 1976. The chemical composition of the atmospheric aerosol. Technical Report, Graduate School of Oceanography, University of Rhode Island, Kingston, RI.

(15) Rahn, K.A. 1981. The MN/V ratio as a tracer of large-scale sources of pollution aerosol for the Arctic. Atmos. Env. **15:** 1457–1464.

(16) Settle, D.M., and Patterson, C.C. 1982. Magnitudes and sources of precipitation and dry deposition fluxes of industrial and natural leads to the North Pacific at Enewetak. J. Geophys. Res. **87:** 8857–8869.

(17) Sittig, M. 1975. Environmental Sources and Emissions Handbook. Park Ridge, NJ: Noyes Data Corporation.

(18) Walsh, P.R.; Duce, R.A.; and Fasching, J.L. 1979. Tropospheric arsenic over marine and continental regions. J. Geophys. Res. **84:** 1710–1718.

Changing Metal Cycles and Human Health, ed. J.O. Nriagu, pp. 43–69.
Dahlem Konferenzen 1984. Berlin, Heidelberg, New York, Tokyo: Springer-Verlag.

FLUXES OF METALS THROUGH THE ATMOSPHERE AND OCEANS

P.E. Buat-Ménard
Centre des Faibles Radioactivités, Laboratoire mixte CNRS-CEA
91190 Gif-sur-Yvette, France

Abstract. The tropospheric transport of metals from the continents to the oceans is an important pathway of their biogeochemical cycle. There is now strong evidence that the long-range transport of anthropogenic aerosols constitutes a major source for the atmospheric concentrations of many heavy metals and metalloids, at least in the Northern Hemisphere. The magnitude of the anthropogenic input from the atmosphere to the ocean exhibits a strong geographical variability which depends primarily on meteorological factors. Within a few years, at most, atmospheric inputs of trace metals are transferred to the deep ocean by the removal of these elements from the surface layers of the ocean by settling particles, primarily of biogenic origin. There is therefore a fast coupling between the atmospheric inputs of trace metals and their transport to the deep ocean. Even with the limited data available, it appears that the deep ocean is no longer excluded from anthropogenic inputs from the atmosphere, especially in the case of Pb.

INTRODUCTION

The presence of metallic compounds from natural and man-made sources in the marine atmosphere raises the question of how global geochemical cycles of metals at the surface of the earth depend on atmospheric chemistry and how these cycles are being perturbed by human activities (6,26,30,31,37). The oceans receive from the continents sedimentary material, nutrients, and other biologically useful elements, as well as increasing amounts of pollutants via atmospheric deposition. At present, the open ocean is probably more affected by pollution inputs through tropospheric transport than through riverine transport. We need therefore to know the magnitude of the anthropogenic input from the atmosphere to the oceans and how this input affects the composition of the oceans. As will be emphasized later, since there is a considerable amount of recycling of trace substances across the air-sea interface, the evaluation of the net input of these substancess from the continents to the oceans via atmospheric deposition is often extremely difficult.

In the marine atmosphere, direct flux measurements are scarce, and in many cases, fluxes are calculated indirectly from atmospheric concentration data. For some elements, the source functions are known with enough certainty so that the right order of magnitude of net air-to-sea and sea-to-air exchange rates can be obtained. Some insight into the variability of these fluxes with space and time will also be given. Finally, when enough information is available, geochemical implications of the calculated fluxes will be emphasized.

SOURCES OF TRACE METALS IN MARINE AEROSOLS

Among the trace constituents of the marine atmosphere, metals and especially heavy metals and metalloids have been extensively studied in the last ten years, primarily because of concern about atmospheric pollution input to the ocean and its possible effect on marine life. With the exception of Hg, these elements are found primarily associated with atmospheric particulate matter. It has been shown that atmospheric particulate matter in remote areas is strongly enriched in trace elements such as Pb, Cd, Hg, Se, Sb, As, Zn (Fig. 1), i.e., the atmospheric concentrations of these metals are much higher than expected from the "unfractionated" dispersions of soil material (quartz, aluminosilicate particles) and seawater (sea-salt aerosol) which are the major sources in term of mass of particulate matter in the marine atmosphere (8,14–16,39).

Many studies have focused on the causes of these apparent "enrichments" in marine aerosols. The long-range transport of anthropogenic aerosols was first recognized as a potential source for these anomalous elements. Because of their volatility (4,16,53), certain heavy metals can apparently be injected into the atmosphere in the vapor phase by high temperature combustion processes (fossil fuel combustion, incineration of wastes, metallurgy, etc.). This will lead to their enrichment in the air, compared to the concentration of more refractory elements. This emission in the vapor phase is followed by homogeneous or, more probably, heterogeneous condensation.

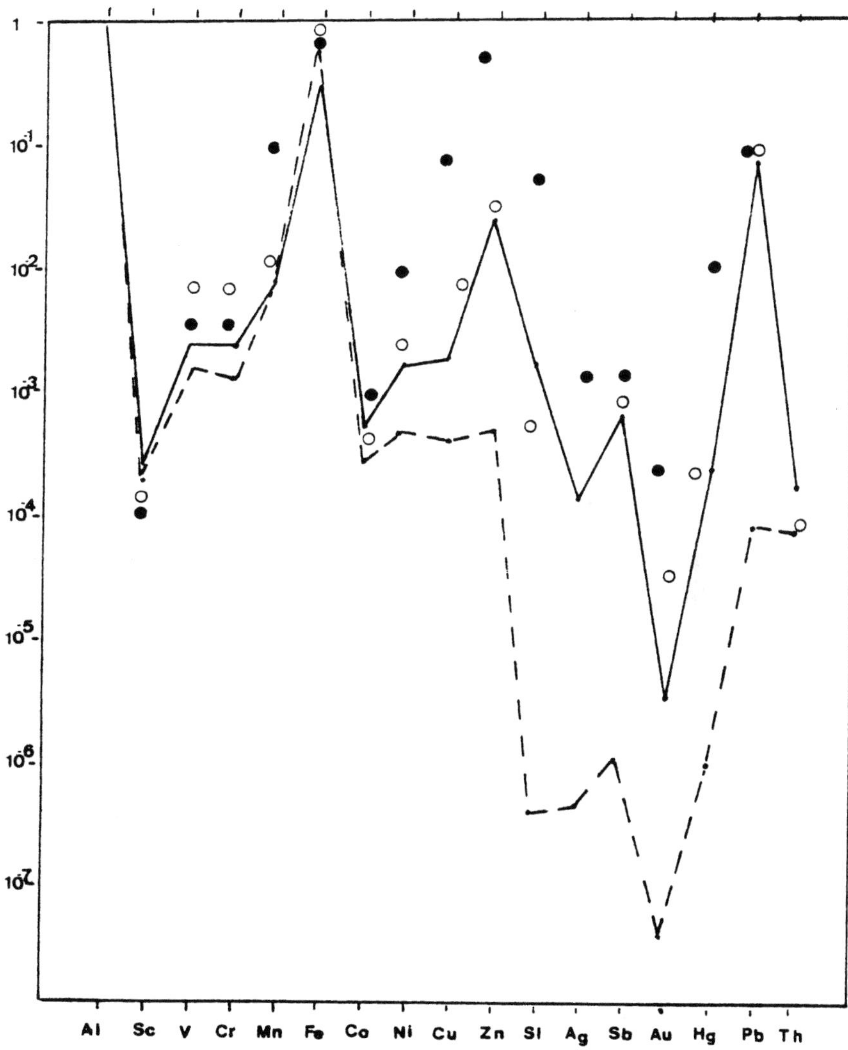

FIG. 1 – Abundances of trace metals in North Atlantic Aerosols, – , (8); Average Crustal Material (49), – – –; Urban Aerosols, ○, (39); and Aerosols collected in the plume of Mount Etna Volcano, ●, (7). All data are normalized to Al = 1.

As a consequence, the enriched elements will occur preferentially in the submicrometer fraction of the aerosol, whereas refractory elements occur with much larger particles. This has been shown to be the case for metals such as Cu, Zn, Sb, Cd, and Pb, not only in urban air (39) but also in the marine atmosphere. For example, at Bermuda (15) and over the Western Mediterranean Sea (2) results of the analysis of high volume cascade impactor samples do indicate that these enriched elements are concentrated in the submicrometer size-range, whereas elements derived from oceanic and crustal sources are concentrated in much larger particles.

It has been realized, however, that such an observation in the marine atmosphere does not equivocally indict pollution sources (16). Natural processes also volatilize elements into the atmosphere. These processes are numerous, but poorly documented. Among them are high temperature volatilization by volcanic activity (7,36), crustal degassing (25), and biological mobilization through methylation by land plants and marine life (3,52). Hence, by itself, the determination of elemental concentrations as function of particle size only suggests the existence of primary and secondary sources for the marine aerosol. Such an approach, although not sufficient, is nevertheless a first important step in source identification.

A further explanation arises from the fact that enriched heavy metals in the marine aerosol, such as Cu, Zn, Pb, have a small but sizable large-particle component (6,15) which could be derived from the sea (50). The source region for sea-derived aerosols is a very thin layer at the air/sea interface, the sea-surface microlayer, together with the material, particulate or dissolved, scavenged by rising bubbles. The now well-documented enrichment of many trace substances (organic compounds, trace metals, P, I, etc.) in the sea-surface microlayer suggests strongly that these substances should be enriched, relative to the composition of bulk surface-seawater, on the particles ejected into the atmosphere when bubbles burst at the air/sea interface. The big question is the exact magnitude of this enrichment. Because it depends on, among other factors, the biological activity in the surface waters, the enrichment of trace substances in sea-derived aerosols is likely to vary with space and time (6). Thus, we cannot identify an oceanic source for atmospheric particulates if we rely only upon the knowledge of the bulk chemical composition of the marine aerosol. Together with concentration data as a function of particle size, other criteria must be used that are more specific for the substance and source being investigated. These other criteria can include: chemical analysis of individual particles, measurements of the isotopic composition of some elements and radioisotopes, analysis of specific "source markers" such as organic compounds or classes of compounds, chemical speciation studies, and carefully designed laboratory and "in-situ" experiments.

CALCULATION OF ATMOSPHERIC DEPOSITION TO THE OCEAN

Owing to inherent sampling difficulties, direct measurements of dry deposition and wet deposition over the ocean are scarce compared to aerosol particle concentration data. As a consequence indirect flux calculations have most often been based on the latter type of data. It is therefore important to consider if aerosol concentration data can provide the basis for reliable dry and wet deposition estimates. With a few examples, I will present different approaches to this problem.

Dry Deposition: Problems

As recently shown by McDonald et al. (34), dry deposition rates of sea-salt aerosol particles may be satisfactorily estimated from measured concentrations and theoretical dry deposition velocity, provided extreme care is taken to obtain air samples that represent the true sea-salt particulate mass and size distribution. Even giant particles (radii >20 μm) must be efficiently collected. These results show the strong wind speed dependence of the sea-salt deposition rate (Table 1). When the wind speed increases from 3.4 to 10 m s^{-1}, the salt deposition rate increases 50-fold, compared to a 7–10-fold increase in atmospheric salt concentration. This results almost uniquely because large particles dominate the deposition, and the mass concentration of such particles increases much faster with increasing wind speed than that of smaller particles. Even at the low wind speed of 3.4 m s^{-1}, 10μ radius sea-salt particles consitiute only 13% of the total sea-salt mass, yet contribute 70% of the deposition. Therefore, for any element which in marine aerosols has a sizable sea-derived component, the dry deposition flux may be mostly due to the recycled component of that flux. This has been shown to be the case for Pb and Zn over the Tropical North Pacific (18). While only 5% of atmospheric Pb and Zn is present on salt particles with radii >10 μm, these particles account for 90% of the total Pb and Zn associated with dry deposition. Obviously, in the lower marine atmosphere, direct dry deposition measurements cannot provide any reliable estimate of the net dry flux of submicron anthropogenic lead aerosol (41). On the other hand, such measurements may provide the best data to quantify the flux of lead from the sea surface to the atmosphere. The best approach for calculating the net dry deposition of Pb to the ocean therefore seems to be to consider only that fraction of Pb not recycled (i.e., the submicron fraction) and to use predicted dry deposition velocities.

Wind Speed $m\ s^{-1}$	Atmospheric salt concentration $\mu g\ m^{-3}$	Salt flux $ng\ cm^{-2}hr^{-1}$
3.4	2.7	8
6.5	14	170
10	18	410

TABLE 1 – A comparison of sea-salt concentrations and fluxes as a function of wind speed.

However, deposition velocities for submicron aerosols over the ocean surface are not yet known with great accuracy (42). In the 0.1 to 1 μm radius interval, there seems to be a large range of uncertainty in data reported, which vary from 0.02 cm s^{-1} to 1 cm s^{-1}. Currently used cascade impactors do not discriminate well in this size range, so that, in my opinion, predicted values probably give only the correct order of magnitude of the net flux. Such sampling devices can also bias the true size distribution of atmospheric particles. For high volume cascade impactors, it seems that particle "bounce" effects, especially for dry particles such as aluminosilicate minerals, can result in a size distribution shifted to smaller sizes. This seems to be a serious problem for such particles not only in continental air (48) but also in the remote marine atmosphere ((9), and see Fig. 2).

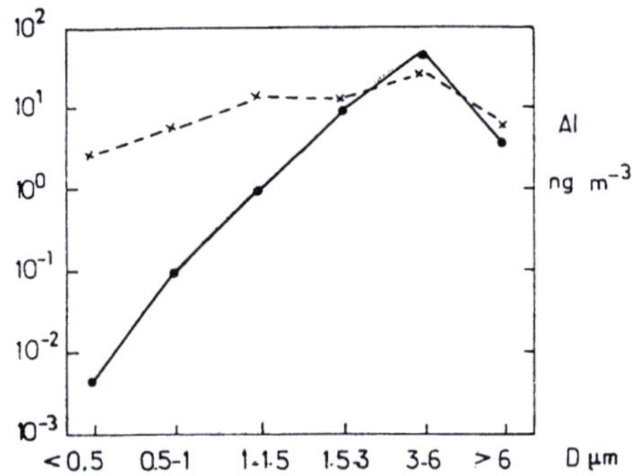

FIG. 2 – Aluminum concentration as a function of particle size in Enewetak air as found with a high-volume cascade impactor ---- (14) and as found by low volume bulk filtration and individual analysis of aluminosilicate particles by analytical scanning electron microscopy — (9).

Wet Deposition: Problems

We will consider first those elements that exist only in the particulate form in the marine atmosphere. On a global scale, wet removal rates can be assessed if we have representative rain concentration data and rain precipitation statistics over the ocean. This is not yet the case. At a given site, sampling over long time periods and integrating different meteorological regimes is also required. Great care has to be taken during collection to sample the entire rain episode. Indeed, sequential analysis of rains has shown that the concentration of trace materials is substantially greater during the early period of rainfall. This may be caused by partial evaporation of drops en route to the earth's surface through air not yet saturated, and/or by the first use of the largest condensation nuclei in the cloud (42).

In general, as described by Slinn (42), it is not easy to predict accurately the wet removal rates of elements by using washout factors (or scavenging ratios) and aerosols particle concentration data from a few meters above the surface of the ocean. An example of such a problem is given by recent results of the mineralogical study of aluminosilicate particles in Tropical Pacific air and rain (9). The mineralogy of particles in one air sample collected over a one-week sampling period is different from that from a one-hr rain sample collected during these three weeks. Whether this difference reflects different sampling time scales or different source regions for the dust in air and rain requires further investigation. It is interesting to note that at the same sampling site, lead-isotope measurements (41) do indicate that the lead in rain does not reflect directly the lead aerosol content of the lower marine troposphere. It is clear that data on the geochemistry of marine aerosols as a function of altitude are urgently needed if we want to solve such problems.

Another important question is that of the scavenging efficiency of aerosol particles by cloud and rain droplets as a function of aerosol particle size. Gatz (24) and Duce et al. (17) have computed the washout factor or scavenging ratio, W, for certain trace metals at urban and continental sites,

$$W = \rho_a C_r / C_a ,$$

where C_r is the concentration of metal in rain ($\mu g\ kg^{-1}$), C_a is the concentration of material in ground level air ($\mu g\ m^{-3}$), and ρ_a is the density of air.

Values for W ranged from a few hundred to a few thousand. Results from both studies indicate a tendency for W to increase with the mass-mean radius of the aerosol particles containing the element. If such results apply to the marine atmosphere, they would imply, as is the case for the dry deposition flux, that the relative contribution of the recycled component to elemental concentrations in rain might be much higher than the contribution of the recycled material to the marine aerosol. Again we will take lead as as example. Assume that 10% of the aerosol lead is recycled lead associated with large sea-salt particles (with a mean washout factor W = 2,000) and that 90% of aerosol lead is not recycled and is submicrometer in size (with a mean washout factor of two hundred). It follows that the percentage of recycled lead in rain will be 55% of the total lead in rain. Such a possibility suggests that we must be cautious in interpreting total, wet-removal fluxes as representing a net input to the ocean. Also, we must be careful when considering the use of some tracers to derive deposition rates to the ocean surface. For example, ^{210}Pb flux meaurements may to some extent be biased by the existence of a recycled component. If this isotope behaves like stable lead, an overestimation by a factor of two in net flux assessment due to wet deposition could result.

However, very recent data obtained over the Tropical North Pacific do not seem to indicate a particle-size dependence of W in the remote marine atmosphere. Indeed, at Enewetak Atoll, it has been found that W for Pb is slightly greater for continental dust (38), despite the fact that the mass-mean radius of Pb, about 0.2 μm, is one fourth that of dust, about 0.8μm (19). Data based on the analysis of individual aluminosilicate particles by scanning electron microscopy (9) also show no tendency for W to increase with particle size at that site. Such results suggest that in relatively clean remote areas (as opposed to polluted urban areas) clouds use essentially all available particles as cloud condensation nuclei, so that little or no particle-size dependence of W would be expected (42). In any event, these observations show again the need for additional studies, including measurements of the distribution of the aerosol with altitude. Thus, effects of size, particle chemistry, and other factors make it difficult to predict W more accurately than to within a factor of ten.

Despite these uncertainties, washout factors can be used to assess the relative importance of wet and dry removal of marine aerosols. For small particles, it seems clear that (in areas of intense precipitation) dry deposition fluxes are almost negligible compared to wet deposition. On the other hand, dry and wet deposition of large particles over the ocean appear, on the average, to be of comparable importance (37,42). From these considerations, and taking into account what we know about the marine versus nonmarine components of elemental concentrations as a function of particle size, we can infer that atmospheric net fluxes from the continents to the oceans will primarily be due to wet deposition in the marine atmosphere. This is particularly true for the anomalously enriched elements which are generally found on submicron sized aerosols. Dry deposition, on the other hand, is certainly significant for sea-derived aerosols, as well as for continental dust from soil erosion.

So far, we have only considered the case of conservative particulate elements in aerosols; i.e., those which do not undergo appreciable gas-to-particle conversion. It is interesting to consider the removal of those elements that have both a vapor-phase component and a particle component in the marine atmosphere. Recent calculations indicate that for Hg, the scavenging efficiency of the gas phase by rain is much smaller than the scavenging efficiency of the aerosol phase (21). The Hg content of marine rain can be explained entirely by the scavenging of the particulate phase, despite the fact that particulate Hg is at least one hundred times less abundant in the marine atmosphere than vapor-phase Hg. Therefore, the impact of the atmospheric input of anthropogenic mercury on the marine chemistry of that element is probably more dependent on the atmospheric cycle of particulate Hg than on the atmospheric cycle of vapor-phase Hg, although the two cycles may be interdependent.

These considerations of the present state-of-the art of the removal processes for aerosols may appear pessimistic. They show how difficult it is to make precise assessments of particulate fluxes to and from the ocean. However, with reasonable care and provided a good data base is used, our present knowledge does permit first approximations of such fluxes and their geochemical significance.

GLOBAL FLUXES OF METALS THROUGH THE ATMOSPHERE

Global fluxes of metals have been estimated in different ways: from emission source strengths, from removal rates by rain and dry deposition over land and ocean, and from concentration data and mean residence times for aerosols in the troposphere. As indicated previously, only first approximations are possible from such estimates. The correct magnitude of such estimates can be checked, however, if one looks for agreement among fluxes calculated in different ways.

An example of the state-of-the-art in this context is the new global atmospheric cycle of arsenic (47), summarized in Table 2. Source strengths and removal rates have been calculated separately for the Northern and the Southern Hemispheres. Based on the most reliable estimates, source strengths calculations indicate that 80% and 35% of the emissions are of anthropogenic origin in the Northern and Southern Hemisphere, respectively. The most significant anthropogenic sources appear to be volatilization from metal processing and

	NORTHERN HEMISPHERE	SOUTHERN HEMISPHERE
1 - Source strengths		
A - Natural sources (particles d<5 μm)		
Ocean		
Bubble bursting	0.012	0.016
Gas exchange	0.05	0.06
Earth's crust		
Particle weathering	0.16	0.077
Direct volatilization	0.0005	0.0002
Volcanoes	3.5	3.5
Forest wildfires	0.11	0.05
Terrestrial biosphere	0.17	0.09
Natural sources: total	4.0	3.8
B - Anthropogenic sources (particles d<5 μm)		
Copper production	12.	1.
Iron/steel production	3.8	0.4
Lead/zinc production	2.0	0.2
Arsenic agriculture	1.7	0.2
Other sources	2.1	0.2
Anthropogenic sources: total	21.6	2.0
Natural and anthropogenic sources: total	25.6	5.8
2 - Removal$^+$		
Over land	20.0	7.6
Over the ocean	2.2	0.56
Global removal	22.2	8.2

$^+$Calculated assuming the tropospheric burden of arsenic is removed 40 times per year.

TABLE 2 – The global atmospheric mass balance of arsenic, unit of 10^9 grams per year, adapted from Walsh et al. (47).

from agricultural chemical usage. The most important natural source may be volcanoes which also involve a volatilization process. The total arsenic emission from all of these volatilization processes is found to be about 120 times greater than primary particle sources, i.e., normal crustal weathering and sea-salt production. This estimate is in agreement with the order-of-magnitude arsenic enrichments relative to crustal or bulk seawater composition in remote aerosol particles. Walsh et al. (47) suggest that arsenic concentrations are predominantly industrial in the Northern Hemisphere and predominantly natural in the Southern Hemisphere.

Removal rates of arsenic by dry and wet deposition were calculated by Walsh et al. (47) from representative global surface concentration data in three ways: a) by assuming that tropospheric aerosols are washed out 40 times per year; b) by using total washout factors for arsenic from continental areas and regional precipitation volumes; and c) by using the mean dry deposition velocity of arsenic over England, assuming that over land the removal by rain is twice the removal by dry deposition. All three estimates have considerable uncertainties. Assumption a) implies an average residence time of aerosols of nine days. Since arsenic is primarily present on submicron aerosols, this value is probably correct within a factor of two, considering predicted residence times for submicron radioactive aerosols (29,37). Assumptions b) and c) are probably correct within a factor of 3–5 (see earlier in this section). Nevertheless, the three estimates of arsenic removal rates are all within a factor of two of each other and with the total arsenic source estimates. This mass balance model indicates that the ocean is a net sink for arsenic and that arsenic inputs in the Northern Hemisphere are predominantly anthropogenic. This geochemical conclusion is not in agreement with that reached by Lantzy and Mackenzie (28). This latter work only considered data on arsenic in rain compiled by Wedepohl (49). The large calculated output of arsenic from the atmosphere by precipitation could not be matched by known inputs, so that Lantzy and Mackenzie postulated a large source of arsenic by biological volatilization from the surface ocean. Although such a source has been postulated by Wood (52), its strength is probably largely overestimated by Lantzy and Mackenzie for the following reasons: a) Rain data used by Lantzy and Mackenzie were found to be much higher than recent data on rain samples from remote marine areas as reported by Andreae (1). Andreae's data are comparable to those predicted by Walsh et al. (47) using washout factors. b) Chemical speciation studies by Andreae (1) of arsenic in rain indicated that methylated forms of arsenic contributed a negligible fraction to the total concentration.

Order-of-magnitude estimates of global fluxes through the atmosphere are given for some metals in Table 3. This table is based on the computations of Weisel et al. (50). Sea-to-air fluxes take into account enrichment of these elements in sea-derived aerosols, as estimated from the results of the analysis of aerosol particles artificially produced from bursting bubbles. From this table, one can conclude that, except for the alkali and alkaline earth metals, the ocean is a sink for trace metals. Two major types of input dominate: mineral dust from soil erosion and man-made inputs. However, the ocean could be a significant source of V and perhaps Cd to the atmosphere.

Element	Natural Sources				Anthropogenic Sources	
	Ocean	Crust	Volcanoes	Vegetation	Fossil fuel	Other
	Mean					
Al	150	20,000	700	40	2,000	2,000
Cd	0.5	0.05	0.1	0.05	0.2	5
Cu	4.0	10	6	0.9	2	50
K	4,000	6,000	200	100	200	200
Mg	10,000	3,000	80	20	300	200
Fe	50	10,000	300	20	2,000	4,000
Mn	5	200	9	5	8	400
Pb	5	3	0.4	0.2	4	400
V	20	30	0.7	0.2	20	2
Zn	5	80	10	10	80	200

TABLE 3 — Source strengths (x 10^9 grams per year) for some trace metals on atmospheric particulate material according to Weisel et al. (50).

METAL FLUXES FROM THE ATMOSPHERE TO THE OCEAN: GEOGRAPHICAL VARIABILITY

The net atmospheric input, per unit area of the ocean surface, will depend on the remoteness from continental areas and on the initial load of continental air masses. Table 4 shows that the rate of net input of metals decreases by many orders of magnitude from coastal areas to the remote regions of the Tropical North Pacific. Flux values shown for the North Sea and the Tropical North Pacific are based on direct measurements. The other values given in this table are based on mean atmospheric concentration data. The fluxes have been derived assuming either that tropospheric aerosols are removed 40 times per year (15), or that the total deposition velocity of continental aerosols in the marine atmosphere is 1 cm s^{-1} (2) based on the ^{210}Pb total deposition data (45,46). One can note that for the North Atlantic, there is relatively good agreement between the values calculated by Duce et al. from concentration data obtained during 1973 and 1974 at Bermuda, and those calculated by Buat-Ménard and Chesselet from atmospheric samples collected during four cruises in the Tropical North Atlantic in 1974 and 1975.

For coastal and semi-remote areas, it is interesting to compare the relative importance of atmospheric and river inputs (and waste discharges), especially for those metals that are enriched in marine aerosols and may therefore have a polution origin. Such a comparison is presented for some metals in Table 5.

From this table, it appears that the atmospheric input of elements is comparable to, if not greater than, the estimated dissolved riverine input. But, such a comparison must be considered with caution, since the riverine flux is for soluble fraction only. Essentially most of the suspended sediments in rivers are at the present time trapped in estuaries. Atmospheric fluxes are for total particulates. Except for crustal elements, however, a significant percentage of the trace metals on atmospheric particles may dissolve upon entering seawater. The fate of the dissolved flux is complex, since recycling between dissolved phase and particulate mineral and organic phases will occur. Nevertheless, for elements such as Zn, Cd, and Pb, the atmospheric flux will effectively compete with the direct dissolved input from rivers and coastal discharges. It appears, therefore, that atmospheric deposition contributes significantly to marine pollution in coastal and semi-remote marine areas. The influence of atmospheric inputs will depend on vertical and horizontal mixing processes (solar, tidal, wind driven) as well as on biological and chemical recycling.

Element	New York Bight (20)	North Sea (13)	Western Mediterranean (2)	South Atlantic Bight (51)	Bermuda (15)	Tropical North Atlantic (8)	Tropical North Pacific (18)
Al	6,000	30,000	5,000	2,900	3,900	5,000	1,900
Sc	-	5	1	-	0.6	1.1	0.4
V	-	480	-	-	5 ?	1.7	2.8
Cr	-	210	49	-	9	14	6
Mn	-	920	-	60	45	70	18
Fe	5,700	25,500	5,100	5,900	3,000	3,200	1,300
Co	-	39	3.5	-	1.2	2.7	0.6
Ni*	-	260	-	390	3	20	-
Cu*	-	1,300	96	220	30	25	2
Zn*	1,400	8,950	1,080	750	75	130	13
As*	-	280	54	45	3	-	-
Se*	-	22	48	-	3	14	4.5
Ag*	-	-	3	-	-	0.9	~ 0.2
Cd*	30	43	13	9	4.5	5	~ 0.5
Sb*	-	58	48	-	1.0	3.5	0.1
Au*	-	-	0.05	-	-	0.1	-
Hg*	-	-	5	24	-	2.1	-
Pb*	3,900	2,650	1,050	660	100	310	7
Th	-	4	1.2	-	-	0.9	0.9

* designates elements generally enriched in marine aerosols.

TABLE 4 – Estimated mean fluxes of trace metals from the atmosphere to the sea surface (ng $cm^{-2}yr^{-1}$).

Element	South Atlantic Bight (51)	New York Bight (20)	North Sea (13)	Western Mediterranean Sea (2)
Arsenic	2.1	1.0	1.7	-
Cadmium	2.7	3.1	1.1	-
Copper	1.9	-	1.9	-
Iron	5.8	6.4	1.7	-
Manganese	0.6	-	0.8	-
Mercury	22	-	2.1	0.8
Nickel	1.7	-	1.3	-
Lead	9.5	20	6.8	6.2
Zinc	2.3	3.1	1.9	0.8

TABLE 5 – Ratio of atmospheric flux to riverine flux (dissolved) of some trace elements in coastal and semi-remote oceanic areas.

Metallic pollutants present in coastal waters (from either direct injection in the water or from anthropogenic inputs from the atmosphere) can be recycled back to land and can thus contribute to coastal atmospheric pollution under sea breeze conditions. The proper assessments of such contributions should be made in the near future. The most convincing evidence of the magnitude of this recycling process for trace metals comes from results of atmospheric measurements of radionuclides previously discharged to sea (12,23,33). In coastal areas, close to the nuclear plants at La Hague, France, and Windscale, UK, elevated concentrations of radioactive plutonium, americium, antimony, cesium, ruthenium, and cerium have been found in aerosols, dry and wet deposition samples, grasses, and lichens. These excess concentrations decrease inland with increasing distance from the coast. The radionuclides most likely come with sea spray, probably associated with the largest particles. Therefore, the fraction of sea-derived aerosols deposited on land, mainly in coastal regions, may contribute significantly to gross river inputs, not only for sea-salt, but also for the enriched elements and other substances in such aerosols. This may explain why observed fluxes of some heavy metals transported by rivers are higher than theoretical predictions derived from rock weathering (32).

ATMOSPHERIC METAL FLUXES TO THE OPEN OCEAN: GEOCHEMICAL IMPLICATIONS

If we assume that enriched elements in marine aerosols are predominantly of anthropogenic origin, it can be seen from Table 4 that the North Atlantic waters are potentially more influenced by pollution than those of the Tropical North Pacific. Up to now, few data are available for the Southern Hemisphere with the exception for Pb (41). We would expect that atmospheric net fluxes of pollutant elements should be, on the average, up to one order of magnitude lower in the Southern than in the Northern Hemisphere. The question now is: can we get an insight into the importance of the atmospheric input as a net source for these elements in the ocean?

One approach is to search for an effect on the dissolved elemental concentrations in seawater. Such an effect has been unequivocally shown for Pb. The reasons for this are found in the very distinctive features of the Pb cycle in the atmosphere and in the ocean at the present time (10,22,40,41). Lead has a very fast cycle in the ocean, since its residence time is estimated to be about two years in surface waters and only about one hundred years in the deep ocean. At present, the atmospheric lead burden (at least in the Northern Hemisphere) is entirely dominated by industrial inputs, which have increased dramatically in the last two hundred years. The present lead cycle in the ocean is therefore not at steady state with respect to atmospheric inputs. Since the residence time of lead in the ocean is about 10 times smaller than the turnover time of the deep waters, different vertical distributions of lead can be expected from one oceanic region to another, depending on the local atmospheric source strength. This is illustrated by the data presented in Fig. 3. It can be seen that lead concentrations in the surface waters are always higher than in deep waters, which illustrates the importance of the present atmospheric input rates. Also, the magnitude of lead concentrations in the waters are related to the level of the atmospheric input rate. Recent data reported by Byrd and Andreae (11) indicate that the vertical distribution of tin in the northwestern Atlantic is somewhat similar to that of lead and would also reflect the fact that the present fluxes of tin through the atmosphere and the ocean are dominated by anthropogenic effects.

A second approach to the assessment of the importance of the atmosphere as a source of metals in the ocean has been to compare the atmospheric input with the inputs of metals to the ocean from all sources. Total inputs have been estimated on the basis of the known elemental composition of deep-sea sediments and the measured sediment accumulation rates. Such a comparison has been made for the North Atlantic (15,37) and the Tropical North Pacific (18). Over the North Pacific, the total deposition of many airborne trace metals (e.g., Al, Fe, V, Sc, Cr, Cs, Th, Zn, and Cu) to the ocean was within a factor of two to three of their estimated rates of accumulation in marine sediments, suggesting that atmospheric transport plays a significant role in supplying these elements to the ocean. The atmospheric deposition of other elements (e.g., Pb, Zn, Cd,

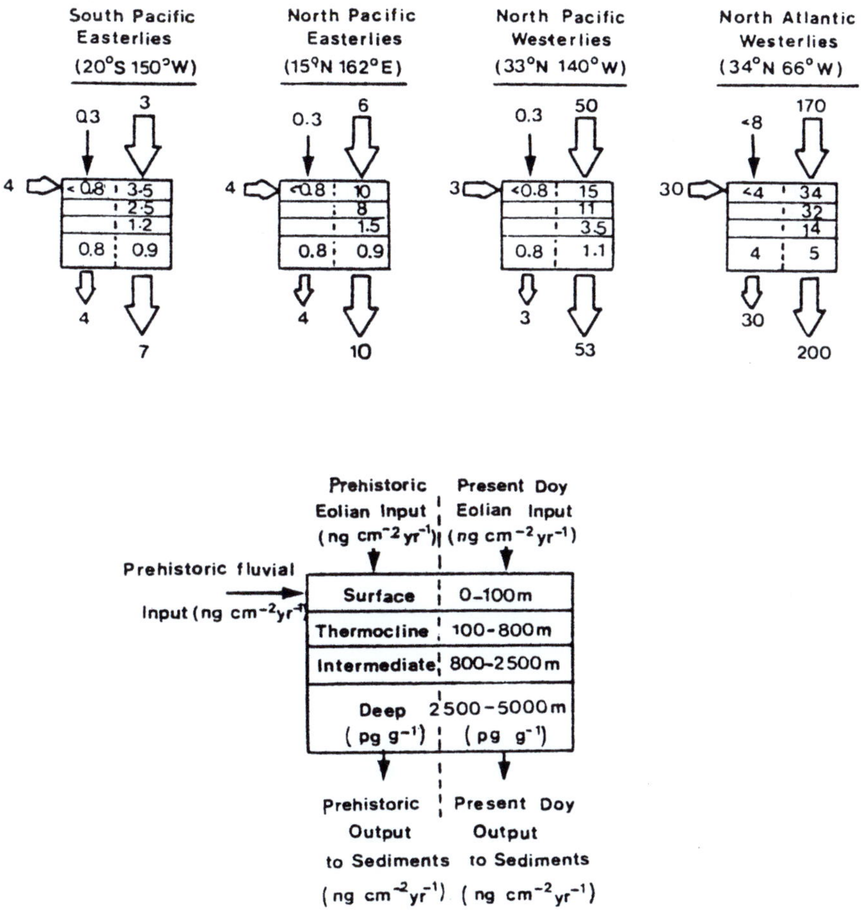

FIG. 3 – Relationships among prehistoric lead and present-day eolian inputs, vertical concentration profiles and sediment outputs of lead on a global scale (Patterson, personal communication; data can be found in (22) and (41)).

Se, and Sb) was considerably greater than their sediment accumulation rates, suggesting that either these elements have been recycled from the ocean surface so that their deposition does not represent a net input to the ocean, or that the present-day atmospheric input of these elements is elevated considerably above the mean for the past 10–15,000 years. It is of interest that the latter elements are those which are enriched in marine aerosols and may therefore have a pollution origin.

Such an approach unfortunately does not take into account the transport processes of metals through the oceanic water column. Consider first the atmospheric flux of crustal material (quartz and aluminosilicate minerals) which is essentially insoluble in seawater. Most of the vertical particulate flux of terrigenous dust in the ocean is accelerated because of the attachment of these small dust particles with fast sinking, large biogenic particles which reach the ocean floor in a few weeks at most (6,18,28). This would mean that atmospheric dusts are deposited within months in deep-sea sediments and undergo little horizontal transport (perhaps less than 100 km). However, in some oceanic areas, resuspension and relatively fast (a few cm s^{-1}) horizontal transport of recently deposited sediments does occur. This means that in such areas the comparison between the atmospheric dust flux and the sedimented flux is not valid. This has been shown likely to be the case in the Tropical North Atlantic, west of the Mid-Atlantic Ridge (5), based on measurements of rubidium and strontium isotopes in clay minerals.

Consider now the fate of atmospheric inputs of metals that may dissolve partially or totally upon entering seawater. Examples would include most metallic pollutants (e.g., Hg, Cd, Pb, As, Se, Cu, Zn, etc.) which are known to be potentially toxic to marine life. Their cycle in the ocean (Fig. 4) is closely linked to the biological cycle which involves primarily active or passive scavenging of metals by particles (6,8,44). In surface waters where most of the biological activity takes place, metals can be, like nutrients, recycled many times between the dissolved phase and the particulate phase. At depths greater than about one hundred meters, biogenic detritus and other particles are less frequently recycled so that they leave the surface water compartment by sedimentation. In deep waters and at the sediment seawater interface, the biogenic particulate flux is known to be primarily recycled back to the water column. As is the case for nutrients, many metals can be remobilized in the dissolved form. They may therefore cycle one or several times in the ocean before being finally buried in sediments. At each cycle (~ 1,000 yrs), only a small fraction of the vertical particulate flux from surface waters would be trapped in sediments, as suggested by studies of bomb-produced metallic radionuclides (27). This, added to the fact that the present rate of the net atmospheric input of "enriched" metals may represent a recent phenomenon, precludes drawing any meaningful geochemical conclusion from a comparison between the

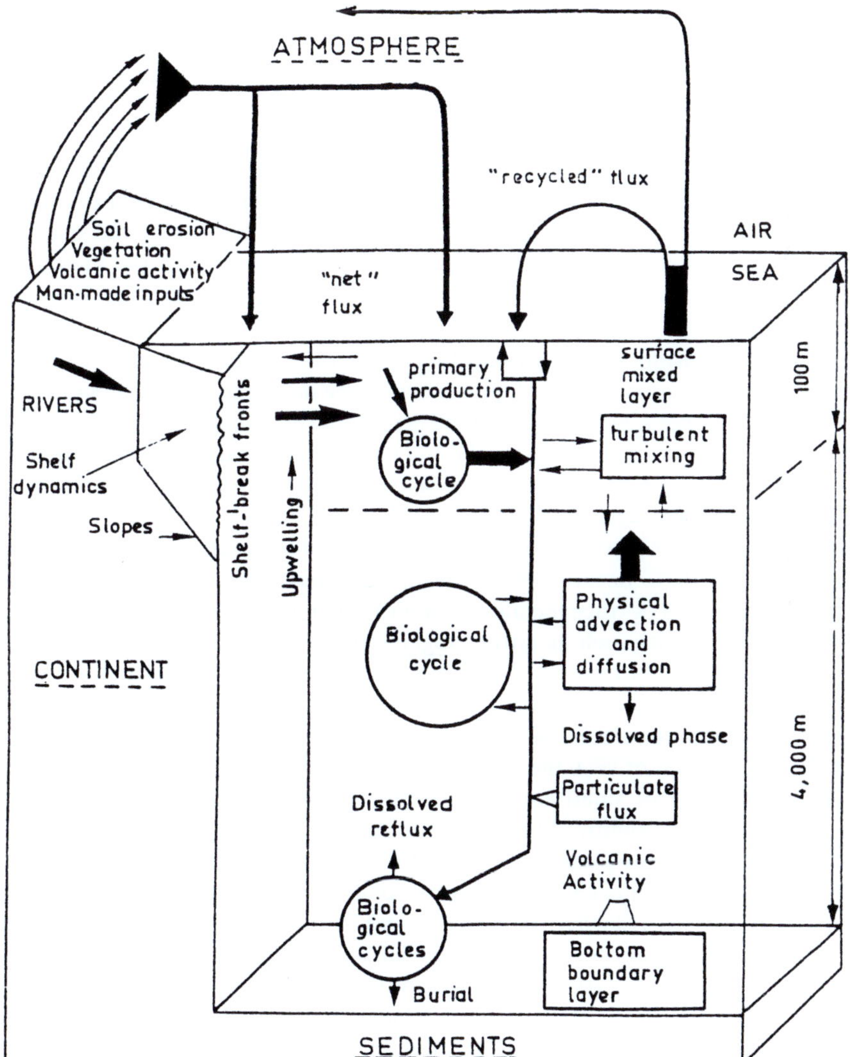

FIG. 4 – Schematic diagram of major ocean processes for reactive elements (from Buat-Ménard (6)).

atmospheric deposition and the rate of accumulation of these elements in deep-sea sediments. One may infer, however, that if the present rate of the atmospheric input of enriched metals to the ocean is higher than the average clay- sedimentation rate for these elements (as is the case for Cd, Zn, Se, and Pb in the North Atlantic (37)), then this is evidence of the importance of the atmosphere as a source of these elements for the ocean. The restriction here is that deep-sea clays are not the only sink for trace elements in the ocean. Major sinks for nutrient elements are probably to be found in continental rise and shelf areas. As stated by Turekian (44), elements involved in the biological cycle of the ocean are sphinx-like in their silence about their origin and depositories.

From these considerations, it seems that a better approach to assess the importance of current atmospheric deposition of trace metals in the open ocean waters is to consider a mass balance between input from the atmosphere and removal from the surface layers by sinking particles. Indeed, since the residence time of such elements in the surface-water compartment (a few months to a few years) is considerably shorter than that of the water itself (about twenty years), their major removal process is particle controlled rather than due to mixing processes (6,8). Because of this fast geochemical coupling between the atmospheric inputs of trace metals and their transport to the deep ocean, recent atmospheric inputs such as anthropogenic inputs are partially or totally recorded in the flux associated with particles sinking from surface waters. Therefore, induced chemical changes in the deep water column will be first observed in that particulate flux. Figure 5 presents, for some trace metals, the relative contribution of (F_a), the net atmospheric flux to (F_s), the flux associated with the settling of small and large particles from the surface layers of the Tropical North Atlantic. Since we can assume a steady state between the atmospheric input (F_a) and the removal flux (F_s), the ratio between these two fluxes provides an insight into the atmospheric source function to ocean waters. As discussed elsewhere (8), both F_a and F_s are calculated values based on a number of assumptions which have significant uncertainties. Nevertheless, it appears from this figure that present atmospheric inputs to the North Atlantic represent, depending on the trace element being considered, between 5% and 100% of total inputs to surface waters. Thus atmospheric deposition of trace metals into the ocean is a significant source in the North Atlantic.

Since the geochemical conclusions which can be drawn from the information in this figure have been discussed extensively (6,8), we will consider here only those elements which are anomalously enriched in the marine aerosol (Ni, Cu, Zn, Se, Sb, Au, Hg, Pb). Among these elements only Se seems to come mostly

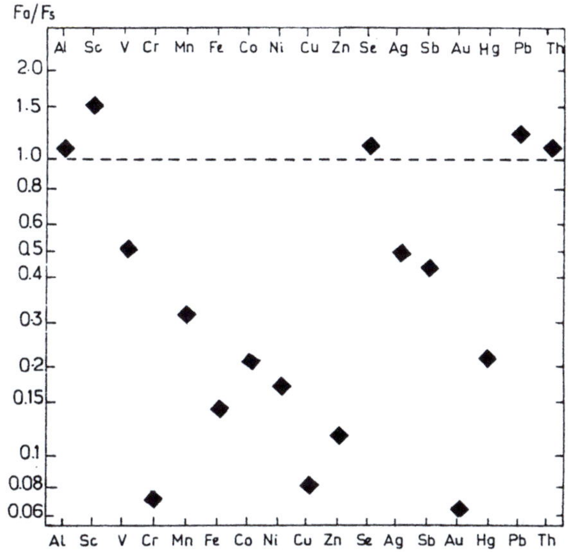

FIG. 5 – Ratio of flux from atmosphere, Fa, to downward particulate flux from surface layers, Fs, in the Tropical North Atlantic. Adapted from Buat-Ménard and Chesselet (8).

from natural sources such as vegetation, volcanic activity (7), or the ocean surface (35). If volcanic activity is the dominant source, Fig. 5 would imply that the atmospheric input of Se to the ocean is the major natural source of this element to ocean surface waters. With a marine source, on the other hand, the atmospheric input of Se to the ocean would simply be a recycled flux.

For those airborne trace metals which can be derived from anthropogenic sources (notably Ni, Cu, Zn, Sb, Ag, Au, Hg, Pb), Fig. 5 indicates that the anthropogenic flux is generally smaller than natural fluxes in the ocean, with the exception of Pb for which the particulate flux in the ocean is at present primarily anthropogenic. However, one can speculate that if in the future, atmospheric deposition of Ni, Cu, Zn, Sb, Ag, Au, Hg, is to be one order of magnitude higher than the present rate over the North Atlantic, then detectable effects may be recorded in oceanic suspended matter and conceivably will alter the oceanic cycle of such elements. Such effects are nevertheless extremely difficult to predict at present. Indeed, changes in the rate of input from the atmosphere will be cycled biologically at the same rate only if such changes do not affect the biological cycle itself. For example, increased levels of toxic metals in surface waters could reduce the biological activity. However, if at the same time increased nutrient availability were to occur, it would counteract the first effect. Although we know that appreciable changes in availability of nutrients by large-scale changes in water circulation and trace element fluxes have occurred in the past, the sedimentary record has not yet been able to tell us precisely how nature takes care of such stresses.

CONCLUDING REMARKS

It seems that our understanding of the cycling of metals through the marine atmosphere has progressed significantly during the last few years. Sources and sinks and transport processes can now be investigated on regional scales. This is necessary because global scale models can only give a crude idea of the various source and sink functions, mean residence times, etc. Residence times may vary geographically and temporally, depending, for example, on the amount of precipitation.

In many cases, our knowledge is still restricted to the Northern Hemisphere. Obviously, reliable data are needed for the Southern Hemisphere. At the tropospheric level, interhemispheric mixing time is long (about six months) compared to the residence time of aerosol particles of a few days. Also, continental and anthropogenic sources are less important in the Southern Hemisphere than in the Northern Hemisphere. Therefore, the Southern Hemisphere appears to be the best place for natural background assessments. Furthermore, most of our knowledge is based on data obtained either from remote islands or from ships. There is a lack of concentration data as a function of altitude, especially below and above clouds. Such data are needed for the understanding of cloud and rain chemistry.

REFERENCES

(1) Andreae, M.O. 1980. Arsenic in rain and the atmospheric mass-balance of arsenic. J. Geophys. Res. **85:** 4513–4518.

(2) Arnold, M.; Seghaier, A.; Martin, D.; Buat-Ménard, P.; and Chesselet, R. 1982. Géochimie de l'aérosol marin au-dessus de la Méditerranée Occidentale. Workshop on Pollution of the Mediterranean, Cannes, December 2–4. Monaco: C.I.E.S.M., in press.

(3) Beauford, W.J:; Barlser, J.; and Barringer, A.R. 1977. Release of particles containing metals from vegetation into the atmosphere. Science **195:** 571–573.

(4) Bertine, K.K., and Goldbert, E.D. 1971. Fossil fuel combustion and the major sedimentary cycle. Science **173:** 233–235.

(5) Biscaye, P.E.; Chesselet, R.; and Prospero, J.M. 1974. Rb-Sr, ^{87}Sr / ^{86}Sr isotope system as an index of provenance of continental dust in the open Atlantic Ocean. J. Rech. Atmos. **8:** 819–829.

(6) Buat-Ménard, P. 1983. Particle geochemistry in the atmosphere and oceans. **In** Air-Sea Exchange of Gases and Particles, eds. P.S. Liss and W.G.N. Slinn, Ch. 10. Dordrecht, Holland: Reidel, in press.

(7) Buat-Ménard, P., and Arnold, M. 1978. The heavy metal chemistry of atmospheric particulate matter emitted by Mount Etna volcano. Geophys. Res. Lett. **5:** 245–248.

(8) Buat-Ménard, P., and Chesselet, R. 1979. Variable influence of the atmospheric flux on the trace metal chemistry of oceanic suspended matter. Earth Planet. Sci. Lett. **42:** 399–411.

(9) Buat-Ménard, P.; Ezat, U.; and Gaudichet, A. 1983. Size distribution and mineralogy of auminosilicate dust particles in Tropical Pacific air and rain. **In** Proceedings of the 4th International Conference on Precipitation Scavenging, Dry Deposition and Resuspension (1982). Amsterdam: Elsevier, in press.

(10) Burnett, M.W., and Patterson, C.C. 1980. Perturbation of natural lead transport in nutrient calcium pathways of marine ecosystems by industrial lead. **In** Isotope Marine Chemistry, eds. E. Goldberg, Y. Horibe, and K. Saruhashi, pp. 413–438. Tokyo: Uchida Rokakuo Publishing Co.

(11) Byrd, J.T., and Andreae, M.O. 1982. Tin and methyltin species in seawater: concentrations and fluxes. Science **218:** 565–569.

(12) Cambray, R.S., and Eakins, J.D. 1982. Pu, ^{241}Am and ^{137}Cs in soil in West Cumbria and a maritime effect. Nature **300:** 46–48.

(13) Cambray, R.S.; Jefferies, D.F.; and Topping, G. 1975. An estimate of the input of atmospheric trace elements into the North Sea and the cycle Sea (72–73). U.K. A.E.A., Harwell Report 7733.

(14) Duce, R.A.; Arimoto, R.; Ray, B.J.; Unni, C.K.; and Harder, P.J. 1982. Trace elements in marine aerosols from Enewetak Atoll: I. Concentrations, Sources, and Temporal Variability. J. Geophys. Res., in press.

(15) Duce, R.A.; Hoffman, G.L.; Ray, B.J.; Fletcher, I.S.; Wallace, G.T.; Fasching, J.L.; Piotrowicz, S.R.; Walsh, P.R.; Hoffman, E.J.; Miller, J.M.; and Heffter, J.L. 1976. Trace metals in the marine atmosphere: sources and fluxes. **In** Marine Pollutant Transfer, eds. H. Windom and R. Duce, pp. 77–119. Lexington, MA: D.C. Heath and Co.

(16) Duce, R.A.; Hoffman, G.L.; and Zoller, W.H. 1975. Atmospheric trace metals at remote northern and southern hemisphere sites: pollution or natural? Science **197:** 551–557.

(17) Duce, R.A.; Unni, C.K.; Harder, P.J.; Ray, B.J.; Patterson, C.C.; Settle, D.M.; and Fitzgerald, W.F. 1979. Wet and dry deposition of trace metals and halogens in the marine environment. Abstract. CACGP Symposium, Boulder, Colorado.

(18) Duce, R.A.; Unni, C.K.; Ray, B.J.; and Arimoto, R. 1982. Atmospheric deposition of trace metals to the Tropical North Pacific. EOS **63:** 987.

(19) Duce, R.A.; Unni, C.K.; Ray, B.J.; Prospero, J.M.; and Merrill, J.T. 1980. Long-range atmospheric transport of soil dust from Asia to the Tropical North Pacific: temporal variability. Science **209:** 1522–1524.

(20) Duce, R.A.; Wallace, G.T.; and Ray, B.J. 1976. Atmospheric trace metals over the New York Bight. NOAA Technical Report ERL 361–MESA4.

(21) Fitzgerald, W.F.; Gill, G.A.; and Hewitt, A.D. 1982. Air-sea exchange of mercury. **In** Trace Metals in Seawater, eds. C.S. Wong, J.D. Burton, E. Boyle, K. Bruland, and E.D. Goldberg. New York: Plenum Press, in press.

(22) Flegal, A.R., and Patterson, C.C. 1983. Vertical profiles of lead in the central Pacific. Earth. Planet. Sci. Lett., in press.

(23) Fraizier, A.; Masson, M.; and Guary, J.C. 1977. Recherches preliminaires sur le rôle des aerosols dans le transfert de certains radio-éléments du milieu marin en milieu terrestre. J. Rech. Atmos. **11:** 49–60.

(24) Gatz, D.F. 1977. Scavenging ratio measurements in Metromex. **In** Precipitation Scavenging (1974), Conference 741003, pp. 71–87. Spring Field, VA: N.T.I.S.

(25) Goldberg, E.D. 1976. Rock volatility and aerosol composition. Nature **260:** 128–129.

(26) Goldberg, E.D. 1976. The Health of the Oceans. Paris: The UNESCO Press.

(27) Labeyrie, L.D.; Livingston, H.D.; and Bowen, V.T. 1976. Comparison of the distribution in marine sediments of the fall-out derived nuclides ^{55}Fe and $^{329,\,240}Po$. **In** Transuranium Nuclides in the Environment, pp. 121–137. Vienna: I.A.E.A.

(28) Lambert, C.E.; Jehanno, C.; Silverberg, J.S.; Brun-Cottan, J.C.; and Chesselet, R. 1981. Log-normal distributions of suspended particles in the open ocean. J. Mar. Res. **39:** 77–98.

(29) Lambert, G.; Ardouin, B.; and Polian, G. 1982. Volcanic output of long-lived radon daughters. J. Geophys. Res., in press.

(30) Lantzy, R.J., and Mackenzie, F.T. 1979. Atmospheric trace metals: global cycles and assessment of man's impact. Geochim. Cosmochim. Acta **43:** 511–525.

(31) Li, Y.H. 1981. Geochemical cycles of elements and human perturbation. Geochim. Cosmochim. Acta **45:** 2073–2085.

(32) Martin, J.M., and Meybeck, M. 1979. Elemental mass-balance of material carried by major world rivers. Mar. Chem. **7:** 173–206.

(33) Martin, J.M.; Thomas, A.J.; and Jeandel, C. 1981. Transport atmosphérique des radionucleides artificiels de la mer vers le continent. Oceanolog. Acta **4:** 263–266.

(34) McDonald, R.L.; Unni, C.K.; and Duce, R. A. 1982. Estimation of atmospheric sea-salt dry deposition: wind speed and particle size dependence. J. Geophys. Res. **87:** 1246–1250.

(35) Mosher, B.W., and Duce, R.A. 1982. Vapor phase and particulate selenium in the marine atmosphere. J. Geophys. Res., in press.

(36) Mroz, E.J., and Zoller, W.H. 1975. Composition of atmospheric particulate matter from the eruption of Heimaey, Iceland. Science **190:** 461–464.

(37) National Academy of Sciences. 1978. The Tropospheric Transport of Pollutants and Other Substances to the Oceans. Washington, D.C.: Printing and Publishing Office, National Academy of Sciences.

(38) Ng, A., and Patterson, C.C. 1981. Natural concentrations of lead in ancient Arctic and Antarctic ice. Geochim. Cosmochim. Acta **45:** 2109–2121.

(39) Rahn, K.A. 1976. The Chemical Composition of the Atmospheric Aerosol. Technical Report, Graduate School of Oceanography, University of Rhode Island.

(40) Settle, D.M., and Patterson, C.C. 1980. Lead in Albacore: a guide to lead pollution in America. Science **207:** 1167–1176.

(41) Settle, D.M., and Patterson, C.C. 1982. Magnitude and sources of precipitations and dry deposition fluxes of industrial and natural leads to the North Pacific at Enewetak. J. Geophys. Res. **87:** 8857–8869.

(42) Slinn, W.G.N. 1983. Wet and dry deposition of aerosol particles. **In** Air-Sea Exchange of Gases and Particles, eds. P.S. Liss and W.G.N. Slinn, Ch. 8. Dordrecht, Holland: Reidel, in press.

(43) Taylor, S.R. 1964. Abundance of chemical elements in the continental crust: a new table. Geochim. Cosmochim. Acta **28:** 1273–1285.

(44) Turekian, K.K. 1977. The fate of metals in the oceans. Geochim. Cosmochim. Acta **41:** 1139–1144.

(45) Turekian, K.K., and Cochran, J.K. 1981. ^{210}Pb in surface air at Enewetak and the Asian dust flux to the Pacific. Nature **292:** 522–524. Corrigenda. Nature **294:** 670.

(46) Turekian, K.K.; Nozaki, Y.; and Benninger, L.K. 1977. Geochemistry of atmospheric radon and radon products. Ann. Rev. Earth Planet. Sci. **5:** 227–255.

(47) Walsh, P.R.; Duce, R.A.; and Fasching, J.L. 1979. Considerations of the enrichment, sources and flux of arsenic in the troposphere. J. Geophys. Res. **84:** 1719–1726.

(48) Walsh, P.R.; Rahn, K.A.; and Duce, R.A. 1978. Erroneous elemental mass-size fractions from a high-volume cascade impactor. Atmos. Env. **12:** 1493–1495.

(49) Wedepohl, K.H., ed. 1969. Handbook of Geochemistry, vol. II / 3. Berlin: Springer-Verlag.

(50) Weisel, C.P.; Duce, R.A.; Fasching, J.L.; and Heaton, R.W. 1983. Estimates of the transport of trace metals from the ocean to the atmosphere. J. Geophys. Res., in press.

(51) Windom, H.L. 1981. Comparison of atmospheric and riverine transport of trace elements to the continental shelf environment. **In** River Inputs to Ocean Systems, pp. 360–369. New York: U.N.E.P. and U.N.E.S.C.O.

(52) Wood, J.M. 1974. Biological cycles for toxic elements in the environment. Science **183:** 1049–1052.

(53) Zoller, W.H.; Gladney, E.S.; and Duce, R.A. 1974. Atmospheric concentrations and sources of trace metals at the South Pole. Science **183:** 198–200.

Changing Metal Cycles and Human Health, ed. J.O. Nriagu, pp. 71–94.
Dahlem Konferenzen 1984. Berlin, Heidelberg, New York, Tokyo: Springer-Verlag.

METAL POLLUTION OF TERRESTRIAL WATERS

U. Förstner
Technische Universität Hamburg-Harburg
2100 Hamburg 90, F.R. Germany

Abstract. Metal pollution in surface waters, which caused considerable public concern during the 1970s, seems to have peaked in critical examples (e.g., in the Rhine River), partly as result of recent improvements in wastewater treatment. Problems with metal pollution, however, still exist both on a local and regional scale: Waters of large areas in southern Scandinavia and eastern North America are severely endangered by the combined effects of airborne pollutants and acidity. Increasing quantities of dredged materials represent long-term reservoirs for the release of metals possibly involving detrimental effects on the quality of both groundwater and agricultural products. During the last decade, the major objectives of research on metal-polluted waters have changed from the initial surveys of sources and pathways to more detailed investigations of the mechanisms controlling the mobility and bioavailability of different metal species.

METAL CONCENTRATIONS IN WATER SAMPLES

Due to the variety of rock formations and the fluctuations in water discharge, it is even more difficult to establish natural background values for metals in inland waters than in seawater (26). As a result of the findings of Patterson et al. (51) on lead values from oceanic samples, it has furthermore been argued that the same contamination effects associated with marine sampling and analysis also exist in fresh and estuarine waters, and "it is safe to assume that many of the natural water values quoted in the literature are high solely for this reason (45)." Another problem is the effect of colloidal metal species in water samples, and many difficulties are associated with the discrimination of dissolved and particulate metal concentrations by filtration, centrifugation, etc. (36). Clearly, just relying on 0.45 micron physical filters cannot ensure that all microbiological and inorganic colloidal materials are removed. Centrifugation can impose similar difficulties if indigenous organisms in natural water samples are disrupted to "leak" out bioconcentrated metals into the bulk water sample. Finally, it is also clear that preservation of water samples by quick-freezing can cause cellular disruption also leading to metal "leakage," whereas not stabilizing or sterilizing samples can permit continued microbial transformations of analyte metals to occur.

For a general overview, metal data from less contaminated and polluted water samples are given in Table 1. Compared with estimated background values, all metals — except arsenic — are enriched by one to two orders of magnitude in the rivers Rhine and Ruhr. The data for the Rhine River show a significant decrease in the dissolved Cd and Pb concentrations between 1973 and 1978; these changes as well as the different metal sources in the Ruhr River catchment will be discussed later. It should be noted here that much higher concentration factors for dissolved metals have been analyzed in inland waters affected by acidic mine effluents; levels of dissolved Ni, Mn, and Co may exceed the background values for surface waters by a factor of more than 10,000; values of zinc, chromium, lead, arsenic, and cadmium are often enhanced by factors of 100–1,000 in the waters compared to "normal" surface waters (23).

Metal	Background data (ref)		Mississippi R. (68)	L. Michigan (16,19)	Rhine River 1973 (18)**	Rhine River 1978 (18)**	Ruhr River (17)***
As	1	(16)	3	1	0.7	1.2	5.9
Cd	0.02	(58)	0.1	0.3	1.0	0.2	2.3
Cr	0.5	(68)	0.5	1.7	2	2	20
Cu	1	(5)	2	5	15	10	44
Hg	0.01	(26)	0.1	0.03	0.2	0.2	0.07
Ni	0.3	(29)	1.5	3	8	10	62
Pb	0.05	(61)	0.2	1.5	15	3	17
Zn	0.5-4	(37,61)	10	16	65	50	226

* Different filtration techniques and filter pore sizes are used
** Station Cologne
*** Station Wetter

TABLE 1 — Dissolved metal concentrations (μg/l) in inland waters.*

Lowering of pH-values on a regional scale by acid precipitation is probably the most important single factor affecting metal mobility in surface waters, particularly for Al, Cd, Mn, and Zn (Table 2). Acidity of surface waters imposes problems in all aspects of metal enrichment, ranging from the toxification of drinking water to problems concerning the growth and reproduction of aquatic organisms, the increased leaching of nutrients from the soil and the ensuing reduction of soil fertility, the increased availability and toxicity of metals, and finally to the undesirable acceleration of Hg methylation in sediments (20).

DISSOLVED VS. PARTICULATE METAL CONCENTRATION

It has been suggested that the source of trace metals in aquatic systems significantly determines their distribution ratio between the aqueous and solid phases. For example, the bulk of detrital trace-element particulates never leaves the solid phase from initial weathering to ultimate deposition. Similarly, metal dust particles (e.g., from smelters) and effluents containing heavy metals associated with inorganic and inorganic matter undergo little or no change after being discharged into a river. In natural systems, such as in the Amazon River and to a large extent also in the Mississippi River, more than 90% of the metal load is transported by particulate matter (Table 3). A similar

	Al	Cd	Mn	Zn
Lakes with pH 6.0 - 7.8	<50	<0.1	<100	<30
Lakes with pH 4.1 - 5.3	up to 600	0.6	400	120

TABLE 2 — Metal concentrations in acidified and nonacidified lakes in Scandinavia and North America (32). Data in μg/l.

	Amazon R. (29)	Mississippi River (68)	Polluted Rivers in U.S.A. (40)	Polluted Rivers in F.R.G. (34)
Cadmium	-	88.9%	-	30%
Zinc	-	90.1%	40%	45%
Manganese	83%	98.5%	-	8 - 97%
Chromium	83%	98.5%	76%	72%
Copper	93%	91.6%	63%	55%
Lead	-	99.2%	84%	79%
Iron	99.4%	99.9%	98%	98%

TABLE 3 — Particulate-bound metal as percentage of total discharge.

sequence of the ratios between particulate and dissolved heavy metals has also been found for polluted systems; typically, however, the dissolved fractions in polluted waters are significantly higher than in less polluted systems, particularly for metals such as Cd, Zn, and Cu. This can possibly be attributed to the larger portion of dissolved or organically complexed metals from anthropogenic sources.

Geographical-climatical Effects

The geographical variability of elements in the suspended matter differs widely (46). The most constant elements are the rare earths, Si and Th. Al, Fe, K, Ga, Hf, Sc, Ta, Ti, and U do not vary much. On the other hand, As, Cr, Mg, and Na vary moderately, and Ca, Cs, Cu, Li, Mo, Ni, Pb, Sr, and Zn are highly variable. The more variable elements are those either carried mainly in solution (Ca, Na, Mg, As, Sr, Li), or significantly enriched in the suspended material, for example, by anthropogenic influences (Zn, Pb, Cu). There appear to be two different types of rivers according to the intensity of the weathering action (46): In tropical rivers the contents of Al, Fe, and Ti in the suspended matter are the highest, while contents of Ca and Na are the lowest; for this type of river the particulates mainly originate from soil material enriched in insoluble rather than the more soluble elements, as these have been leached away. On the other hand, contents of Al, Fe, and Ti in temperate and arctic rivers are much lower; the suspended load of these types of rivers is mainly derived from rock debris or poorly weathered particles.

Seasonal Variations

Studies by Grimshaw et al. (31) on the River Ystwyth in mid-Wales, where strong metal pollution from past mine operations is still obvious, indicate that metal concentrations in solution are highest during low flow periods. This suggests a dilution effect which has been found by many other investigators, in particular in less polluted rivers, but also in polluted ones. For brief periods during initial stages of storm runoff, there is a very significant increase in the metal concentrations in solution, apparently due to a flushing effect.

Studies on suspended sediments from the middle Rhine River (59) reveal a clear dependency of Cd concentrations on river water flow (Fig. 1). However, even the moderate cadmium contents in the winter period exceed the maximum values for the spring/summer, despite the much higher water discharge. (None of the cadmium values, from either the summer or the winter period, fell below approximately 4 mg Cd/kg, which is more than ten times greater than the tentative natural background concentration of cadmium in the

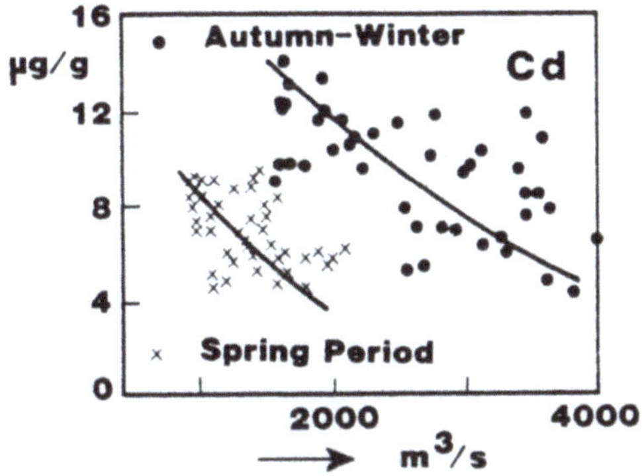

FIG. 1 — Water discharge vs. cadmium concentration in suspended sediments of the Rhine River near Koblenz (59) (from (55)).

suspended solids of the Rhine River system.) Such a development is due to the suspended sediments, rich in cadmium, being held back in periods of poor flow, for example, mainly during the summer, in the lock-regulated Neckar and Main rivers. In autumn/winter the tendency is reversed, the material is carried by high water flows into the Rhine River in increased quantities.

Data on the temporal variability of metal transport from the Susquehanna River to the Chesapeake Bay (12) show that the trace metal concentrations correlate well with the amount of solids discharged. When data are calculated for weight concentrations of metals in the solid fraction, it is found that all metals generally peak during December and January, and secondary peaks occur for Co, Cr, Ni, Cu, and Mn in July. Troup and Bricker (69) suggest that this effect is due to decaying organic matter which is abundant in the Susquehanna River during these two periods. Examples on the seasonal effect of planktonic material to adsorb metals from solution have been described from Bristol Channel, U.K. (1) and Monterey Bay, California (39). Investigations of Baccini (3) on the temporal variations in copper distribution in biomass, allochthonous particulates, and dissolved phase of the mesotrophic Lake of the Four Cantons in Switzerland reveal that the "biomass" portion of the total particulate phase for Cu increases during periods of high organic production. It is suggested that Cu (and Zn) introduced in soluble form is transferred into the particulate matter by plankton and that the Cu from decomposed sedimentary plankton is partially returned to the solution, whereas the allochthonous particles are deposited on the lake bottom relatively unchanged. From measurements of the seasonal effects of Cu, Zn, Cd, Pb, and Hg on the photosynthetic activity of phytoplankton in two pre-Alpine lakes, it has been indicated by Gächter (27) that increased metal loading would primarily change the phytoplanton composition in lakes which in turn could influence the balance between photosynthesis and respiration (64).

SOURCES, FLUXES, LOADS, AND SINKS OF METALS IN INLAND WATERS

Evaluation of natural versus anthropogenic fluxes of metals into aquatic systems is increasingly using data from sediment core studies. Such calculations, however, have to consider changing environmental conditions — such as deforestation and agricultural changes — and their effects on the sedimentation rate in a particular basin. The data given by Hamilton-Taylor (33) from Lake Windermere in England show that the present-day natural input rates are four to five times higher than the precultural influxes from the same catchment area (Table 4); it is assumed that this is simply a function of the increased overall rate of deposition. Anthropogenic metal fluxes into Lake Windermere are highest for mercury, followed by lead, whereas the cultural effects on fluxes of zinc and copper are much lower (Table 4).

Metal pollution in lakes originates from tributary inflow, coastal runoff, industrial and domestic effluent discharges, atmospheric fallout, shipping operations, and the recreational use of the lake. The quantitative inventory of sources and sinks of metals in Lake Erie by Nriagu et al. (49) revealed that atmospheric inputs account for 20%, 35%, and 50% of the Cu, Zn, and Pb discharges, respectively, originating from the lake's catchment. The contributions of Cu, Zn, and Pb from sewage effluents are 45%, 30%, and 20%, respectively (Table 5).

Type of flux	rate, $\mu g \cdot cm^{-2} \cdot yr^{-1}$			
	Zinc	Lead	Copper	Mercury
Precultural	1.9	0.7	0.3	0.001
Present-day natural	7.5	2.6	1.2	0.005
Present-day anthropogenic	16.9	9.9	1.8	0.049
Maximum anthropogenic	47.9	25.1	2.8	0.049

TABLE 4 — Fluxes of heavy metals into Lake Windermere (33).

	Copper	Lead	Zinc
Detroit River (import from upper lakes)	1640	630	5220
Tributaries, U.S.A.	100	52	271
Tributaries, Ontario	31	19	140
Sewage discharges	448	283	759
Dredged spoils	42	56	175
Atmospheric inputs	206	645	903
Shoreline erosion	190	221	308
Total, all sources	2657	1906	7776
Export, Niagara River and Welland Canal	1320	660	4400
Retained in sediments (t/a)	1337	1246	3776
Retained in percent of input	50%	65%	35%
Metal concentrations (μg/l)	1.5	2.0	6.0
Mean residence time (days)[a]	104	180	152

[a] Residence time was calculated from the volume of the lake (469 km^3) and from the input data of the present table.

TABLE 5 — Inventory of sources and sinks of heavy metals in Lake Erie (49). Flux rate in tonnes/year.

	Copper	Lead	Cadmium	Zinc
Lake Constance				
Atmospheric input $ng \cdot cm^{-2} \cdot yr^{-1}$	714	11000	20	8400
Deposition $ng \cdot cm^{-2} \cdot yr^{-1}$	6500	9500	100	36000
Metal concentration (ng/l)	300-800	50-100	6-29	1000-4000
North Atlantic Ocean				
Atmospheric input* $ng \cdot cm^{-2} \cdot yr^{-1}$	25	310	-	130
Vertical transitional flux $ng \cdot cm^{-2} \cdot yr^{-1}$	234	330	-	1040
Metal concentration (ng/l)	100 (30-300)	3 (1-15)	10 (1-120)	100 (10-600)

* From Buat-Mênard and Chesselet (9); see also Buat-Mênard, this volume.

TABLE 6 — Inputs, desposition rates, and concentrations of heavy metals in Lake Constance compared with data for the North Atlantic (61).

Concentration profiles of dissolved and particulate trace metals in Lake Constance (61) reveal similar curves as obtained from oceanic systems. This implies a removal of trace metals in the surface layers by biological processes and adsorption onto allochtonous particles, and subsequent partial release due to decomposition of the authigenic particles during settling to the lake bottom. Although the input of trace metals into Lake Constance is one to two orders of magnitude higher than that of the ocean, the metal concentrations are on the same order of magnitude, apparently due to the highly efficient removal processes operating in these systems (Table 6).

An example of the heavy metal budget in a densely populated region is given for the 4488 km^2 large Ruhr River catchment area, a tributary to the Rhine River, by Imhoff et al. (35). The Ruhr River is the main source of drinking water for a population of about 6 million people and process water for the industries. At the same time the Ruhr and its tributaries receive treated wastewaters from 2.2 million people and 2 million industrial population equivalents. Among other industries, about 300 plating plants have settled in the Ruhr basin; these plants discharge a considerable amount of heavy metals, thereby increasing metal concentrations in the river water up to critical levels (e.g., nickel, which appears in Ruhr water in higher concentrations than in other rivers). Figure 2 shows that approximately 55% of the heavy metals (Pb, Cu, Zn, Ni, Cr, Cd) in the Ruhr River are discharged from municipal and industrial wastewater treatment plants; 45% are of natural origin. The municipal treatment plants receive 31% of the heavy metals from domestic and 69% from industrial wastewater.

The origin of the elements and their fraction in relation to the total load varies considerably. For example, 90% of the chromium in the Ruhr River is discharged by industries; nickel and cadmium derive up to 66% from industries, while only about 50% of the copper and zinc originate from industrial wastewater (35).

Due to the combined effects of the economic crisis, the reduction of the metal inputs by enhanced wastewater treatment, and the partial replacement of metals in critical applications (e.g., mercury in chlorine-alkali-electrolysis), the metal loads in many polluted rivers are decreasing after having reached a maximum in the early 1970s. From a review of data from the Rhine River at the German-Dutch border, Malle and Müller (43) showed that the concentrations of suspended solids hardly changed since 1973, while the contents of Cr, Cu, Zn, Cd, Hg, and Pb are retrograde (Fig. 3). The decrease in the Cr and Hg contents mainly occurred in the suspended solids, and Zn and Cd concentrations were found to be reduced in the liquid phase. (In addition, there are also effects of high water flow in "wet" years.)

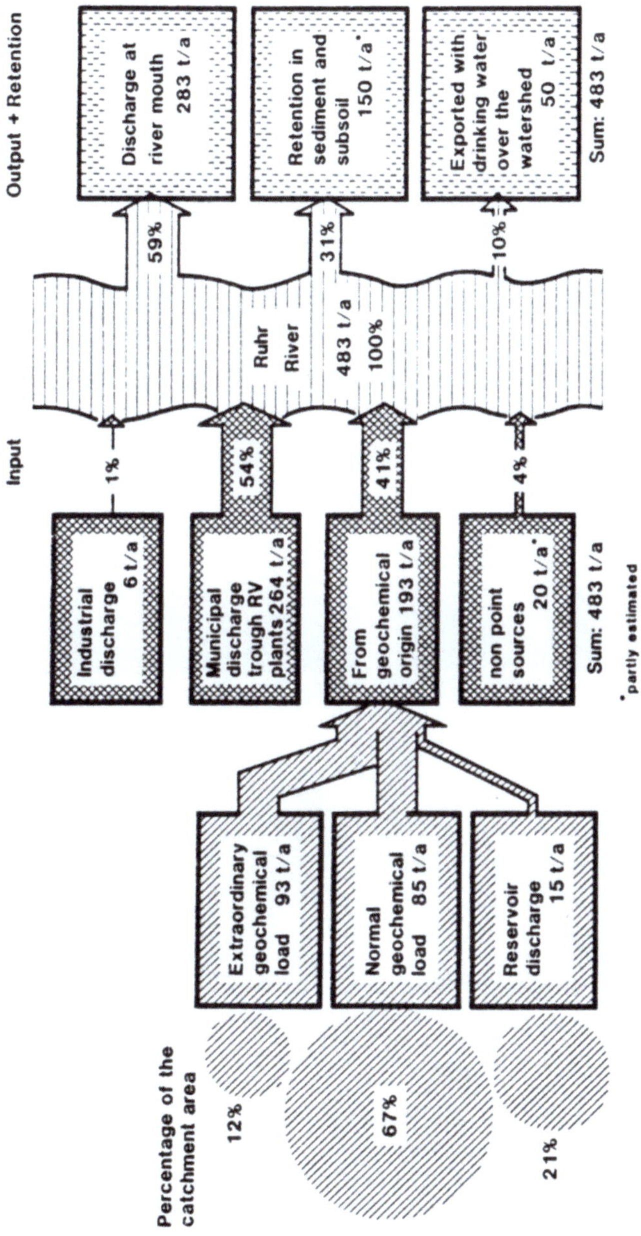

FIG. 2 — Balance of heavy metal loads in the Ruhr River catchment (35) (from (55)).

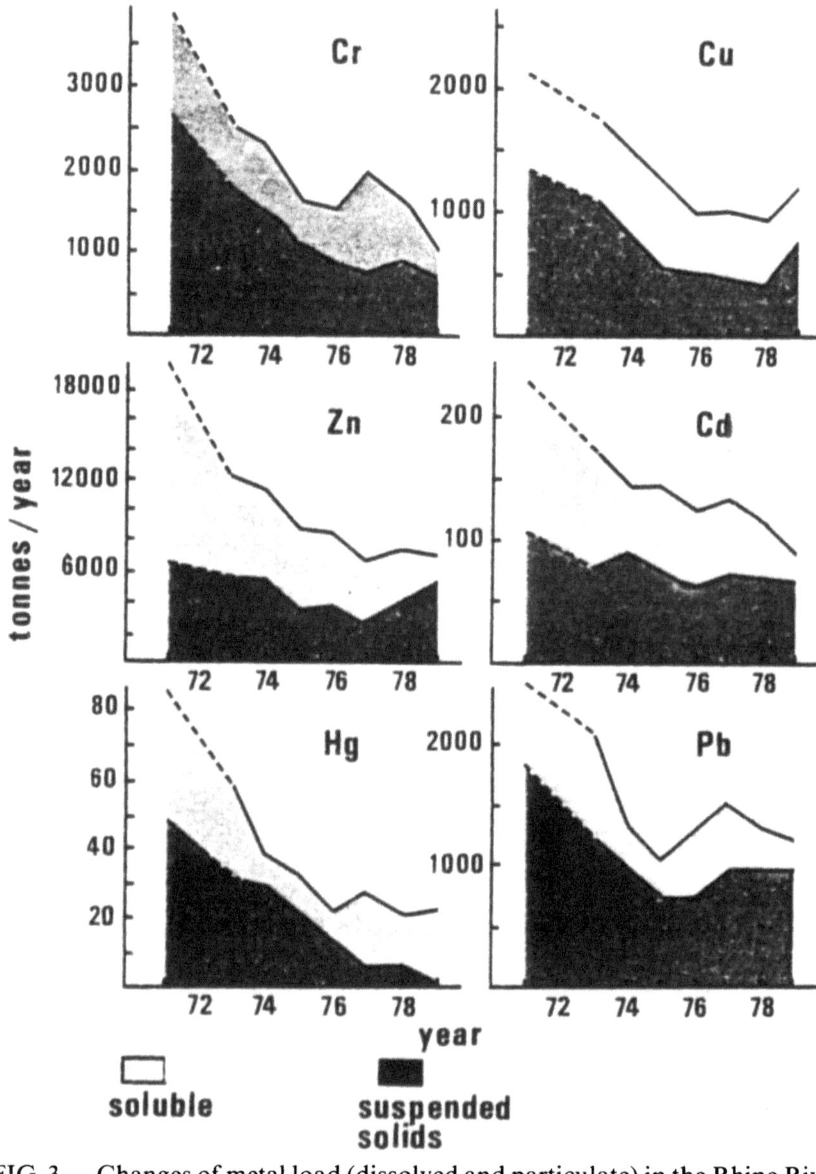

FIG. 3 — Changes of metal load (dissolved and particulate) in the Rhine River at the Dutch-German border from 1971 to 1979 (43) (from (55)).

The recent development of metal loads in the suspended matter is reflected in the sediments of the Rhine River in the Netherlands (samples partly originate from 15th and 18th century polders and from archives). Figure 4 (Salomons et al. (56)) indicated a definitive decrease of mercury (and arsenic), while cadmium concentrations are still increasing and are presently about 100-fold above natural background values. The contents in the 1920 sample were already fifteen times the baseline level of cadmium in the Rhine River.

Of the total amount of metals transported by the Rivers Rhine and Meuse, about two-thirds accumulates in the Netherlands, which, in fact, acts as an effective "treatment plant" for these river discharges (56). This situation may be regarded as beneficial for the North Sea, but it poses problems for the management of inland waters and the landfill areas in the Netherlands. The problems with polluted sediments are intensified by the closing of lagoons, the extension of harbor areas, the deepening of navigation channels, and the channelization of rivers. The total amount of dredged material from the Rotterdam

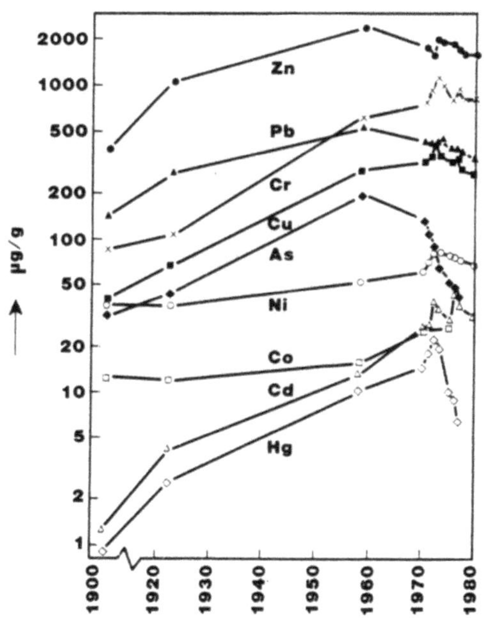

FIG. 4 — History of trace metal pollution in the Rhine River in the Netherlands as reflected in its sediments (56) (from (55)).

Harbor is presently about $20 \cdot 10^6$ m^3, which represents a 20–25-fold increase compared to 70 years ago. The most severely contaminated spoil from the easterly harbors, with high admixtures of fluvial sediment, is dumped on land. Thus the landfill areas are an important sink for fluvial (and partly marine) particulate metals. A tentative balance for trace metals of fluvial origin in spoils from the Rotterdam Harbor is presented in Table 7.

Problems arise from the metal concentration in sediments both with land disposal and marine dumping. From pot experiments under controlled conditions and by surveying the contents of contaminants in the crops grown on polluted sediments under field conditions, it has been shown that agricultural use of the landfill areas is restricted due to cadmium accumulation in crops, particularly in the green, leafy vegetables, wheat grain and other cereals, and in most green fodders. With respect to the contamination of groundwater in disposal areas, the transport of pollutants was calculated with a one-dimensional dispersion program (38). Chloride, showing conservative behavior, will become apparent in the adjacent polders in approximately the year 2100, the heavy metals one to three centuries later, and pesticides several thousand years later.

MOBILIZATION OF METALS FROM FRESHWATER SEDIMENTS

The sediment approach, which has only recently become a major subject of interest in the research of aquatic systems, has a number of perspectives: First, sediments are an expression of the condition of a water system; they can reflect the current quality of the system as well as the historical development of certain hydrologic and chemical parameters, e.g., metal anomalies in zones of mineralization and from pollution sources (26). Sediment data play an increasing role within the framework of environmental forensic investigations (48),

	Zinc	Copper	Lead	Cadmium
Input Rhine + Meuse	10051	1215	1707	195
Transport to landfill	1269	223	280	24.3
Percent of input	13%	18%	16%	13%
Transport to North Sea	462	78	116	7.6
Percent of input	5%	6%	7%	4%

TABLE 7 — Tentative metal balance for the Rivers Rhine and Meuse and of dredged material of fluvial origin from the Rotterdam Harbors in 1977 (56). In tons/year.

particularly in those cases in which a short-term or past pollution event is not or only insufficiently traceable from water analysis. Here, however, the effects of different grain size must be especially considered (26). In addition, sediments are increasingly recognized as a pollutant proper and as a carrier and possible source of contaminants in aquatic systems a well as with respect to the biological effects of polluted solid materials on agricultural land (see above). Metals are not necessarily fixed permanently by the sediment but may be recycled via biological and chemical agents, both within the sedimentary compartment and also back to the water column. Metals temporarily immobilized in the suspended matter and in bottom sidiments of aquatic systems may be released, for example, as a result of physicochemical changes, such as (a) lowering of pH (see introductory section on metal concentrations in water samples), (b) increased salinity, (c) an alteration in the redox conditions, and (d) increased input of organic chelators. With respect to the effects of higher salinities it has been demonstrated by Salomons et al. (56) that already at chloride levels of 200 mg/l — which are "normal," for example, in the lower Rhine River — the adsorption of cadmium on the suspended matter is adversely affected. Remobilization experiments simulating the situation when anoxic sediments deposited at low salinity regions in the harbors are brought into an oxic, marine environment, indicate the significant release of cadmium, approximately 50% of the original sediment concentration. The rate of release, however, is slow and takes about four to five weeks to level off (56). This effect, which seems to be related to the formation of relative stable cadmium-chloro-complexes, has recently been evidenced from field data in the Weser River estuary in northern Germany (24).

It was found from isotope studies with Mississippi River sediments that exchangeable ^{109}Cd levels are significantly increased as cadmium-contaminated sediment is transferred from a near-neutral pH, reducing environment to a moderately acid, oxidizing environment (28). Under these conditions, levels of cadmium (and probably other trace metals) of subsurface water from upland disposal of dredged materials and metal availability to plants growing on the material may be enhanced (28).

A considerable impact on heavy metal mobility may result from the growing use of synthetic complexing agents, e.g., nitrilotriacetic acid in detergents to replace polyphosphates. Data given by Salomons and Van Pagee (57) indicate that the effect of reduced adsorption of metals is more significant than the possible (active) remobilization of metals and could already occur at NTA-levels of 50–200 μg/l for metals such as zinc, nickel, and cadmium.

CHEMICAL SPECIATION OF TRACE METALS IN FRESHWATER SYSTEMS

The availability of trace metals for metabolic processes is closely related to their chemical species both in solution and in particulate matter. Following the earlier mercury pollution incidents such as that at Minamata Bay, it was shown that the toxicity and fate of waterborne metal contaminants was dependent on chemical forms, and that the quantification of these forms would be more meaningful than measurement of total metal concentrations (4).

Biological Uptake of Dissolved and Particulate Trace Metals

Jointly with the geochemical studies on the Rhine River in the Netherlands (56), an active biological monitoring program has been carried out using Dreissena Polymorpha (44). This bivalve lives attached to solid surfaces in the

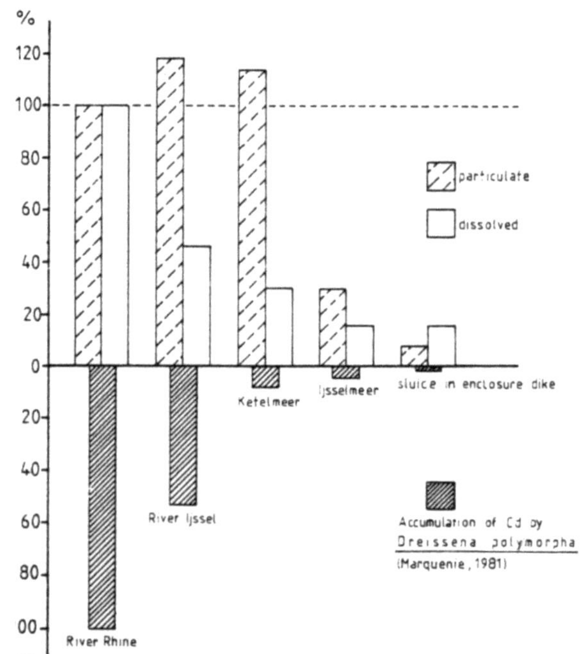

FIG. 5 — Cadmium contents in dissolved and particulate phases in the Rhine River, IJssel River, the Ketelmeer, the IJsselmeer, and sluices in the enclosure dike (values in percent of the Rhine River data), and cadmium in Dreissena polymorpha (44,52,56).

water and is exposed to trace metals in the dissolved and particulate phases. Data in Fig. 5 demonstrate that cadmium concentrations in the organism reflect the dissolved cadmium contents which decreases from the River Rhine via River IJssel to Ketelmeer (and are even lower in the IJsselmeer), rather than the increasing Cd-contents in the suspended matter. Apparently cadmium present in the suspended matter is less bioavailable.

Metal Species in Particulate Matter

Only part of the metals in solid matter may take part in short-term geochemical processes and/or are bioavailable (25). For the differentiation of the solid metal species — e.g., cation exchangeable forms, carbonate associations, easily and moderately reducible phases, associations with organic substances and sulfides, and the relative inert "residual" sediment fractions — sequential extraction procedures (10, 54, 65) have been developed which — other than direct measurements, e.g., by scanning electron microscopy/EDAX analysis — are not phase-specific or "selective," as has sometimes been claimed. These data can at best be used for the estimation on the potential remobilization of metals under changing environmental conditions, while the unsatisfactory situation with respect to the assessment of the quantitative extent of bioavailable element concentrations from solid speciation data can now be explained as being due mainly to the effect of competition between more or less strong adsorption sites in the solid substrate and selective mechanisms of the metal uptake by different organisms (42,66). A promising approach to bridge the gap between the chemical speciation data of the solid matter and the biological uptake values is the application of pore water analysis, since it has been demonstrated that the nondetrital phases of the sedimentary material constitute "the reservoir for potential subsequent release of contaminants into the water column and into new interstitial waters" (15).

Metal Speciation in Aqueous Phases

Several methods are used to determine the speciation of trace metals in solution (Fig. 6). For example, with such physical separation methods as filtration, ultrafiltration, dialysis, centrifugation, or ion-exclusion chromatography, metal species in solution are classified according to molecular size. With the chemical methods it is possible to classify the metals according to their interaction with adsorbing substances, e.g., cation-exchange resins or their behavior at the mercury electrode during voltammetric measurements (55). Usually, these methods are combined into speciation schemes (22). A summary on the aspect of "trace element speciation in surface waters" has recently been published (41), including contributions on physical separation techniques (63),

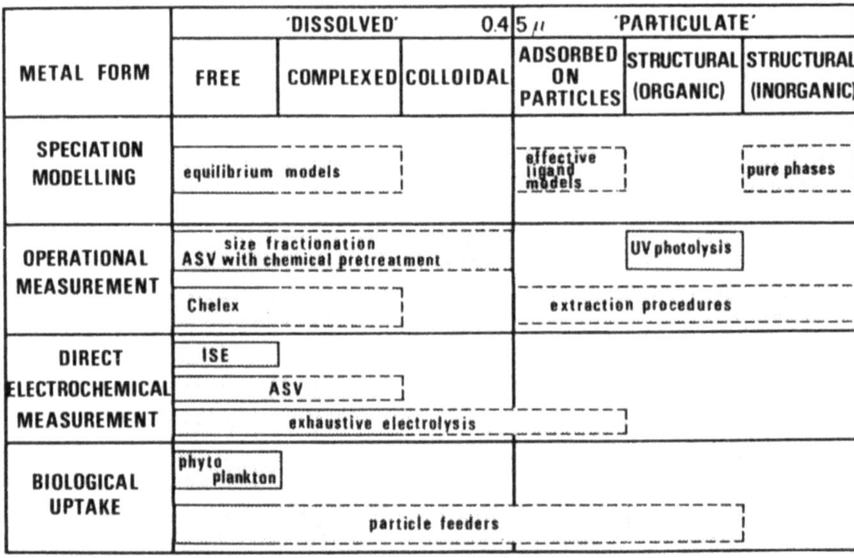

FIG. 6 — Classification of metal species in water (70) (from (55)).

voltammetric studies (50), and biological control of metal species in the aqueous phase (30,62).

There now exist a number of well-defined operational procedures for the quantification of selected groups of metal species, and considerable data for a range of natural water systems have now been accumulated (4,22). Significant among these findings is the existence of high proportions of the total metal concentrations in complexed or colloidally-associated forms; however, "the ways in which sampling and measurement procedures perturb the natural species have not yet been fully investigated" (4). The major differences between freshwater and seawater milieux are (60): (a) lower concentrations of major cations and anions, (b) lower ionic strengths, (c) higher content of adsorbing surfaces, and (d) higher concentrations of organic ligands in freshwater systems.

A summary of the biological consequences of trace element speciation in the aquatic environment has been given by Smies (62). With respect to the bioavailability/apparent toxicity, inorganic species in the dissolved phase do not seem to have significant effects, with the exception of chloride and carbonate complexing; biological control of inorganic speciation is only of interest for the geochemical cycles of As and Se (2,47). Inorganic ionic (mainly free ionic) species of metals are usually taken up by phytoplankton, but there is evidence that organically bound metals cannot be assimilated. Thus, natural and artificial chelating substances (e.g., EDTA, NTA, humic and fulvic acids) can reduce metal toxicity (53).

It is clear that the picture now based on the arbitrary borderline at 0.45 μm between "dissolved" and particulate metal-containing species will be altered with respect to the question of both chemical reactivity and bioavailability by the application of newer speciation methods involving gas or liquid chromatography coupled to such element-specific detectors as AA, MS, flame emission, etc. (7, 8, 13, 21). In this respect a new horizon of knowledge has been opened up: "One which focusses on the formation, pathways and fate of organometals and organometalloids in ecosystems, phenomena which are difficult to predict by conventional chemistry theories" (14). As seen from the work of Cantillo and Segar (11), Braman and colleagues (6), and Brinckman and co-workers (7), this field has made particularly strong advances during the last few years, e.g., in the separation and determination of organic and inorganic species of mercury, tin, lead, and arsenic. The data show that many of these compounds have serious impacts on the environment because of their modified toxicological properties (14,67).

Acknowledgements. Thanks are due to P. Buat-Ménard for the critical review of the manuscript and to F. Brinckman for detailed contributions to the discussions on the chemical speciation of colloidal metals. The comments of F. Baldi, H. Heinrichs, E.K. Silbergeld, and H.J. Tobschall are gratefully appreciated. W. Salomons (Delft Hydaulics Laboratory) kindly provided material for the present text and for Figs. 5 and 6.

REFERENCES

(1) Abdullah, M.I., and Royle, L.G. 1974. A study of the dissolved and particulate trace element in the Bristol Channel. J. Mar. Biol. Assoc. U.K. **54**: 581–597.

(2) Andreae, M.O. 1978. Distribution and speciation of arsenic in natural waters and some marine algae. Deep-Sea Res. **25**: 391–402.

(3) Baccini, P. 1976. Untersuchungen über den Schwermetallhaushalt der Seen. Schweiz. Z. Hydrol. **38**: 121–158.

(4) Batley, G.E. 1983. The current status of trace element speciation studies in natural waters. **In** Trace Element Speciation in Surface Waters and Its Ecological Implications, ed. G.G. Leppard, pp. 17–36. New York: Plenum Press.

(5) Boyle, E.A. 1978. Trace element geochemistry of the Amazon and its tributaries. EOS **59**: 276.

(6) Braman, R.S.; Johnson, D.L.; Foreback, C.C.; Ammons, J.M.; and Bricker, J.L. 1977. Separation and determination of nanogram amounts of inorganic arsenic and methylarsenic compounds. Anal. Chem. **49**: 621–627.

(7) Brinckman, F.E.; Blair, W.R.; Jewett, K.L.; and Iverson, W.P. 1977. Application of a liquid chromatograph coupled with a flameless atomic absorption detector for speciation of trace organometallic compounds. J. Chromatogr. Sci. **15**: 493–505.

(8) Brinckman, F.E.; Jackson, J.A.; Blair, W.R.; Olson, G.J.; and Iverson, W.P. 1983. Ultratrace speciation and biogenesis of methyltin transport species in estuarine waters. **In** Trace Metals in Sea Water, eds. C.S. Wong et al., pp. 39–72. New York: Plenum Press.

(9) Buat-Ménard, P., and Chesselet, R. 1979. Variable influence of the atmospheric flux on the trace metal chemistry of oceanic suspended matter. Earth Planet. Sci. Lett. **42**: 399–411.

(10) Calmano, W., and Förstner, U. 1983. Chemical extraction of heavy metals in polluted river sediments in Central Europe. **In** Biological Availability of Trace Metals, eds. R.E. Wildung and E.A. Jenne, pp. 77–90. Amsterdam: Elsevier.

(11) Cantillo, A.Y., and Segar, D.A. 1975. Metal species identification in the environment — a major challenge for the analyst. Proceedings of the International Conference on Heavy Metals in the Environment, Toronto, Canada, vol. I, pp. 183–204.

(12) Carpenter, J.H.; Bradford, W.L.; and Grant, V. 1975. Processes affecting the composition of estuarine waters (H_2CO_3, Fe, Mn, Zn, Cu, Ni, Cr, Co, and Cd). **In** Estuarine Research, ed. L.E. Cronin, vol. 1, pp. 137–152. New York: Academic Press.

(13) Chau, Y.K., and Wong, P.T.S. 1977. An element- and speciation-specific technique for the determination of organometallic compounds.

In Environmental Analysis, ed. G.W. Ewing. New York: Academic Press.

(14) Chau, Y.K., and Wong, P.T.S. 1983. Direct speciation analysis of molecular and ionic organometals. **In** Trace Element Speciation in Surface Waters, ed. G.G. Leppard, pp. 87–103. New York: Plenum Press.

(15) Chen, K.Y.; Gupta, S.K.; Sycip, A.U.; Lu, J.C.S., Knezevic, M.; and Choi, W.-W. 1976. Research study on the effect of dispersion, settling, and resedimentation on migration of chemical constituents during openwater disposal of dredged materials. Dredged Material Research Program, Rept. D-76-1. Vicksburg, MS: U.S. Army Corps of Engineers.

(16) Copeland, R.A., and Ayers, J.C. 1972. Trace Element Distributions in Water, Sediment, Phytoplankton, Zooplankton, and Benthos of Lake Michigan. Ann Arbor, MI: Environmental Research Group, Inc.

(17) Dietz, F., and Koppe, P. 1982. Schwermetalle in der Ruhr. **In** Schadstoffe im Wasser: Metalle, eds. J. Reichert and U. de Haar, vol. 1, pp. 180–256. Boppard: Harald Boldt Verlag.

(18) Dittrich, V., and Klein, T. 1982. Schwermetalle im Rhein bei Köln. **In** Schadstoffe im Wasser: Metalle, eds. J. Reichert and U. de Haar, vol. 1, pp. 97–113. Boppard: Harald Boldt Verlag.

(19) Edgington, D.N., and Robbins, J.A. 1976. Records of lead deposition on Lake Michigan sediments since 1800. Env. Sci. Technol. **10**: 266–273.

(20) Fagerström, T., and Jernelöv, A. 1972. Aspects of the quantitative ecology of mercury. Water Res. **6**: 1193–1202.

(21) Fernandez, F.J. 1977. Metal speciation using atomic absorption as a chromatography detector. At. Absorpt. Newsl. **16**: 33–35.

(22) Florence, T.M., and Batley, G.E. 1980. Chemical speciation in natural waters. CRC Crit. Rev. Analyt. Chem. **9**: 219–296.

(23) Förstner, U. 1981. Trace metals in fresh waters (with particular reference to mine effluents). **In** Handbook of Strata-Bound and Stratiform Ore Deposits, ed. K.H. Wolf, vol. 9, pp. 271–303. Amsterdam: Elsevier.

(24) Förstner, U., and Calmano, W., 1982. Bindungsformen von Schwermetallen in Baggerschlämmen. Vom Wasser **59**: 83–92.

(25) Förstner, U., and Salomons, W. 1983. Trace element speciation in surface waters: Interactions with particulate matter. **In** Trace Element Speciation in Surface Waters and Its Ecological Implications, ed. G.G. Leppard, pp. 245–273. New York: Plenum Press.

(26) Förstner, U., and Wittmann, G. 1981. Metal Pollution in the Aquatic Environment, 2nd ed. Berlin: Springer.

(27) Gächter, R. 1976. Untersuchungen über die Beeinflussung der planktonischen Photosynthese durch anorganische Metallsalze im eutrophen Alpnachersee und der mesotrophen Horwer Bucht. Schweiz. Z. Hydrol. **38**: 97–119.

(28) Gambrell, R.P.; Khalid, R.A.; Verloo, M.G.; and Patrick, W.H. 1977. Transformations of heavy metals and plant nutrients in dredged sediments as affected by oxidation reduction potential and pH. II. Materials and methods/results and discussion. Dredged Material Research Program, Rept. D-77-4. Vicksburg, MS: U.S. Army Corps of Engineers.

(29) Gibbs, R.J. 1977. Transport phases of transition metals in the Amazon and Yukon Rivers. Geol. Soc. Am. Bull. **88**: 829–843.

(30) Giesy, J.P. 1983. Biological control of trace metal equilibria in surface waters. **In** Trace Element Speciation in Surface Waters, ed. G.G. Leppard, pp. 195–210. New York: Plenum Press.

(31) Grimshaw, D.L.; Lewin, J.; and Fuge, R. 1976. Seasonal and short-term variations in the concentration and supply of dissolved zinc to polluted aquatic environment. Env. Pollut. **11**: 1–7.

(32) Haines, T.A. 1981. Acidic precipitation and its consequences for aquatic ecosystems: A review. Trans. Am. Fisheries Soc. **110**: 669–707.

(33) Hamilton-Taylor, J. 1979. Enrichments of zinc, lead, and copper in recent sediments of Windermere, England. Env. Sci. Technol. **13**: 693–697.

(34) Heinrichs, H. 1975. Die Untersuchung von Gesteinen und Gewässern auf Cd, Sb, Hg, Tl, Pb und Bi mit der flammenlosen Atomabsorptions-Spektralphotometrie. Thesis, University of Göttingen, Göttingen, F.R. Germany.

(35) Imhoff, K.R.; Koppe, P.; and Dietz, F. 1980. Heavy metals in the Ruhr River and their budget in the catchment area. Prog. Water Technol. IAWPR **12**: 735–749.

(36) Jenne, E.A.; Kennedy, V.C.; Burchard, J.M.; and Ball, J.W. 1980. Sediment collection and processing for selective extraction and for total trace element analysis. **In** Contaminants and Sediments, ed. R.A. Baker, vol. 2, pp. 169–191. Ann Arbor: Ann Arbor Science Publ.

(37) Kennedy, V.C., and Sebetich, M.J. 1976. Trace elements in northern California streams. **In** Geological Survey Research 1976, pp. 208–209. Washington, D.C.: U.S. Government Printing Office.

(38) Kerdijk, H.N. 1981. Groundwater pollution by heavy metals and pesticides from a dredge spoil dump. In Quality of Groundwater, eds. W. van Fuyvenboden, P. Glasbergen, and H. van Lelyveld, pp. 279–286. Amsterdam: Elsevier.

(39) Knauer, G.A., and Martin, J.H. 1973. Seasonal variations of cadmium, copper, manganese, lead, and zinc in water and phytoplankton in Monterey Bay, California. Limnol. Oceanogr. **18**: 597–604.

(40) Kopp, J.F., and Kroner, R.C. 1968. Trace Metals in Waters of the United States. Washington, D.C.: Federal Water Pollution Control Administration, Division Pollution Surveillance.

(41) Leppard, G.G., ed. 1983. Trace Element Speciation in Surface Waters and Its Ecological Implications. Nato Conference Series I: Ecology. New York: Plenum Press.

(42) Luoma, S.N., and Bryan, G.W., 1979. Trace metal bioavailability: Modeling chemical and biological interactions of sediment-bound zinc. **In** Chemical Modeling in Aqueous Systems, ed. E.A. Jenne, pp. 577–609. American Chemical Society Symposium Series, vol. 93. Washington, D.C.: American Chemical Society.

(43) Malle, K.-G., and Müller, G. 1982. Metallgehalt und Schwebstoffgehalt im Rhein. Z. Wasser Abwasser Forsch. **15**: 11–15.

(44) Marquenie, J.M. 1981. The freshwater mollusc Dreissena Polymorpha as a potential tool for assessing bio-availability of heavy metals in aquatic systems. **In** Proceedings of the International Conference on Heavy Metals in the Environment, Amsterdam, pp. 409–412. Edinburgh: CEP Consultants.

(45) Martin, J.H.; Knauer, G.A.; and Flegal, A.R. 1980. Distribution of zinc in natural waters. **In** Zinc in the Environment, ed. J.O. Nriagu, vol. 1, pp. 193–197. New York: John Wiley & Sons.

(46) Martin, J.-M., and Meybeck, M. 1979. Elemental mass balance of material carried by major world rivers. Mar. Chem. **7**: 173–206.

(47) Measures, C.I., and Burton, J.D. 1978. Behaviour and speciation of dissolved selenium in esturarine waters. Nature **273**: 293–295.

(48) Meiggs, T.O. 1980. The use of sediment analysis in forensic investigations and procedural requirements for such studies. **In** Contaminant and Sediments, ed. R.A. Baker, vol. 1, pp. 297–308. Ann Arbor, MI: Ann Arbor Science Publ.

(49) Nriagu, J.O.; Kemp, A.L.W.; Wong, H.K.T.; and Harper, N. 1979. Sedimentary record of heavy metal pollution in Lake Erie. Geochim. Cosmochim. Acta **43**: 247–258.

(50) Nürnberg, H.W. 1983. Voltammetric studies on trace metal speciation in natural waters. Part II: Application and conclusions for chemical oceanography and chemical limnology. **In** Trace Element Speciation in Surface Waters, ed. G.G. Leppard, pp. 211–230. New York: Plenum Press.

(51) Patterson, C.C.; Settle, D.; Schaule, B.; and Burnett, M. 1976. Transport of pollutant lead to the ocean and within ocean ecosystems. **In** Marine Pollution Transfer, ed. H.L. Windom and R.A. Duce, pp. 23–38. Lexington: D.C. Heath Co.

(52) Ravera, O. 1982. Ecological aspects of heavy metal speciation. **In** Ecological Effects on Heavy Metals Speciation in Aquatic Ecosystems, Ispra Courses, 15–19 November 1982, Ispra/Italy.

(53) Ravera, O.; Cartesano, A.; de Bernardi, R.; and Guzzi, L. 1973. Effects of chelating agents (EDTA and NTA) on the incorporation of radionuclides by freshwater filter feeding organisms (codepods and lamellibranch). Atto 5th Colloquium International Oceanology of the Mediterranean, Messina/Italy, pp. 437–448.

(54) Salomons, W., and Förstner, U. 1980. Trace metal analysis on polluted sediments. II. Evaluation of environmental impact. Env. Technol. Lett. **1**: 506–517.

(55) Salomons, W., and Förstner, U. 1984. Metals in the Hydrological Cycle. Berlin: Springer, in press.

(56) Salomons, W.; van Driel, W.; Kerdijk, H.; and Boxma, R. 1982. Help! Holland is plated by the Rhine. **In** Effects of Waste Disposal on Groundwater and Surface Water, Proceedings of the Exeter Symposium. IAHS-Publ. **139**: 255–269.

(57) Salomons, W., and Van Pagee, J.A. 1981. Prediction of NTA levels in river systems and their effects on metal concentrations. Proceedings of the International Conference on Heavy Metals in the Environment, Amsterdam, pp. 694–697. Edinburgh: CEP Consultants.

(58) Schell, W.R., and Nevissi, A. 1977. Heavy metals from waste disposal in central Puget Sound. Env. Sci. Technol. **11**: 887–893.

(59) Schleichert, U. 1975. Schwermetallgehalte der Schwebstoffe des Rheins bei Koblenz im Jahresablauf — eine gewässerkundliche Interpretation. Dtsch. Gewässerkundl. Mitt. **19**: 150–157.

(60) Sibley, T.H., and Morgan, J.J. 1977. Equilibrium speciation of trace metals in fresh water: sea water mixtures. Proceedings of the International Conference on Heavy Metals Environment, Toronto 1975, vol. 1, pp. 319–338.

(61) Sigg, L.; Sturm, M.; Stumm, W.; Mart, L.; and Nürnberg, H.W. 1982. Schwermetalle im Bodensee — Mechanismen der Konzentrationsregulierung. Naturwissenschaften **69**: 546–548.

(62) Smies, M. 1983. Biological consequences of trace element speciation in the aquatic environment. **In** Trace Element Speciation in Surface Waters and Its Ecological Implications, ed. G.G. Leppard, pp. 177–193. New York: Plenum Press.

(63) Steinnes, E. 1983. Physical separation techniques in trace element speciation studies. **In** Trace Element Speciation in Surface Waters, ed. G.G. Leppard, pp. 37–47. New York: Plenum Press.

(64) Stumm, W., and Baccini, P. 1978. Man-made chemical perturbation of lakes. **In** Lakes — Chemistry, Geology, Physics, ed. A. Lerman, pp. 91–126. New York: Springer.

(65) Tessier, A.; Campbell, P.G.C.; and Bisson, M. 1979. Sequential extraction procedure for the speciation of particulate trace metals. Anal. Chem. **51**: 844–851.

(66) Tessier, A.; Campbell, P.G.C.; Auclair, J.C.; Bisson, M.; and Boucher, H., 1982. Evaluation de l'Impact des Rejets Miniers sur des

Organismes Biologiques. Rapp. Sci. Université Quebec, INRS-Eau No. 146.

(67) Thayer, J.S., and Brinckman, F.E. 1982. The biological methylation of metals and metalloids. Adv. Organomet. Chem. **20**: 313–356.

(68) Trefry, J.H., and Presley, B.J. 1976. Heavy metal transport from the Mississippi River to the Gulf of Mexico. **In** Marine Pollution Transfer, eds. H.L. Windom and R.A. Duce, pp. 39–76. Lexington: D.C. Heath Co.

(69) Troup, B.N., and Bricker, O.P. 1975. Processes affecting the transport of materials from continents to the oceans. **In** Marine Chemistry in the Coastal Environment, ed. T.M. Church. American Chemical Society Symposium Series, vol. 18, pp. 133–151.

(70) Whitfield, M., and Turner, D.R. 1980. Theoretical studies of the chemical speciation of lead in sea water. **In** Lead — Occurrence, Fate and Pollution in the Marine Environment, eds. M. Branica and Z. Konrad. Oxford: Pergamon Press.

Changing Metal Cycles and Human Health, ed. J.O. Nriagu, pp. 95–111.
Dahlem Konferenzen 1984. Berlin, Heidelberg, New York, Tokyo: Springer-Verlag.

POLLUTION OF SOILS BY CADMIUM

T. Asami
Faculty of Agriculture, Ibaraki University
Ami-machi, Ibaraki, Japan

Abstract. Cadmium is the most important metal in the soil pollution problem, because human beings consume Cd through many crops, and there is a small margin between present intake and the intake that may cause a significant effect on human health. In Japan, rice is the main Cd source and 4,816 ha of paddy fields have been designated as the "contaminated" land under the Agricultural Land Soil Pollution Prevention and Others Law. Furthermore, about 9.5% of Japanese paddy soil is more or less polluted by Cd. Cadmium is adsorbed by soil and remains there for a long period. In polluted soil it is mainly in exchangeable or humus complex form and is available to plants. Absorption of soil Cd by the rice plant is high in the oxidized condition and decreases in the reduced condition because of the formation of insoluble CdS. Cadmium accumulated in a plant at levels dangerous to the animal or human consumer without any inhibition of plant growth. The only method for improving the polluted paddy fields is by covering the polluted surface soils with a 30 cm deep layer of unpolluted soil. Liming and P applications are not effective in paddy soils. Japanese rice has a higher Cd concentration than the rice of other countries because of the pollution in a large area of Japan. When sewage sludge is applied to agricultural soil, attention must be paid to possible increases in the Cd concentration of the soil.

INTRODUCTION

Environmental pollution by metals has now become one of the most important problems in the world, especially in Japan. Environmental heavy metal pollution has a long history in Japan, and in recent years, it has spread all over the country. Health injury due to heavy metals has been recognized officially by the Pollution-related Health Damage Compensation Law, e.g., the cases of Itai-itai disease due to Cd, chronic As poisoning, and Minamata disease due to methyl Hg compounds (Table 1).

Of these diseases, only the Itai-itai disease was derived from soil pollution and was mainly caused by intake of Cd in the rice grain from Cd-polluted fields. The other two diseases were not caused by soil pollution. Minamata disease was caused by the intake of methyl-Hg compounds in fish and shellfish and chronic As poisoning by the intake of As in air and drinking water.

In general, rice is the main Cd source for Japanese people in polluted and unpolluted areas, and > 60% of Cd has been consumed from rice in unpolluted

Disease	Designated Area	No. of Certified Persons	No. of Deaths from the Disease	No. of Suspected Persons
Itai-itai Disease	Lower Jinzu River Basin	106	66	65*
Minamata Disease	Lower Agano River Basin	684	96	90**
	Coastal Area of Minamata Bay	1,737	437	5,624**
Chronic Arsenic Disease	Sasagadani District	21	9	
	Toroku District	129	20	

In addition to these specific diseases, nonspecific diseases such as chronic bronchitis, bronchial asthma, asthmatic bronchitis, pulmonary emphysema, and their complications are certified under the Pollution-related Health Damage Compensation Law, and the number of certified persons is 79,166 in 41 areas.

* Number of suspected patients living in the designated region under medical care and observation.

**Number of persons awaiting certification.

TABLE 1 — Designated areas and numbers of certified persons, etc., with metal poisoning in Japan by the end of March 1981 (based on (11)).

areas (25). As has generally been recognized, there is only a small margin between present intake and the intake that may cause a significant prevalence of renal tubular damage to the general population. Therefore, Cd is the most important metal in the soil pollution problem in Japan. As it is absorbed more by such upland crops as wheat, barley, corn, and vegetables than by rice (9,27), the resulting soil pollution will be the most important problem in other countries in connection with human health as well. I will therefore limit my discussion to the soil pollution problem by Cd.

POLLUTED AREAS OF AGRICULTURAL SOIL IN JAPAN

Maximum allowable limits of heavy metals in soil have been set for Cd, Cu, and As. The requirements for the designation of agricultural land soil pollution policy areas under the Agricultural Land Soil Pollution Prevention and Others Law are based on the maximum allowable limit of Cd (1.0 ug/g in unpolished rice) in rice, and on the concentration of Cu (125 ug/g by 0.1 N HCl extraction) and of As (15 ug/g by 1 N HCl extraction) in soils. The maximum allowable limit of Cd in soil is determined based on the concentration of Cd in rice, set so as not to increase the Cd concentration in rice to a level hazardous to human health, and the relationship between the Cd concentration in rice and that in soil can be changed by many factors. Areas in which rice contaminated with over 1.0 ug/g of Cd is produced, or which lie in the vicinity of some such area and thus are highly probable of producing rice contaminated with the same amount, are designated agricultural land soil pollution policy areas (Fig. 1). One ug/g of Cd in unpolished rice is the maximum allowable limit as set by the Ministry of Health and Welfare in 1970 (for review see (6)).

By the end of January 1982, a total of 48 areas covering about 4,816 ha were designated under the Law; 41 for Cd, 11 for Cu, and 5 for As. The countermeasures are formulated for 41 areas (3,015.1 ha), which include the covering of the polluted surface soil with a layer of unpolluted soil and in some cases the use of different water sources for irrigation. The countermeasures have been completed in 21 areas (about 960 ha). In addition, a total of 42 other areas (about 330 ha) have been budgeted as independent prefectural projects (12).

In addition to the detailed survey, a general survey for soil pollution by Cd and other metals has been conducted at the same points in agricultural areas all over Japan. The paddy fields cover about 3,000,000 ha and the upland fields including orchard fields about 2,500,000 ha in Japan. Sampling has been made on every 1,000 ha of paddy fields and on every 2,000 ha of upland fields. Results of the general survey proved that 9.5% of paddy soil, 3.2% of upland soil,

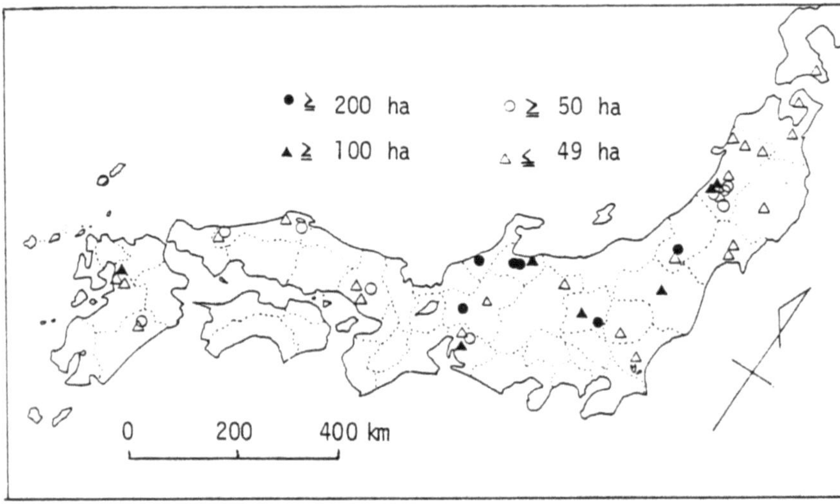

FIG. 1 — Agricultural land soil pollution policy areas in Japan (based on (12)).

and 7.5% of orchard soil are polluted by Cd. These soils are also polluted by Zn and Cu (3).

As the pollution sources such as factories, incinerators,and motor vehicles increase, there is also the possibility of an increase in pollution of the agricultural soils all over Japan by Cd and other metals.

BEHAVIOR OF CADMIUM IN PADDY SOIL AND ABSORPTION BY RICE PLANT

Soil Accumulation of Cadmium from Polluted Irrigation Water

Various concentrations of Cd in solution (0.001–10 ug/ml) were prepared using river water (pH was 7.3) containing 18 ug/ml Ca, 3.38 ug/ml Mg, 0.62 ug/ml K, and 4.2 ug/ml Na. One liter of the solution was added to 10 g of soil and shaken for 1 hr. Nearly all of the Cd was absorbed by the soils taken from Hokuriku and Toyama Agricultural Experiment Stations, the textures of which are light clay and loam, respectively. The adsorption of Cd at equilibrium fits Freundlich's adsorption isotherm; the slope of the regression straight line is about 0.9 as shown in Fig. 2 (16). The background concentration of Cd in unpolluted streams is about 0.1 ng/ml (15). When the concentration of the equilibrium solution exceeds this concentration, the Cd concentration in soils should become higher than the natural level.

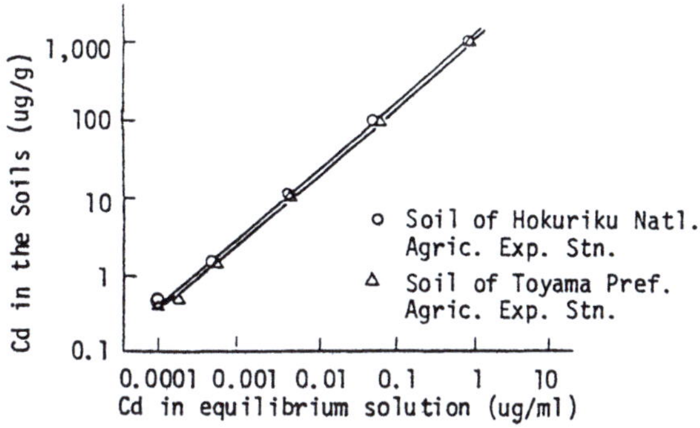

FIG. 2 — Adsorption equilibria of Cd between soil and solution (16).

Iimura (16) conducted a column experiment in which river water containing 0.01 ug/ml Cd was leached through a 5 cm thick soil column at the rate of 5 cm (50 ml) per day. More than 90% of the Cd was adsorbed, and consequently the concentration of leachates was lowered to several tenths of 1 ng/g, which is near the concentration level of clean water (15) and unpolluted soil leachates. From these results it was concluded that if the irrigation water is polluted by Cd, so also is the paddy soil. Cadmium remains in paddy soils for a long time and is absorbed by rice plants.

Cadmium Status in Soil

Factors affecting the sorption of metals by soils include: exchange adsorption on the surface of clay and humus, complex formation with humus, adsorption and occlusion by the hydrated oxides of Al, Fe, and Mn, etc., and formation of insoluble compounds (e.g., CdS) from soluble metals, especially under reduced condition.

Many data suggested that the main forms of Cd in soils were exchangeable and humus complex. Table 2 shows the result of systematic fractionation of Cd in air-dried polluted soils obtained by the McLaren and Crawford method (4). About half of Cd exists in exchangeable form and most of the remaining half is in specific sorption in organic form, while according to McLaren and Crawford more than half of soil Cu is in the residual fraction which is extractable by HF treatment, followed by the fraction which is especially sorbed by organic materials, and finally the fraction occluded by oxides. A fair amount of Cd is extracted by 0.1 N HCl. These data for Cd show that Cd in soils is highly available to plants.

Soil	Total Cd ug/g	0.1 N HCl sol. Cd ug/g	Cd in the individual fractions (%)			
			CA	AAC	PYR	OX
A	2.19	1.57	50	5	39	8
B	4.59	4.27	59	11	49	2
C	5.12	3.93	54	20	49	4
D	7.47	7.20	56	13	51	7

CA, soil solution and exchangeable – extracted by 0.05 M $CaCl_2$; AAC, bound by inorganic sites (specific sorption) – extracted by 2.5% CH_3COOH; PYR, bound by organic sites (specific sorption) – extracted by 1 M $K_4P_2O_7.3H_2O$; OX, occluded by free oxides – extracted by acid oxalate (4).

TABLE 2 — Fractionation of cadmium in paddy soils.

Absorption of Cadmium by Rice Plants from Paddy Soil

The most characteristic difference between paddy soil and upland soil is the drastic change in the oxidation-reduction state. Absorption of heavy metals by paddy rice is influenced seriously by water management or the oxidation-reduction state of the paddy soil. Figure 3 shows that the Cd concentration ratio (= Cd conc. in unpolished rice/Cd conc. in soil) was less than 0.01 in the minimum case when the rice was grown under submerged conditions throughout the whole period of growth (16). While in the maximum case, in relatively coarse textural Toyama soil, the ratio exceeded two when the submerging

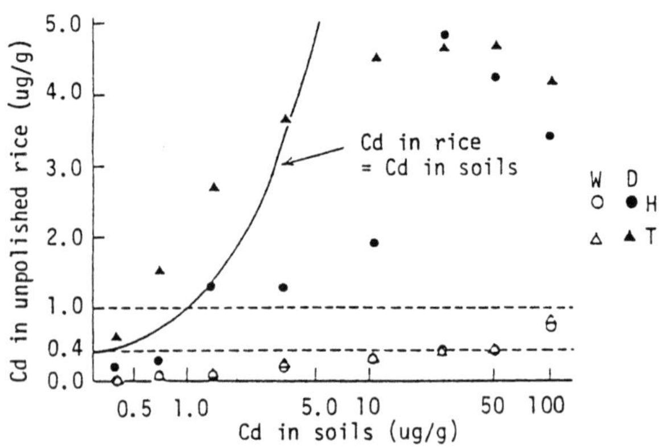

FIG. 3 — Influence of water management of paddy soils on Cd contents in unpolished rice in pot culture experiment. D, drained after tillering stage; W, submerged during whole growth period; H, soil of Hokuriku Natl. Agric. Exp. Stn.; T, soil of Toyama City (16).

water was drained after the tillering stage. Soil texture does not affect the Cd absorption in reduced condition, but it is significantly affected under oxidized condition. The increase in Cd absorption by rice due to draining of the submerged water after the panicle formation stage is most prominent.

Change in Solubility of Cadmium Accompanying Change in Redox Potential of Paddy Soil

In upland (oxidized) condition of soils, Cd availability to plants is high, while Cd absorption by rice is much decreased under submerged (reduced) condition. The most acceptable explanation for the decrease is considered to be a decrease in solubility due to the formation of CdS. Figure 4 shows the results of an experiment to verify this hypothesis in which three types of soil were incubated under submerged conditions avoiding the exposure of the submerged water to the atmosphere, and the Eh and the solubilities of Cd in 1.15 N CH_3COONH_4 solution (pH 5.5) were determined intermittently (16). The curve of the solid line in the figure is the theoretical curve for the $SO_4^{2-} \rightleftarrows H_2S$ oxidation-reduction system.

RELATION BETWEEN THE CADMIUM CONCENTRATION OF UNPOLISHED RICE AND RICE YIELD

Cadmium accumulates in a plant at levels dangerous to the animal or human consumer without any inhibition of plant growth. Figure 5 is an example of a water culture experiment of rice (18). Cadmium concentration in unpolished rice far exceeded 1.0 ug/g before the decrease in yield. Nearly the same tendency was found in the soil culture experiments (18).

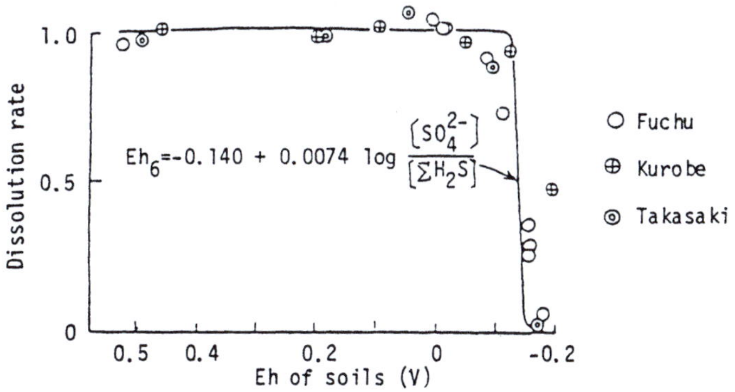

FIG. 4 — Dissolution of Cd from soils influenced by Eh (16).

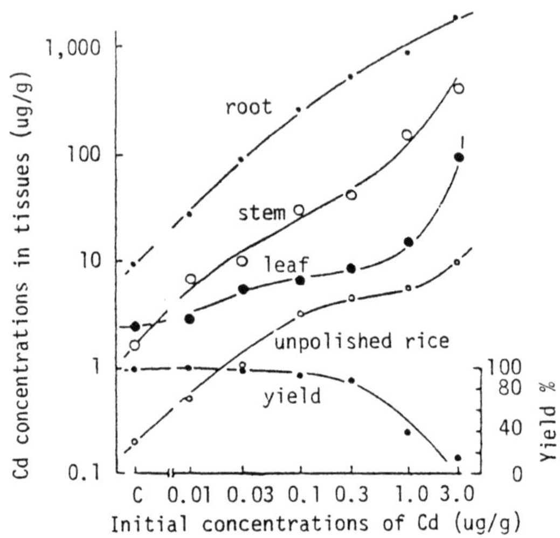

FIG. 5 — Absorption of Cd by rice plant and yield of unpolished rice (water culture experiment) (18).

EFFECT OF ZINC ON THE ABSORPTION AND TRANSLOCATION OF CADMIUM BY RICE PLANT

Cd-polluted soils are also polluted by Zn in most cases. Two characteristics of interactions among the metals are the competition or antagonism among the metals and the co-absorption or synergism. Japanese scientists showed that the addition of Zn to the soil increases Cd absorption by rice plants. They assumed that Zn dissociates Cd fixed to the binding sites in the soil due to the competition for the sites and increases the solubility of Cd. This phenomenon is also observed in rice plants grown in culture solution.

Honma and Hirata (14) found that Zn increased the translocation of Cd from the roots to tops but decreased the Cd incorporation into the roots. They postulated that Zn may have interfered in absorption into the vacuole and increased translocation. Furthermore, they found that the ratio of Cd to Zn in the nutrient solution controlled the occurrence of the synergism and antagonism between the two elements. For example, when the Cd concentration in the solution was 0.5 ug/ml, the addition of 0.5–5 ug/ml of Zn increased the Cd concentration in the tops, but the addition of 25 ug/ml Zn decreased it.

REHABILITATION OF CADMIUM POLLUTED PADDY SOILS

The improvement of the polluted paddy soils was conducted mainly by covering the polluted surface soil with a 30 cm deep layer of unpolluted soil, thus reducing practically the Cd concentrations of unpolished rice below 0.1 ug/g in every field and every year (if Cd was not re-added to the paddy soils) (7,8).

In upland fields, liming decreases Cd absorption by crops by raising the pH of the soil. But in paddy fields, soil pH is usually higher than 6.0 under submerged condition because of the formation of soluble Fe^{2+}. Thus, liming is not as effective as would be expected in upland fields. The effectiveness of adding P to polluted paddy soils is also limited (1–3).

CADMIUM CONCENTRATION IN UNPOLISHED RICE IN JAPAN

Polluted Areas

Cadmium polluted areas in which rice with over 1.0 ug/g has been produced or in which there is a high possibility of production of such rice are designated

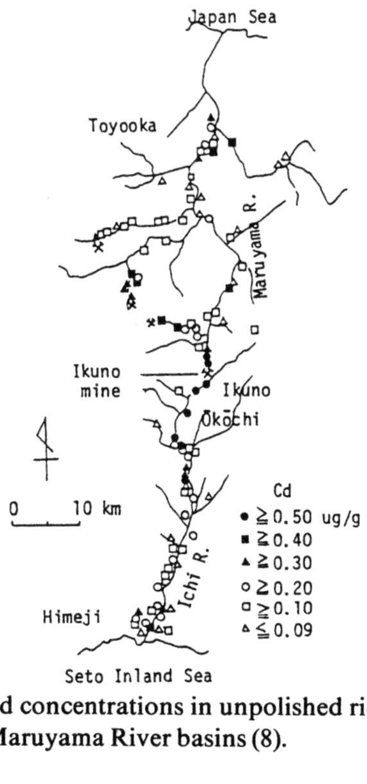

FIG. 6 — Average Cd concentrations in unpolished rice produced in each village of the Ichi and Maruyama River basins (8).

as the agricultural land soil pollution policy areas as shown in Fig. 1. Saito et al. (23) stated that an effect of health was observed in inhabitants usually consuming rice with over 0.3 ug/g Cd. According to Nogawa (this volume), Cd concentration in rice, which would not cause renal tubular damage, may be set under 0.3–0.4 ug/g Cd wet weight. Figure 6 shows the average Cd concentrations in unpolished rice produced in each village in the Ichi and Maruyama River basins (8). While some paddy fields in Ikuno and Okochi Town were improved based on the Law, unimproved paddy fields in many villages have been producing Cd polluted rice with over 0.3 ug/g every year. Therefore, Cd polluted rice has been produced every year all over Japan in the paddy fields around the agricultural land soil pollution policy areas and in the polluted areas which are not officially designated.

Unpolluted Areas

According to the general survey of 1971–73 (1–3), the average Cd concentration of unpolished rice in the general area (unpolluted area) was 0.09 ug/g (N = 8,125). In a general survey conducted in Chiba Prefecture near Tokyo, 6% of the rice contained above 0.3 ug/g Cd and 2% of the rice had above 0.4 ug/g Cd (20). Ito and Harada (19) analyzed the unpolished rice sent to Tokyo for 1973–80 and found that the average concentration of Cd was 0.105 ± 0.142 ug/g, and 3.3% of the samples contained above 0.4 ug/g. Suzuki et al. (24) studied the Cd concentration in rice from the data of the literature as well as their own analysis and concluded that Japanese rice has consistently two or three times higher Cd content than rice produced in other countries.

Table 3 shows the Cd concentration of unpolished rice all over Japan as analyzed by the National Institute of Hygienic Sciences and many other local hygienic institutes (26). These data show that Cd concentration of Japanese rice

Year	Sample No.	Min.	-10%-	-50%-	-90%-	Max.	Average	SE
1971	412	1.0	20.0	110.0	350.0	3,010	186.2	295.7
1972	200	0.5	10.0	75.0	260.0	550	112.7	101.9
1973	457	0.5	20.0	160.0	580.0	3,040	239.4	271.3
1974	514	5.0	20.0	80.0	260.0	1,710	124.3	156.2
1975	346	1.0	10.0	50.0	230.0	2,300	121.7	238.2
1976	256	1.0	20.0	70.0	206.0	868	99.0	103.5
1977	417	0.5	20.0	63.0	460.0	1,330	150.0	231.2
1978	333	2.5	10.0	60.0	210.0	640	89.5	90.7
1979	242	2.5	30.0	90.0	200.0	770	106.7	87.7
1980	202	1.0	5.0	50.0	240.0	800	96.6	110.2

TABLE 3 — Cadmium concentrations in unpolished rice (ng/g) (26).

is elevated by a slight pollution of paddy soils in so-called unpolluted areas. Cadmium in rice rarely exceeds 0.1 ug/g in unpolluted paddy soils under usual cultivating conditions.

These results in Table 3 show that, in spite of the countermeasures against Cd polluted rice, Cd in Japanese rice is still higher than that produced in other foreign countries. Therefore, Cd pollution of rice is a severe problem not only in the polluted areas, but also in so-called unpolluted areas.

In the areas without specific pollution sources, the main Cd sources for soil pollution are atmospheric fallout and fertilizer application (5). Table 4 shows the amounts of Cd in atmospheric fallout in some places of Japan and Sweden (5,13,25). Even in such urban areas as the Wards of Tokyo, Kyoto City, or Osaka City, the amount of atmospheric fallout of Cd was much less than that around the Annaka Zn smelter. According to Hirata (13), only 33% of the Cd emitted from the chimneys fell out around the Zn smelter, and the rest of it was carried to greater distances. The amount of atmospheric fallout of Cd in the Wards of Tokyo, Kyoto City, or Osaka City was two to four times as much as that in each of the control areas. The amount of atmospheric Cd in Uppsala was less than that in the three control areas of Japan. Nriagu (21) presented a list of the amounts of deposition of Cd in atmospheric fallout in the UK, the USA, and some other countries.

Location	Cd Fallout Rate ($mg/m^2/year$)	Sampling Period	Reference
Uppsala	0.11	May 1976 - May 1978	(5)
Wards of Tokyo	0.76	Jul. 1976 - Jun. 1979	(25)
Okutama, Tokyo**	0.20	Jul. 1976 - Jun. 1979	(25)
Kyoto City	0.31	Aug. 1982 - Dec. 1982	*
Seki, Kyoto**	0.14	Aug. 1982 - Dec. 1982	*
Osaka City	0.94	Aug. 1981 - Dec. 1981	*
Toyono, Osaka**	0.22	Aug. 1981 - Dec. 1981	*
Annaka City			
500 m***	6.6	Oct. 1971 - Apr. 1978	(13)
1,000 m***	4.0	Oct. 1971 - Apr. 1978	(13)
2,000 ***	2.2	Oct. 1971 - Apr. 1978	(13)

* Onozuka, H., Asami, T., and Homma, S., unpublished data.
** Control locations of the Wards of Tokyo, Kyoto City, and Osaka City, respectively.
*** Distance from the chimney of Annaka Smelter of Toho-Aen Co., Ltd. (Zn smelter).

TABLE 4 — Deposition of cadmium in atmospheric fallout.

LAND APPLICATION OF SEWAGE SLUDGE CONTAINING HEAVY METALS

Sewage sludge has been applied to agricultural soils in many countries. If it is used as compost, the most important problem is heavy metal pollution of soils. As shown in Table 5a, the legal regulation of concentrations of Hg, Cd, and As is established by the Fertilizer Control Law in Japan to prevent pollution. With regard to solubility of toxic matters in the applied sewage sludge, the legal regulation was made on the critical concentrations of toxic substance as shown in Table 5b. The sewage sludge for land application is required to meet both regulations for toxic matters (17).

"Recommended maximum cumulative sludge metal applications for privately-owned croplands" in the USA are shown in Table 6 (10). Under maximum applications, Cd in soils exceeded 5, 10, and 20 ug/g, respectively, assuming the plow layer is 10 cm and the apparent soil density is one. These values are found in the heavily polluted soils in Japan. According to "Recommended maximum permissible addition of elements in sewage sludge to uncontaminated noncalcareous soils over a period of 30 years or more" in the UK (22), maximum permissible addition of Cd is 5 kg/ha, which corresponds to 5 ug/g Cd in soil. As mentioned above, the upland crops absorb more Cd than rice.

Figure 7 shows the Cd concentration of edible crops grown on soil treated with a Cd-enriched sewage sludge under greenhouse conditions (9). Ten grams of Cd-enriched sewage sludge were added per 1 kg of soil. From Fig. 7, it is clear that the edible tissues of upland crops, except squash, absorbed more Cd than the rice grain. The amounts of Cd in these upland

(a) Standard of concentration in sewage sludge (mg/kg D.S.).

Components	Hg	Cd	As
Concentration	2.0	5.0	50.0

(b) Standard of concentration in a leachate water (mg/l).

Components	Cd	Pb	Cr^{6+}	As	Hg	Alk-Hg	CN^-	Org-P	PCB
Concentration	0.3	3.0	1.5	1.5	0.005	ND	1.0	1.0	0.003

TABLE 5 — Legal regulation of toxic matters related to land application of sewage sludge (17).

Metal	Soil Cation Exchange Capacity 0-5	5-10	15 meq/100 g
	maximum application, kg/ha		
Zn	250	500	1,000
Cu	125	250	500
Ni	50	100	200
Cd	5	10	20
Pb	500	1,000	2,000

1. Annual Cd application should not exceed 2 kg/ha from dewatered or composted sludge, or 1 kg/ha from liquid sludge; sludge should not supply more crop available nitrogen than the crop requires.
2. Sludge with Cd 25 ug/g should not be applied unless the Cd/Zn exceeds 0.015; if Cd/Zn exceeds 0.015, an abatement program to reduce sludge Cd should be initiated.
3. These recommendations apply only to soils that are adjusted to pH 6.5 when sludge is applied and are to be managed to pH 6.2 thereafter.
4. Leafy vegetables or tobacco cropland should not receive sewage sludge application.
5. The cation exchange capacity is for unamended soil.

TABLE 6 — Recommended maximum cumulative sludge metal application for privately-owned cropland (10).

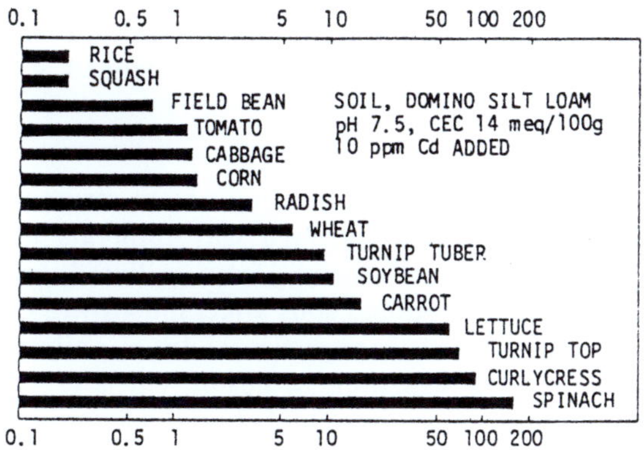

FIG. 7 — Cd concentration of edible tissue grown on a soil treated with a Cd-enriched sewage sludge (based on Bingham et al. (9)).

crops are higher than those in field experiments. Figure 8 shows the results of the survey of Cd concentration in rice grains and wheat grains produced on the same soils in the Annaka district (27), with the concentration of Cd in the wheat grains higher than that of rice grains. Of course, the rice was grown

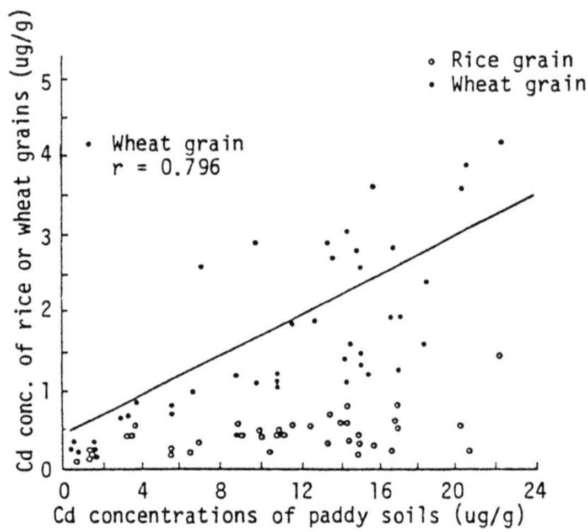

FIG. 8 — Relation between the Cd concentration of paddy soils and that of rice or wheat grains (27).

under submerged conditions and the wheat was grown under upland conditions. According to these data, the recommended values should be lowered both in the USA and UK.

CONCLUSION

Soil pollution by metal is essentially different from air or water pollution, because the soil adsorbs metals through clay minerals and humus and accumulates them. Removal of metals from polluted soils is difficult. Therefore, the metals in the polluted soils affect the growth of crops and the health of animals and humans through the intake of polluted crops over a long period. In connection with human health, Cd is the most important metal in the soil pollution problem. The behavior of Cd in the paddy soil-rice plant-human system is revealed through the studies in Japan.

To protect human beings from toxic elements such as Cd, it is necessary to establish a "pedomedical science" to promote the studies on the soil-plant-(animal)-human relationship of Cd and other metals. Cooperation of the scientists in many fields is now required.

REFERENCES

(1) Agricultural Administration Bureau, Ministry of Agriculture and Forestry. 1972. Results of Survey for Agricultural Land Soil Pollution Policy, 1971. Tokyo: Agricultural Administration Bureau, Ministry of Agriculture and Forestry.

(2) Agricultural Production Bureau, Ministry of Agriculture and Forestry. 1973. Results of Survey for Agricultural Land Soil Pollution Policy, 1972. Tokyo: Agricultural Production Bureau, Ministry of Agriculture and Forestry.

(3) Agricultural Production Bureau, Ministry of Agriculture and Forestry. 1974. Results of Survey for Agricultural Land Soil Pollution Policy, 1973. Tokyo: Agricultural Production Bureau, Ministry of Agriculture and Forestry.

(4) Akita Prefectural Experiment Station. 1978. Studies on the Migration of Damage for Crop by the Heavy Metal in Soil. Akita: Akita Prefectural Agricultural Experiment Station.

(5) Andersson, A. 1981. Cadmium and lead in precipitation and drainage water. Swedish J. Agric. Res. **11**: 119–125.

(6) Asami, T. 1981. Maximum allowable limits of heavy metals in rice and soil. **In** Heavy Metal Pollution in Soils of Japan, eds. K. Kitagishi and I. Yamane, pp. 257–274. Tokyo: Japan Scientific Societies Press.

(7) Asami, T. 1981. The Ichi and Maruyama River basins: Soil pollution by cadmium, zinc, lead, and copper discharged from Ikuno mine. **In** Heavy Metal Pollution in Soils of Japan, eds. K. Kitagishi and I. Yamane, pp. 125–136. Tokyo: Japan Scientific Societies Press.

(8) Asami, T.; Homma, S.; Wada, T.; Nakashima, K.; and Kubota, M. 1983. Pollution of unpolished rice produced in the Ichi and Maruyama River basins by cadmium discharged from Ikuno Mine, Hyogo Prefecture. Jpn. J. Soil Sci. Plant Nutr. **54**: 30–36.

(9) Bingham, F.T.; Page, A.L.; Mahler, R.J.; and Ganje, T.J. 1975. Growth and cadmium accumulation of plants grown on a soil treated with a cadmium-enriched sewage sludge. J. Env. Qual. **4**: 207–211.

(10) Chaney, R.L. 1982. Fate of toxic substances in sludge applied to cropland. **In** International Symposium on Land Application of Sewage

Sludge, October 13–14, 1982, Tokyo, pp. 259–324. Tokyo: Association for Utilization of Sewage Sludge.

(11) Environment Agency, Government of Japan. 1982. Quality of the Environment of Japan, 1981. Tokyo: Health, Welfare and Environment Problems Research Society.

(12) Environment Agency. 1982. White Paper on Environment, 1982. Tokyo: Printing Bureau, Ministry of Finance, Japan.

(13) Hirata, H. 1981. Annaka: Land polluted mainly by fumes and dust from zinc smelter. **In** Heavy Metal Pollution in Soils of Japan, eds. K. Kitagishi and I. Yamane, pp. 149–163. Tokyo: Japan Scientific Societies Press.

(14) Honma, Y., and Hirata, H. 1976. Effects of Cd and/or Zn concentrations in the nutrient medium on the growth and Cd uptake of rice. J. Soc. Soil Manure, Japan **47**: 314–320.

(15) Iimura, K. 1981. Background contents of heavy metals in Japanese soils. **In** Heavy Metal Pollution in Soils of Japan, eds. K. Kitagishi and I. Yamane, pp. 19–26. Tokyo: Japan Scientific Societies Press.

(16) Iimura, K. 1981. Heavy metal problems in paddy soils. **In** Heavy Metal Pollution in Soils of Japan, eds. K. Kitagishi and I. Yamane, pp. 37–50. Tokyo: Japan Scientific Societies Press.

(17) Ishikawa, T., and Dodo, I. 1982. Present status and land application of sewage sludge in Japan. **In** International Symposium on Land Application of Sewage Sludge, October 13–14, 1982, Tokyo, pp. 1–14. Tokyo: Association for Utilization of Sewage Sludge.

(18) Ito, H., and Iimura, K. 1976. The absorption and translocation of cadmium in rice plants and its influence on their growth in comparison with zinc. Bull. Hokuriku Natl. Agric. Exp. Stn. No. **19**: 71–139.

(19) Ito, K., and Harada, H. 1980. Contents of cadmium in brown rice sent in Tokyo. Ann. Rep. Tokyo Metrop. Res. Lab. Publ. Health **31**: 142–148.

(20) Matsumoto, N.; Higurashi, N.; and Miyoshi, H. 1976. Heavy metal pollution in the paddy fields around the city dust incinerators. Bull. Chiba-Ken Agri. Exp. Station **17**: 140–149.

(21) Nriagu, J.O. 1980. Cadmium in the atmosphere and in precipitation. **In** Cadmium in the Environment, ed. J.O. Nriagu, Part 1, pp. 71–114. New York, Chichester, Brisbane, Toronto: John Wiley & Sons.

(22) Ramsay, R.C. 1982. The establishment of guidelines and monitoring system for the disposal of sewage sludge to land. **In** International Symposium on Land Application of Sewage Sludge, October 13–14, 1982, Tokyo, pp. 325–352. Tokyo: Association for Utilization of Sewage Sludge.

(23) Saito, H.; Shitomi, K.; Hagai, K.; and Arikawa, T. 1978. Chronic cadmium poisoning induced by environmental cadmium pollution. **In** Chudokugaku to Eiyogaku (Toxicology and Nutrition Sciences), eds. T. Suzuki, N. Imura, and S. Suzuki, pp. 85–99. Tokyo: Shinohara Publishing Co.

(24) Suzuki, S.; Djuangshi, N.; Hyodo, K,; and Soemarwoto, O. 1980. Cadmium, copper, and zinc in rice produced in Java. Arch. Environ. Contam. Toxicol. **9**: 437–449.

(25) Tokyo Metropolitan Research Institute for Environmental Protection. 1982. Data on Particulate Matter in the Atmosphere. Tokyo: Tokyo Metropolitan Research Institute for Environmental Protection.

(26) Uchiyama, M., ed. 1982. Decenial Summary of Food Contamination Monitoring Data of Japan (1971–1980). Tokyo: National Institute of Hygienic Sciences.

(27) Yamada, K. 1979. Soil pollution by copper mine and zinc smelter. **In** Dojo-osen no Kiko to Kaiseki (Mechanism and Analysis on Soil Pollution), ed. M. Shibuya, pp. 1–37. Tokyo: Sangyo Tosho.

Changing Metal Cycles and Human Health, ed. J.O. Nriagu, pp. 113–141.
Dahlem Konferenzen 1984. Berlin, Heidelberg, New York, Tokyo: Springer-Verlag.

THE CONTAMINATION OF PLANTS AND SOILS WITH HEAVY METALS AND THE TRANSPORT OF METALS IN TERRESTRIAL FOOD CHAINS

A. Kloke*, D.R. Sauerbeck**, and H. Vetter***
*Biologische Bundesanstalt für Land- und Forstwirtschaft
1000 Berlin 33, F.R. Germany
**Institut für Pflanzenernährung und Bodenkunde
Bundesforschungsanstalt für Landwirtschaft
3300 Braunschweig, F.R. Germany
***Landwirtschaftliche Untersuchungs- und Forschungsanstalt
2900 Oldenburg, F.R. Germany

Abstract. The most important sources of heavy metal contamination in their order of significance are air pollution, river sediments, sewage sludges, town waste composts, agricultural chemicals, and industrial waste.

Lead accumulation in plants near highways and roads has decreased considerably since the law controlling the lead content of petrol came into force. Hazardous amounts of lead, however, are still concentrated 10–20 m along the sides of highways and roads, and in the strip of up to 5 m there are also higher concentrations of cadmium.

Soil properties such as texture, pH-value, and ion exchange capacity strongly influence the availability of trace metals to plants. Lead, chromium, and mercury are usually quite tightly fixed in soils, whereas cadmium and zinc can be readily absorbed from there by plant roots. Food and forage contamination with lead is usually due to emission from the air rather than to its uptake from soils.

Guidelines are discussed for critical contents in town waste materials, plants, and soils. Considerable differences in the uptake exist not only between elements and soil types, but also depending on the plant families, plant species, and plant organs concerned. Relatively low concentrations in staple foods may be more hazardous than higher contents in food items consumed in small amounts.

Excess exposure of animals and man by high contents in plants is possible with lead, cadmium, and thallium (and to some extent also with copper), whereas zinc, nickel, chromium, and mercury are more toxic to plants than the hazard which such food plants constitute to their consumers.

Fortunately, critical heavy metal contents in plants and soils have so far been ascertained in small areas only. The areas with the highest accumulation of heavy metals in soils and plants are found in the vicinity of industrial plants, cities, and rivers.

INTRODUCTION

In 1975 the Deutsche Forschungsgemeinschaft published a map of the world entitled "Worldwide Shortage of Raw Materials Makes Increased Efforts in Scientific Research Imperative" (2) which shows which countries export which mineral substances and metals to the Federal Republic of Germany (Fig. 1), and every year the Metallgesellschaft AG, Frankfurt, compiles statistics (4) which give information about the amounts of nonferrous metals being produced and used in the FRG and other countries.

These studies, however, do not inquire as to the whereabouts of these heavy metals after they have done their job in the different products. They get into the waste from which they may then pass into arable soil via town waste compost and sewage sludges or via effluents and dust from waste dumps and landfills. Such places may also be grassed over after their closure (104). Figure 2 summarizes the cycle of heavy metals in industrial societies.

After being processed by man in industry, heavy metals are transported into the soil via polluted air, liquid and solid waste (sewage sludge and compost), and by some agricultural chemicals (1,13,19,20,25,32–34,45,48,55,65,76,

A

B

CANADA
Iron
Zinc
Titanium
Copper
Lead
Nickel
Molybdenum
Nb-Ta-V

MEXICO
Zinc

PERU
Zinc
Lead
Copper

BOLIVIA
Zinc
Lead
Tin
Tungsten
Antimony

KOREA
Lead
Tungsten

JAPAN
Manganese

CHINA
Tungsten
Aluminum

EEC
Iron
Copper
Aluminum
Lead
Tin
Kaolin
Fluorite
Uranium

VENEZUELA
Iron

BRAZIL
Iron
Manganese
Nb-Ta-V
Tungsten

CHILE
Copper
Molybdenum

NEW GUINEA
Copper

POLAND
Sulfur
Copper

MALAYSIA
Tin

INDONESIA
Tin

SCANDINAVIA
Iron
Pyrites
Titanium
Zinc
Lead
Copper
Nickel
Cobalt

SPAIN
Iron
Copper
Fluorite
Tungsten

USSR
Chromium
Pyrites
Manganese
Copper
Nickel

SE-EUROPE
Aluminum
Copper

AUSTRALIA
Iron
Aluminum
Manganese
Titanium
Lead
Zircon
Nickel

USA
Copper
Molybdenum
Tungsten
Zircon
Cobalt
Nb-Ta-V

NICARAGUA
Copper

GUAYANA
Aluminum

MOROCCO
Manganese
Phosphate
Copper
Lead

MAURITANIA
Iron

SIERRE LEONE
Iron
Aluminum

LIBERIA
Iron

IRAN
Chromium
Zinc

PHILIPPINES
Copper

SOUTH AFRICA
Chromium
Manganese
Copper
Nickel
Zircon

MADAGASCAR
Graphite

NIGERIA
Tin

GABUN
Manganese

ZAIRE
Copper
Cobalt

ANGOLA
Iron

ZAMBIA
Copper

CYPRUS
Copper

TURKEY
Chromium
Antimony

THAILAND
Tin
Tungsten

FIG. 1 — (A) Map illustrating flow of mineral substances and metals from supplier countries into the Federal Republic of Germany. (B) Supplier countries and the respective mineral substances and metals they furnish (cf. (2,3)).

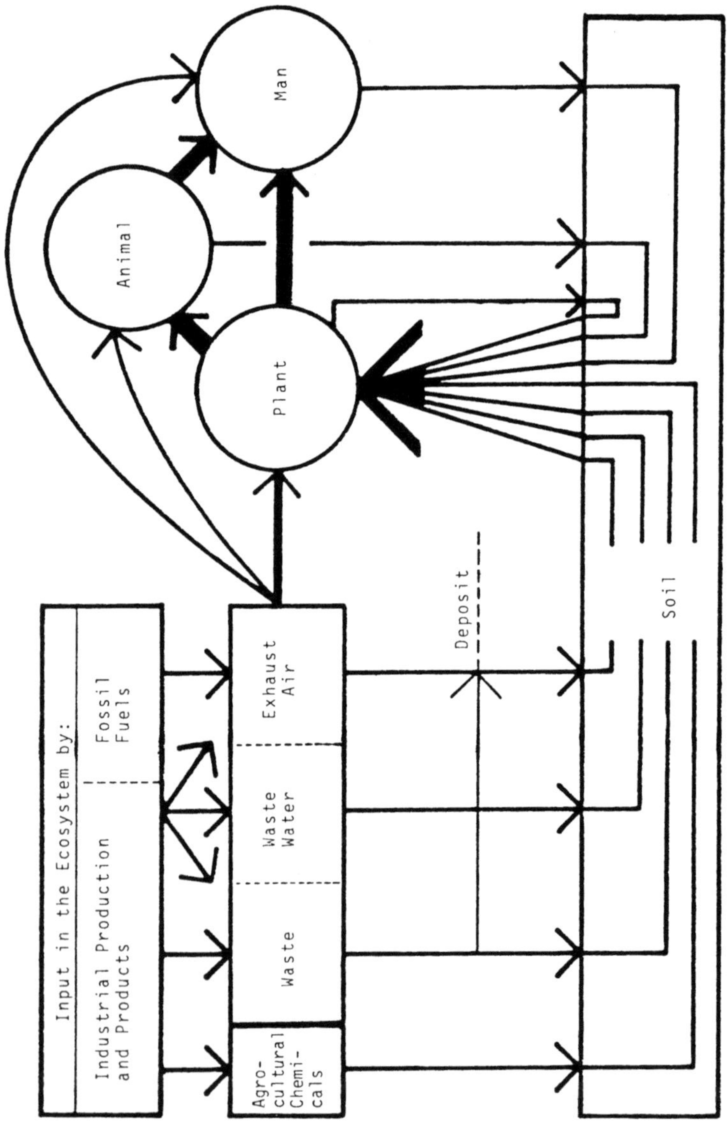

FIG. 2 — Cycle of heavy metals in industrial societies (58).

93,96,103). As can be seen from Fig. 2, heavy metals from fossil fuels are also transferred to the soil via exhaust air. Most of them will be absorbed there, but some part may also be transformed into a more mobile form which will enable the metals to be either taken up by plants or washed out from the soil. Immobilization in soil, intake by plants, transfer to the groundwater, as well as effect on fertility, etc., are very different for each heavy metal, each plant, and each soil type (1,21,23–25,28,40,45,61,65,74,82,90,99).

After uptake from the soil by plants, these heavy metals reach man either directly or via animals, then to be reintroduced into the soil, where they again await uptake by plants together with newly transported heavy metals from polluted air, wastewater, and solid waste. Once having entered this cycle most of them can hardly break out of it, except for the biomethylation of elements such as Hg, As, and possibly Sn (1,20,28,45,53–61,65).

AIR POLLUTION

For energy production and combustion, man uses organic materials which originate from plants. Mineral oil, coal, and wood, as well as the fuels made from them, contain practically all elements from aluminum to gold, albeit in very different amounts. Some metals such as Pb and Mn are even intentionally added to liquid fuels for technical reasons. During the burning process some of these elements are released into the air and transported into soil by precipitation. In addition to this, there are also the heavy metals released directly into the air by industry (45).

According to the administrative regulations concerning the amendment of the technical directives for the prevention of air pollution in the FRG, the following limits in Table 1 for lead, cadmium, and thallium have been provided (6):

Metal	Airborne dust $\mu g/m^3$	Dust precipitation $\mu g/m^2/d$	kg/ha/a
Lead	2.000	250	0.913
Cadmium	0.040	5	0.018
Thallium		10	0.036

TABLE 1 — Maximum allowable amounts of some heavy metals in airborne dust (6).

SOIL POLLUTION

Sewage Sludge

The regulations of the FRG concerning sewage disposal require that wastewater be purified before it is fed into rivers and lakes, which has led to a widespread construction of sewage treatment plants. This, however, has only shifted the problem of harmful substances from the aquatic to the terrestrial sphere. In the FRG 8% of the sewage sludge obtained is incinerated (which leads to air pollution), 49% is deposited at dumps (thus occupying land which could be utilized in another way), and 38% is used in agriculture as manure (60,92,93). As some sewage sludges contain high amounts of heavy metals, regulations concerning the use of these wastes in agriculture have been laid down. According to these regulations, heavy metals in sewage sludge used on soils may not exceed the values given in Table 2.

The figures in Table 2 are total contents, although it is known that different chemical bindings and forms of heavy metals exist in soils and have great influence on their availability to plants. Research on metal speciation in soils is, therefore, under way (13,40,45,62,77,84,100), but the results are not yet conclusive enough to be incorporated into legal regulations. As will be explained later, soil properties such as pH and clay and humus content also play a decisive role in heavy metal availability but can hardly be taken into reliable account as far as the generalizing legislation on the agricultural utilization of town waste materials is concerned.

Composts

About 2% of the sewage sludge in the FRG is composted and applied to agricultural and horticultural soils as a portion of the composts which are partly

Element	Sewage sludge (ppm)	Soil (ppm)
Zinc	3,000	300
Copper	1,200	100
Lead	1,200	100
Chromium	1,200	100
Nickel	200	50
Cadmium	20	3
Mercury	25	2

TABLE 2 — Threshold values of certain heavy metals in sewage sludge and soil (dry matter) (5,56,60,68,79).

manufactured from the 24 million tons a year of domestic garbage (42,92,93). There are no precise statistics available concerning the amount of composts being applied or about the size of the area used for food production which is fertilized with this material. It is supposed, however, that such composts are not transported farther than 20–30 km, for otherwise transportation costs would exceed their value. The heavy metal content of composts is sometimes considerable. The range of the heavy metal content of garbage and garbage/sewage-sludge composts is summarized in Table 3.

River Sediments

In stagnant as well as in running waters there is a deposition of solid organic and inorganic materials at their bottom. This process is dependent on many environmental factors. These deposits may contain more or less high amounts of heavy metals which in turn depend on the sewage fed into these waters. In order to maintain shipping it is necessary to dredge waterways periodically. The dredged material is often applied to nearby soils in large amounts. These soils may thus be enriched with considerable amounts of heavy metals (41,58,78,101).

Furthermore, heavy metals and floating particles are deposited when creeks and rivers overflow their banks. In these instances, heavy metals are not only deposited in the soil but there is also some direct contamination of plants (especially those of meadows and pastures). There have been repeated cases of death among cattle after eating plants contaminated with soil containing high amounts of lead (8).

Agricultural Chemicals

Whereas the input of heavy metals into the soil via the atmosphere cannot be termed intentional, and the input via application of garbage/sewage-sludge composts and sewage sludges does not necessarily have to be accepted, the

Heavy metal	Range (ppm)
Lead	20 - 6,000
Copper	50 - 5,000
Zinc	300 - 12,000
Nickel	10 - 500
Cadmium	1 - 40

TABLE 3 — Range of the heavy metal content of garbage and garbage/sewage-sludge composts (42,58,92).

input of certain heavy metals via agricultural chemicals is usually intended. Some of these trace elements are present in agricultural chemicals such as fertilizers in the form of impurities. Iron, copper, zinc, cobalt, manganese, and magnesium are essential nutrients for plants, animals, and man. These are applied, where necessary, under controlled conditions in special fertilizers. Some pesticides in use contain copper, zinc, or manganese as components of their active ingredients. Furthermore, most mineral fertilizers and pesticides — as well as all natural products — contain other elements in more or less minute amounts. The German Fertilizer Legislation (Düngemittelgesetz) has already issued some threshold values for the content of lead and chromium in certain fertilizers (85).

Currently being debated are especially the cadmium contents of phosphate fertilizers and the mercury in seed treatment agents. Rock phosphates used in Europe contain from 0.1 ppm (Kola-Phosphate) up to more than 75 ppm cadmium (phosphates from Western Africa). The phosphate fertilizers manufactured from these raw products in the FRG contain somewhere between 5 and 25 ppm Cd, so that, on the average, 3.5 g Cd/ha/a are introduced by using these fertilizers (46,47,49,81,83,95,96,98). A summary of the cadmium balance of German soils is shown in Fig. 3.

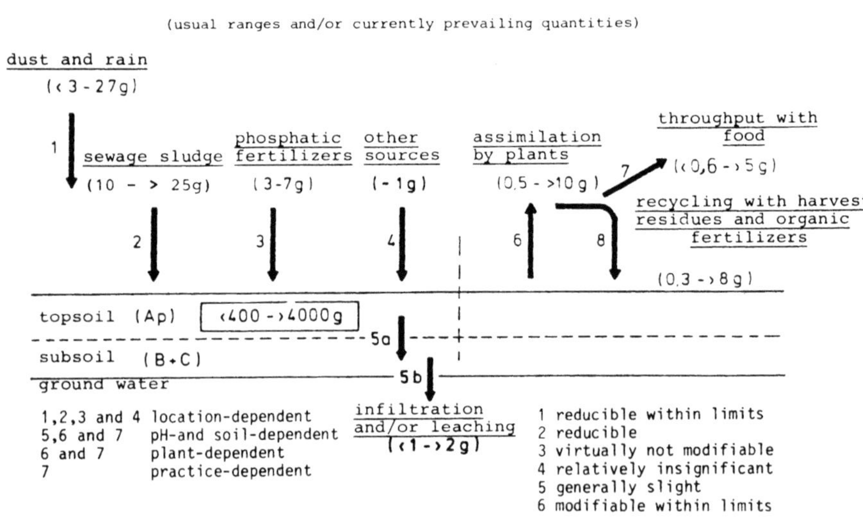

FIG. 3 — Cadmium inputs and throughputs on agricultural land (81).

BEHAVIOR OF HEAVY METALS IN SOILS

When discussing the behavior of heavy metals in soil, the central question is the uptake of these elements by plants, for in the interests of the health of man and animals only a certain level of heavy metals in plants can be tolerated. Guidelines for the maximum allowable amounts in food exist for the elements lead, cadmium, and mercury (16,35,50) (Table 5). These, however, can only be adhered to in plant cultivation if certain corresponding concentrations of

Element		Total amount in air dried soil (mg/kg)		
		Range	Special or contaminated soils	Allowable concentration
As	Arsenic	0.1 - 20	< 8,000	20
B	Boron	5 - 20	< 1,000	25
Be	Beryllium	0.1 - 5	< 2,300	10
Br	Bromine	1 - 10	< 600	10
Cd	Cadmium	0.01 - 1	< 200	3
Co	Cobalt	1 - 10	< 800	50
Cr	Chromium	2 - 50	< 20,000	100
Cu	Copper	1 - 20	< 22,000	100
F	Fluorine	50 - 200	< 8,000	200
Ga	Gallium	0.1 - 10	< 300	10
Hg	Mercury	0.01 - 1	< 500	2
Mo	Molybdenum	0.2 - 5	< 200	5
Ni	Nickel	2 - 50	< 10,000	50
Pb	Lead	0.1 - 20	< 4,000	100
Sb	Antimony	0.01 - 0.5	?	5
Se	Selenium	0.01 - 5	< 1,200	10
Sn	Tin	1 - 20	< 800	50
Tl	Thallium	0.01 - 0.5	< 40	1
Ti	Titanium	10 - 5,000	< 20,000	5,000
U	Uranium	0.01 - 1	< 115	5
V	Vanadium	10 - 100	< 1,000	50
Zn	Zinc	3 - 50	< 20,000	300
Zr	Zircon	1 - 300	< 6,000	300

TABLE 4 — Guidelines for allowable concentrations of certain elements in arable soil (56,60).

these elements in the soil are not exceeded. Some threshold values of the elements most relevant for the environment are suggested for agricultural soils in Table 2. However, research and discussion on tolerable amounts for other elements in soils have only just begun (56,60,89) (Table 4).

Mercury

The native content of mercury in soils is in the range of 0.01–1.0 ppm. It is absorbed both by organic and inorganic soil constituents. Special attention should be called to its methylation (especially in soils rich in organic matter) and to the release of metallic mercury into the atmosphere as a result of its low vapor tension. In undisturbed soil profiles, more mercury is found in the upper layers which are rich in organic matter. Mercury is enriched there by precipitation and shed plant parts which take mercury and its compounds out of the air (48,65). The uptake of mercury by plants is low in uncontaminated soils but may be of some concern in soils which have been fertilized with sewage sludge. However, the use of seed treatment agents containing mercury, which are now prohibited in the FRG, does not seem to have led to an increase of the mercury content in food and fodder crops (31,37,57,59).

Lead

The lead content of soils is normally in the range of 0.1–20.0 ppm. This content, however, can be increased up to 4,000 ppm near industrial plants emitting lead, and to more than 300 ppm in urban soils by the combustion of fuels such as wood, coal, and oil. The exhaust fumes of automobiles can increase the lead content by 100 ppm in the soils at the sides of busy roadways up to a distance of about 20 m (54,58,66,69,101,102). A considerable improvement has been achieved through the reduction of gasoline leading being enforced by law.

Further sources of lead contamination of soils are garbage composts, sewage sludges, and river sediments (particularly when composts are manufactured from garbage containing industrial waste). In soil lead is included in the processes of precipitation and transformation. In the course of time the availability of soil lead to plants usually decreases. This is especially the case when the pH of the soil is about 7.0 and free $CaCO_3$ is available or applied to the soil. On the other hand, fixed lead (as well as most other heavy metals) can be mobilized again by a decrease in the pH. This is especially the case after uncontrolled acidic precipitation for many years and after the regular application of physiologically acidic fertilizers which have not been balanced by soil liming (21,28,43,59,67,74,86).

Cadmium

Of all the elements causing environmental concern, cadmium is at present receiving the largest attention. The normal cadmium content of soils, which is far below 1.0 ppm, can be increased to over 100 ppm by emissions from certain industries, domestic heating, river sediments, and sewage sludges (11, 17,18,29,36,41,52,58,61,65,69,76,78,80,95,96,101,102). The accumulation of Cd near roads from tire wear is usually concentrated in a narrow strip of only 5 m.

The suggested guideline value in Germany for cadmium in agricultural soils is at present 3 ppm (Table 2). When cadmium is transferred to soil, it undergoes processes of transformation and immobilization similar to those mentioned for lead. However, it retains its availability to plants on a much larger scale, and this can be increased even further by only slight decreases in the pH. Its uptake by plants depends very much on the plant species (11,21–23,28,29, 43,59,61,74,82,87,88,90,96,99). Furthermore, the lower the clay and humus contents of the soil, the greater the availability of cadmium to plants. Light and acid soils can immobilize cadmium less readily than heavy soils with a neutral pH. As cadmium is mobilized much more easily by acidic precipitation and regular application of physiologically acidic fertilizers than is lead, limitation of its transfer to soils and regular liming of soils are of utmost importance (11,17,18,21,22,25,29,59,64,90,95,96).

Other Metals of Possible Concern

In Table 4 several other elements of toxicological significance are also listed. These elements enter the soil by different routes and may accumulate there. Particular attention must be paid to As, Be, Cd, Cr, Pb, Ni, Sn, and Se which are themselves carcinogenic, as are their compounds. Given that the threshold limits of the elements as shown in Table 4 are not exceeded and the pH of soil is about 7.0, to our present knowledge harmful effects to the health of man and animals will not result from plants grown on such soil (56,57,59,60).

PLANT POLLUTION

An increasing concentration of heavy metals can decrease plant yields and crop quality, i.e., affect their growth and suitability as food and fodder crops. There are different threshold limits to the heavy metal content prescribed for these uses (15,16,22,37–39,43,50,89,97). The following points are relevant to the determination of such threshold limits:

Critical Contents in Food Plants (Table 5)

If the level of heavy metals in plants is increased by air pollution, damage to plant growth will occur at much higher levels than that which occurs as a result of an uptake of the same heavy metals from the soil alone. Heavy metals deposited on the surface of leaves either do not affect plants or are less detrimental than heavy metals which are actually taken up (45,55,57,82,101,103). In areas heavily contaminated with lead and cadmium these elements affect human health first, and only at higher levels do they affect the health of domestic animals and/or the crop yields (Table 6) (19–21,29,30,82,101,103). A similar situation appears to exist in areas polluted with thallium (44,63,75). In establishing threshold limits for food and fodder crops, more attention should be paid to the fact that a low but constant average intake of heavy metals by man and animals over long periods can affect health more than a much higher short-term exposure. In tests on cattle which were supplied with a dose of 230 ppm Pb in their feed, negative effects on health could be observed after 9 months (101).

Foodstuffs	Guidelines '79 (mg/kg fresh weight)		
	Lead	Cadmium	Mercury
Green vegetables	1.2	0.1	no values given
Sprout vegetables	1.2	0.1	no values given
Fruits	0.2	0.1	no values given
Root vegetables	0.5	0.05	no values given
Pomacious fruits	0.5	0.05	no values given
Stone fruits	0.5	0.05	no values given
Berries	0.5	0.05	no values given
Cereals	0.5	0.1	0.03
Potatoes	0.2	0.1	0.02

* The Ministry of Nutrition, Agriculture, Environment, and Forestry of Baden-Württemberg, Germany, writes on this subject: "These guidelines are not to be exceeded, if possible. Where the content of the foodstuffs of plant origin is twice as high as the 'Guidelines' indicate, the products will be rejected during the ordinary course of the official food quality control, and on a commercial basis are no longer allowed to be brought into circulation" (3).

TABLE 5 — Guidelines '79 for lead, cadmium, and mercury in foodstuffs of plant origin* (16,50).

The heavy metal content of an individual feed or the food intake of one day is not important, but the intake of heavy metals over longer periods of time is. Food of plant and animal origin from polluted areas which is ingested in small amounts must be evaluated differently from that which is consumed constantly in greater quantities. High cadmium content in parsley, e.g., is unimportant, whereas medium concentrations of cadmium in potatoes can be very serious. In this connection the comparatively low guideline values for the heavy metal content of animal livers and kidneys deserve comment (30,50). In polluted areas only 2–4% of the total intake of lead and cadmium by humans are actually due to the consumption of liver and kidney. To be able to meet such low threshold values set for the heavy metal content of livers and kidneys, it was necessary to fix the corresponding threshold values of fodder crops at very low levels, too (15,16,37–39,43,59).

In regions where there are high emissions it is neither necessary nor possible to establish threshold limits for each of the many different plant species growing there. The critical values in areas with air pollution are those which must be kept to with regard to the quality of food and fodder (15,36,55,101–103). The amount in a particular plant species is not decisive for the potential "body burden" of man and domestic animals, but the amount of heavy metals in the total food ingested over long periods of time is. Thus, in high emission areas it may be sufficient to establish guideline values for the heavy metal content of certain standard or test crops such as ryegrass and/or kale, taking into account the relationship between the heavy metal content of these plant species and other food and fodder crops (9,23,36,101,102).

Phytotoxicity

More systematic experiments are necessary to establish the critical concentration of heavy metals for plants' growth (Table 6). Experiments conducted using an addition of heavy metal salts are only of limited relevance as Cd, Ni, and Zn salts applied to soils are taken up by plants 5–7 times more readily than the same heavy metals applied with sewage sludge (21,29,74,88,90,100). Similarly, heavy metals transferred to soil via the air are taken up more easily in most cases than those contained in sewage sludge and river sediments. This can be explained by the fact that not only do smaller particles reach the soil, but also acid-forming compounds are emitted along with the heavy metals which increase the availability of these metals to plants (11,18,55,84, 90,101,103). In areas with heavy soil pollution where the elements copper, zinc, nickel, chromium, and mercury are more likely to affect the plant yield than their nutritional quality, it might be advisable to set corresponding threshold values. A valuable guideline in this case as well would be the determination of the heavy metal content of standard crops such as ryegrass and kale (36,101,102) (Table 6).

The guidelines for permissible concentrations of heavy metals in soil (Tables 2 and 4) are an important indication of the potential burden of plants, but in some cases they need correcting as well as complementing with analyses of plants. A threshold value of 100 ppm for soil copper, e.g., seems to be too high for Northern Germany, for in a light acid soil a copper content of over 30–40 ppm can already affect crop yields. Similar differences may also exist for tropical soils versus those from the temperate climatic zone. On the other hand, an upper limit of 100 ppm for chromium and lead and 300 ppm for zinc is possibly fixed unnecessarily low (20,21,43,58,74,89,102).

Element	transfer coefficients plant/soil[a]	critical concentrations (μg/g d. m.) for plant growth[b]		critical concentrations (μg/g d. m.) in animal feed[c]
Cd	1 - 10	5 - 10	>	0.5 - 1
Co	0.01 - 0.1	10 - 20	<	10 - 50
Cr	0.01 - 0.1	1 - 2	<	10 - 50
Cu	0.1 - 1	15 - 20	≤	30 - 100
Hg	0.01 - 0.1	0.5 - 1		~ 1
Ni	0.1 - 1	20 - 30	<	50 - 60
Pb	0.01 - 0.1	10 - 20	≥	10 - 30
Tl	1 - 10	20 - 30	>	1 - 5
Zn	1 - 10	150 - 200	<	> 500
As	0.01 - 0.1	10 - 20		
Be	0.01 - 0.1	~ 1		
Se	0.1 - 10	10 - 20	>	4 - 5
Sn	0.01 - 0.1	~ 60		

[a] orders of magnitude only; the actual ratios vary depending on plant species and soil properties

[b] threshold values at which growth depressions of sensitive plant species start

[c] farm animals according to (91) and (97)

TABLE 6 — Transfer and critical concentration of trace metals in the soil-plant-animal chain (from (82)).

Heavy Metal Discrimination and Interaction

As already mentioned, most heavy metals, once they enter soil, are strongly fixed there unless low pH values promote their dissolution. This applies mainly to Pb, Cr, and Hg (but also to As, Be, and Sn) so that, contrary to their air deposition on plant surfaces, the carry-over of these elements from soil to plants does not constitute an important pathway to the consumer (Table 6) (19,20,22,28,45,55,57,59,82,93,103).

A second effective barrier for the transfer of most heavy metals is their limited translocation from plant roots to the shoots and fruits. In the plant realm one can roughly distinguish between accumulator, indicator, and excluder plants. The former enrich heavy metals in their tissue as compared with the soil in which they grow, whereas the contents of indicator plants correspond more or less closely to those in the nutrient medium. Excluder plants, on the other hand, can discriminate to some extent against heavy metals by accumulating them primarily in their roots, although above a certain limit the exhaustion of this storage capacity eventually leads to a breakthrough into other plant parts (7,12,73,82,100).

The sensitivity of different plants to heavy metal poisoning depends in part on this behavior and on the particular heavy metal concerned. Pb, Cd, and Tl are known to be tolerated by most plants in greater amounts than what is desirable for their consumption as food and feed. Only at concentrations far above this may toxicity symptoms also develop in the plants themselves. Such symptoms, however, are always indicative of a relatively advanced metabolic injury. Usually long before this an unspecific yield depression, which is just as undesirable, has already occurred (12,21,26,27,43,45,61,72,82).

As far as the elements Cu, Ni, and Zn are concerned, this yield depression starts at considerably lower contents than the tissue concentrations which could be harmful to mammals in edible plants, except perhaps for Cu in sheep (19,20,27,38,59,82). From this it may be concluded that the "passage poisons" Zn, Cu, and Ni in the food chain are more toxic to the plants themselves than to humans and animals which may eventually eat these plants. The "accumulative poisons" Pb and Cd (and probably Tl), on the other hand, are usually tolerated by plants in greater amounts than what is recommended for their long-term utilization as food and feed. Cr and Hg, however (as well as As and Be), are translocated from soils and roots into the plant shoots to such a small extent that toxicologically critical contents in food and feed are hardly to be expected (43,82,101).

The effects of heavy metals on both plants and animals vary not only from element to element but may also be enhanced or counteracted by the existence and metabolic behavior of other trace elements. This interdependency can be antagonistic as well as additive or, in some cases, even synergistic, and the extent and direction of this interrelationship also depends on the actual concentration (10,21,38,45,70,82). The experiments of Oberländer and Roth (70), e.g., have shown that Cu and Fe(III) ions can stimulate the absorption of Cd by plant roots, whereas Fe(II), Zn, and Mn exert inhibitory effect. This seems to have something to do with the electrochemical potential of the particular element concerned (Fig. 4).

However, not only the uptake from the growth medium but also the effect of certain heavy metal concentrations inside the plant tissue depend on the presence of other trace elements. Contrary to the aforementioned decrease of its uptake by increasing Zn additions, for instance, the plant yield depression by Cd is strongly enhanced by an excess of Zn, even though the tissue concentration of each individual metal alone may not yet constitute a significant detriment to plant growth (70).

Of the different plant species and families, the grasses take up much less heavy metals from soils than most dicotyledonous plant species do (23,28,59,

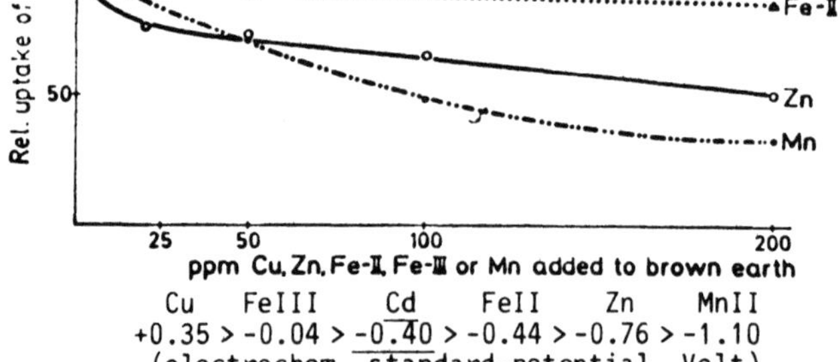

Cu FeIII Cd FeII Zn MnII
+0.35 > -0.04 > -0.40 > -0.44 > -0.76 > -1.10
(electrochem. standard-potential, Volt)

FIG. 4 — Effect of different heavy metals on the uptake of cadmium from soil (1 ppm Cd); 100 = uptake without additions (70).

61,82,90,100). The metabolically most active plant parts, especially the strongly transpiring plant leaves, exhibit considerably higher heavy metal contents than stems and storage organs or fruits and seeds, respectively. Even in cereal grains most heavy metals concentrate in the outer layers, so that the processed flour contains very little as compared with the bran or the inedible vegetative plant parts (Fig. 5) (70,71,90).

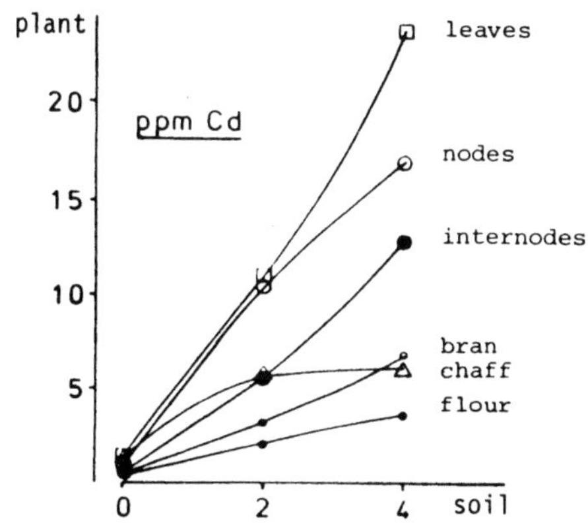

FIG. 5 — Incorporation of Cd into different plant parts and grain fractions of spring wheat (acid cambisol pH 5.60) (90).

high uptake	moderate uptake	low uptake	very low uptake
Lettuce	Kale	Cabbage	Snapbean family
Spinach	Collards	Sweet Corn	Pea
Chard	Beet	Broccoli	Melon family
Escarole	Turnip Roots	Cauliflower	Tomato
Endive	Radish Globes	Brussel Sprouts	Pepper
Cress	Mustard	Celery	Eggplant
Turnip Greens	Potato	Berry Fruits	Tree Fruits
Beet Greens			
Carrot			

TABLE 7 — Relative accumulation of heavy metals into edible plant parts by different crops (99).

On the other hand, some leafy vegetables such as chard, spinach, and lettuce are known to accumulate heavy metals most effectively. This results in a particularly high contamination risk of vegetable plants grown in urban gardens, which have been exposed over long periods to the emissions and wastes of municipal origin (Table 7) (22,53,59,99,102).

WHERE ARE CRITICAL BURDENS REACHED FIRST? WHICH METALS, WHICH SOURCES, AND WHICH AREAS?

So far, critically high concentrations of heavy metals covering large areas have not been ascertained. Looking at the problem on an average scale, the concentration of heavy metals in foodstuffs is far below the permissible amounts. In the case of lead the mean content in food has only reached 20–40% of the allowable amount, and in the case of cadmium 20–50% (15,30,31,51,102). However, more detailed information about these burdens is of crucial importance in polluted areas where the heavy metal content of soil is frequently 20–100 times higher than it is supra-regionally (3,14,36,45,60,66,76,80,102).

The greatest burden and cause of damage to the health of domestic animals locally is lead. Leaving aside the hazards of the work-place and its effects on health, however, only rarely has damage to health in the human population of the FRG been ascertained in spite of high exposure.

There is also some local concern as to the exposure to copper. Although it is a trace element necessary for plant growth, an excess can affect both plant growth and the health of sheep (20,21,37,38,82,97). On a long-term basis cadmium also requires consideration since it is desirable to prevent well in advance possible impairments to man and animals caused by cadmium. Amongst other measures and regulations, it is crucial to prevent its accumulation in the soil, as cadmium — in contrast to lead — can be taken up by some plants in considerable amounts (17,18,28,29,41,61,65,82,90,95,96,100).

Values exceeding the threshold limits for soils have been found in 0.1% of the samples analyzed in Weser-Ems and in Hesse (Table 8) and in 1.6% of the samples analyzed in Baden-Württemberg. As the uptake of cadmium by plants in different soils varies considerably, it is necessary to determine the cadmium content of plants growing on soils rich in cadmium more intensively than has been done up to now (14,29,36,41,43,66,90,102).

Considerable damage to plants and domestic animals by thallium has been observed in the vicinity of some cement factories (38,44,63). Mercury is of great

Element	Hesse approx. 2,000 samples	Baden-Württemberg 3,000 samples	Weser-Ems (Northwest Germany) Cd - 25,000 samples Cu - 50,000 samples
Hg	0	0	-
Cd	0.1	1.6	0.1
Ni	0.1	14.9	-
Cu	0.1	0.4	0.04
Cr	0	1.5	-
Pb	0.2	1.8	-
Zn	0.1	1.9	-

TABLE 8 — Percent of soil samples with heavy metal contents higher than the threshold values (14,102).

importance in the food chain insofar as it accumulates in fish. However, it accumulates much less in plants (20,28,57,82).

As far as nickel, chromium, and zinc are concerned, the task is to determine both whether they are accumulated in amounts hazardous mainly to plants and where they are accumulated. It is probable that at lower concentrations these heavy metals impede plant growth more than they affect the health of man and animals, especially when they are the only metals of concern in soil (Table 6) (21,26,28,43,45,60,82).

REFERENCES

(1) Allaway, W.H. 1968. Agronomic controls over the environmental cycling of trace elements. Adv. Agron. **20**: 235–274.

(2) Anonymous. 1975. Querschnitt: 1. Weltweite Rohstoffverknappung erzwingt verstärkte Forschungsanstrengungen. 2. Denkschrift zur Lage der Lagerstätten-Forschung, von der DFG veröffentlicht. DFG-Mitt. 3/75, pp. 43–45.

(3) Anonymous. 1980. Erlaß des Ministeriums für Ernährung, Landwirtschaft, Umwelt und Forsten des Landes Baden-Württemberg über die Schwermetallbelastung von Böden. Gemeinsames Amtsblatt **28B(39):** 1186–1189.

(4) Anonymous. 1980. Metallstatistik. Frankfurt/Main: Metallgesellschaft AG 67.

(5) Anonymous. 1982. Klärschlammverordnung — AbfklärV vom 25.06.1982. Bundesgesetzblatt, Part I, pp. 734–739. Bonn: Bundesdruckerei.

(6) Anonymous. 1983. Allgemeine Verwaltungsvorschrift vom 23.02.1983 zur Änderung der Ersten Allgemeinen Verwaltungsvorschrift zum Bundesimmissionsschutzgesetz (Technische Anleitung zur Reinhaltung der Luft — TA Luft). Gemeinsames Ministerialblatt **34B(6)**: 94–112.

(7) Baker, A.J.M. 1981. Accumulators and excluders — strategies in the response of plants to heavy metals. J. Plant Nut. **3**: 643–654.

(8) Balks, R. 1961. Untersuchungen über den Bleigehalt des Bodens. Kali-Briefe, Fachgebiet 1, Series 11, vol. 5, pp. 1–7. Hannover: Büntehof, Kali und Salz.

(9) Beckett, P.H.T., and Davis, R.D. 1977. Upper critical levels of toxic elements in plants. New Phytol. **79**: 95–106.

(10) Beckett, P.H.T., and Davis, R.D. 1978. The additivity of the toxic effects of Cu, Ni and Zn in young barley. New Phytol. **81**: 155–173.

(11) Berglund, S. 1982. Plant uptake of cadmium. Summary of Swedish investigations up to and including 1981. Stockholm: The National Environment Protection Board.

(12) Berry, W.L., and Wallace, A. 1981. Toxicity: the concept and relationship to the dose response curve. **In** Trace Element Stress in Plants, Effects and Methodology, eds. A. Wallace and W.L. Berry. J. Plant Nut. **3(Suppl.)**: 13–19.

(13) Brümmer, G. 1981. Funktion des Bodens im Stoffhaushalt der Ökosphäre. **In** Ökologische Grundlagen des Natur- und Umweltschutzes, ed. G. Olschowy, pp. 111–124. Hamburg, Berlin: Parey.

(14) Brüne, H. 1981. Schwermetalle in Klärschlämmen und Böden Hessens. Forum Städtehyg. **32**: 171–173.

(15) Bundesamt für Ernährung und Forstwirtschaft. 1980. Kontamination agrarwirtschaftlicher Produkte mit Fremdimmissionen durch Blei, Cadmium, Quecksilber und Arsen. Arbeitsmaterialien, Bericht des BEF. Frankfurt: Bundesamt für Ernährung.

(16) Bundesgesundheitsamt. 1979. Richtwerte '79 für Blei, Cadmium und Quecksilber in und auf Lebensmitteln. Bundesgesundheitsblatt **22**: 282–283.

(17) Bundesminister des Innern. 1982. Anhörung zu Cadmium. Protokoll der Sachverständigenanhörung Berlin 2.–4. Nov. 1981. Berlin: Umweltbundesamt (BMI).

(18) CEC and GSF. 1981. Commission of the European Communities & Gesellschaft für Strahlen- und Umweltforschung mbH. Ecotoxicology of Cadmium. Brussels, Luxembourg & München-Neuherberg: EUR 7499 EN and GSF-Bericht 0-629.

(19) Chaney, R.L. 1980. Health risks associated with toxic metals in municipal sludge. **In** Sludge — Health Risks of Land Application, eds. G. Bitton et al., pp. 59–83. Michigan: Ann Arbor Science Publishers Inc.

(20) Chaney, R.L. 1981. Health risks associated with toxic metals in municipal sludge. Oak Ridge National Laboratory. Life Science Symposium Series, Gatlingburg, Tennessee, Oct. 4–8.

(21) Chaney, R.L., et al. 1977. Plant accumulation of heavy metals and phytotoxicity resulting from utilization of sewage sludge and sludge composts on cropland. I. National Conference on Composting of Municipal Residues and Sludge Products, Silver Springs, August 23–25, pp. 86–97. Rockville, MD: Information Transfer, Inc.

(22) Chaney, R.L., et al. 1980. Impact of EPA regulations on utilization of sludge in agriculture. Proceedings of the National Conference on Municipal and Industrial Sludge Utilization and Disposal, pp. 16–22. Silver Springs: Information Transfer, Inc.

(23) Cottenie, A., and Camerlynck, R. 1981. Specific behaviour of some monocotyledon and dicotyledon species with regard to excess of trace elements. **In** Trace Elements in Agriculture and in the Environment, ed. A. Cottenie, pp. 43–50. Ghent, Belgium: Laboratory of Analytical and Agrochemistry.

(24) Council for Agricultural Science and Technology. 1976. Applications of sewage sludge to cropland: appraisal of potential hazards of the heavy metals to plants and animals. CAST-Report No. 64. Ames, IA: Council for Agricultural Science and Technology.

(25) Council for Agricultural Science and Technology. 1980. Sewage

sludge applications on agricultural soils: effect of annual and cumulative additions of cadmium and zinc on plants. CAST-Report No. 83. Ames, IA: Council for Agricultural Science and Technology.

(26) Davis, R.D., and Beckett, P.H.T. 1978. Upper critical levels of toxic elements in plants. II. New Phytol. **80**: 23–32.

(27) Davis, R.D.; Beckett, P.H.T.; and Wollan, E. 1978. Critical levels of twenty potentially toxic elements in young spring barley. Plant Soil **49**: 395–408.

(28) Davis, R.D., and Carlton-Smith, C. 1980. Crops as indicators of the significance of contamination of soil by heavy metals. Technical Report TR 140, Stevenage Laboratory, Water Research Centre, Stevenage, England.

(29) Davis, R.D., and Coker, E.G. 1980. Cadmium in agriculture, with special reference to the utilization of sewage sludge on land. Technical Report TR 139, Stevenage Laboratory, Water Research Centre, Stevenage, England.

(30) Diehl, J.F. 1981. Die Belastung des Verbrauchers durch Cadmium — eine kritische Übersicht. Berichte der BFE: 1981/3. Karlsruhe: Bundesforschungsanstalt für Ernährung.

(31) Diehl, J.F. 1983. Schwermetalle in der Nahrung — Werden die Grenzwerte der duldbaren Belastung überschritten. Landwirt. F. **39**: 35–59.

(32) D'Itri, F.M. 1977. Proceedings, "Wastewater Renovation and Reuse." New York, Basel: Marcel Dekker, Inc.

(33) Ditter, P. 1979. Belastung und Belastbarkeit der Landschaft am Beispiel von Schwermetallen. Forschungsbericht des Landes Nordrhein-Westfalen **2828**: 259–314.

(34) Enk, R.H. van. 1979. The pathway of cadmium in the European Community. EUR 6626 EN. Commission of the European Communities, Joint Research Centre, ISPRA.

(35) FAO/WHO. 1978. List of maximum levels recommended for contaminants by the Joint FAO/WHO codex Alimentarius Commission. Third Series. Food & Agriculture Organization of the United Nations and World Health Organization. Rome, Italy: FAO/WHO.

(36) Früchtenicht, K., and Vetter, H. 1983. Charakterisierung der Schwermetallbelastung durch Messung der Schwermetallgehalte in Pflanzen. Landwirt. F. **39**: 154–164.

(37) Gründer, H.-D. 1983. Belastungsgrenzen für Schwermetalle bei Hauswiederkäuern und anderen Haustieren. Landwirt. F. **39**: 60–93.

(38) Hapke, H.-J. 1983. Schwermetalle in der Nahrungskette Belastungsgrenzen für Pflanzen. Landwirt. F. **39**: 1–11.

(39) Hecht, H. 1983. Die verschiedenen Wege und Möglichkeiten des Eintrags von Schwermetallen in die Futtermittel und die dabei zu beachtenden Belastungsgrenzen. Landwirt. F. **39**: 94–107.

(40) Herms, U., and Brümmer, G. 1980. Einfluß der Bodenreaktion auf Löslichkeit und tolerierbare Gesamtgehalte von Nickel, Kupfer, Zink, Cadmium und Blei in Boden und kompostierten Siedlungsabfällen. Landwirt. F. **33**: 408–423.

(41) Herms, U., and Tent, L. 1983. Cadmium-Gehalte in Spülfeldern aus Hafenschlick und in darauf angebauten Kulturpflanzen — eine Felderhebungsuntersuchung. Landwirt. F. **39**: 448–456.

(42) Hoffmann, G. 1980. Richtwerte für Schwermetallkonzentrationen in Müll- und Müllklärschlammkomposten in Deutschland und in der Europäischen Gemeinschaft. Sammelband Informationstagung "Die Verwendung von Müll- und Müllklärschlammkompost in der Landwirtschaft." Gottlieb Duttweiler-Institut, Zürich, November 10, 1980.

(43) Hoffmann, G. 1983. Zusammenhänge zwischen kritischen Schadstoffgehalten im Boden, in Futter- und Nahrungspflanzen. Landwirt. F. **39**: 130–153.

(44) Hoffmann, G.; Schweiger, P.; and Scholl, W. 1982. Aufnahme von Thallium durch landwirtschaftliche und gärtnerisch Nutzpflanzen. Landwirt. F. **35**: 45–54.

(45) Hughes, M.K.; Lepp, N.W.; and Phipps, D.A. 1980. Aerial heavy metal pollution and terrestrial ecosystems. **In** Advances in Ecological Research, ed. A. MacFadyen. London, New York: Academic Press.

(46) Hutton, M. 1983. The environmental implications of cadmium in phosphate fertilizers. Phosphorus & Potassium **123**: 33–36.

(47) Isermann, K. 1983. Enfluß der Phosphatdüngung auf der Cadmiumgehalt des Bodens, ermittelt anhand zahlreicher Dauerversuche in Westeuropa. Landwirt. F. **39**: 283–301.

(48) Jarvis, S.C. 1976. Heavy metal pollution of agicultural soils. Rep. Progr. Appl. Chem. **60**: 375–386.

(49) Jung, J.; Isermann, K.; and Henjes, G. 1979. Einfluß von cadmiumhaltigen Düngerphosphaten auf die Cadmiumanreicherung von Kulturböden und Nutzpflanzen. Landwirt. F. **32**: 262–274.

(50) Käferstein, F.K.; Altemann, H.-J.; Kallischnigg, G.; Klein, H.; Kossen, M.-T.; Lorenz, H.; Müller, J.; Schmidt, E.; and Zufelde, K.P. 1979. Blei, Cadmium und Quecksilber in und auf Lebensmitteln. ZEBS-Berichte 1/1979.

(51) Kampe, W. 1980. Schwermetallkontamination bei Pflanzen. Landwirt. F. **36**: 322–355.

(52) Kick, H.; Bürger, H.; and Sommer, K. 1980. Gesamtgehalte an Pb, Zn, As, Cd, Hg. Cu, Ni, Cr und Co in landwirtschaftlich und gärtnerisch genutzten Böden Nordrhein-Westfalens. Landwirt. F. **33**: 12–22.

(53) Kloke, A. 1972. Die Belastung der gärtnerischen und landwirtschaftlichen Produktion und Erntegüter durch Immissionen. Ber. Landwirt. **50**: 57–68.

(54) Kloke, A. 1974. Beeinträchtigung der Qualität von Nahrungs- und Futterpflanzen durch Umweltchemikalien, insbesondere durch Schwermetalle. Qual. Plant. **24**: 137–157.

(55) Kloke, A. 1980. Immissionsbelastete landwirtschaftliche Standorte. **In** Leistungen von Landwirtschaft und Landschaft zur Wohlfahrt von Verdichtungsräumen, ed. Schriftenreihe für ländliche Sozialfragen. ASD vol. 84, pp. 24–44. Hannover: Schaper.

(56) Kloke, A. 1980. Richtwerte '80: Orientierungsdaten für tolerierbare Gesamtgehalte einiger Elemente in Kulturböden. Mitt. VDLUFA **1-3**: 9–11.

(57) Kloke, A. 1981. Die Anreicherung von Schadstoffen im Getreide. **In** Forschungsbericht "Rückstände in Getreide- und Getreideprodukten," ed. DFG, pp. 30–41. Boppard: Bold-Verlag.

(58) Kloke, A. 1981. Kontamination (des Bodens) durch Schwermetalle. **In** Ullmanns Encyklopädie der technischen Chemie, 4th ed., vol. 6, section F, "Böden," pp. 502–506. Weinheim: Verlag Chemie GmbH.

(59) Kloke, A. 1982. Aufnahme umweltrelevanter Elemente durch die Pflanze. XVII. Vortragstagung der DGQ, pp. 65–90, March 26–27, 1982, Speyer.

(60) Kloke, A. 1982. Re-use of sludges and treated waste water in agriculture. Wat. Sci. Techn. **14**: 61–72.

(61) Kloke, A., and Schenke, H.-D. 1979. Der Einfluß von Cadmium im Boden auf den Ertrag verschiedener Pflanzenarten und deren Cadmiumgehalt. Z. Pflanz. B. **142**: 131–136.

(62) Köster, W., and Merkel, D. 1983. Beziehungen zwischen den Gehalten an Zink, Cadmium, Blei und Kupfer in Böden und Pflanzen bei Anwendung unterschiedlicher Bodenuntersuchungsmethoden. Landwirt. F. **39**: 245–254.

(63) Landesanstalt für Immissionsschutz des Landes Nordrhein Westfalen. 1980. Umweltbelastung durch Thallium. Bonn: Universitätsdruckerei.

(64) Lauwerys, R.R. 1982. The toxicology of cadmium. Report EUR 7649 EN. Luxembourg: Commission of the European Communities.

(65) Lisk, D.J. 1972. Trace metals in soils, plants, and animials. Adv. Agron. **24**: 267–325.

(66) Lux, W., and Hintze, B. 1983. Schwermetallverteilung in Böden und Pflanzen in städtischen Bereichen Hamburgs. Landwirt. F. **39**: 196–201.

(67) Mayer, R. 1981. Ist unser Ökosystem in Gefahr durch Einflüsse der Industrie- und Wohlstandgesellschaft? Schadstoffe außerlandwirtschaftlicher Herkunft. Arbeiten d. DLG, vol. 172, Landbewirtschaftung und Ökologie, pp. 126–130. Frankfurt: DLG-Verlag.

(68) Möller, U. 1981. Anreicherung von Schwermetallen in Klärschlämmen kommunaler Klärwerke. Korrespondenz Abwasser **28**: 176–178.

(69) Oberländer, H.-E. 1981. Die verkehrsbedingte Belastung von Nahrungspflanzen durch Blei und Cadmium in Österreich. Förderungsdienst **2**: 42–46.

(70) Oberländer, H.-E., and Roth, K. 1982. Uptake and distribution of soil-applied labelled heavy metals in cereal plants and products. FAO/IAEA Research Coordination Meeting, "Agrochemical Biota Interactions in Soil and Water Using Nuclear Techniques," Technical Document No. 283, pp. 137–149. Vienna: International Atomic Energy Agency.

(71) Ocker, H.-D.; Seibel, W.; and Brüggemann, J. 1983. Reduktion des Schwermetallgehaltes durch die Getreideverarbeitung. Landwirt. F. **39**: 333–341.

(72) Ohki, K. 1975. Lower and upper critical zinc levels in relation to cotton growth and development. Physiol. Plant. **35**: 96–100.

(73) Pettersson, O. 1976. Heavy-metal ion uptake by plants from nutrient solutions with metal ion, plant species and growth period variations. Plant Soil **45**: 445–459.

(74) Pluquet, E., and Kuntze, H. 1983. Einfluß von Schwermetallen im Boden auf Pflanzenertrag und Gehalte in Abhängigkeit vom Boden-pH und Tongehalt. Landwirt. F., in press.

(75) Prinz, B.; Krause, G.H.M.; and Stratmann, H. 1979. Thalliumschäden in der Umgebung der Dyckerhoff Zementwerke AG in Lengerich, Westfalen. Staub — Reinhalt. Luft **39**: 457–462.

(76) Purves, D. 1977. Trace Element Contamination of the Environment. Science 1. Amsterdam, New York: Elsevier.

(77) Rietz, E., and Söchtig, H. 1982. Extraktionsverhalten und Bindung von Schwermetallen in Böden unterschiedlichen Belastungsgrades. Landwirt. F. **38**: 382–393.

(78) Salomons, W.; Mook, W.G.; and Eysink, W. 1981. Biogeochemical and hydrodynamic processes affecting heavy metals in rivers, lakes and estuaries. Waterloopkundig Laboratorium Delft Hydraulics Laboratory, No. 253.

(79) Sauerbeck, D. 1981. Möglichkeiten und Grenzen der Ausbringung von Siedlungsabfällen auf pflanzenbaulich genutzte Flächen. Ber. Landwir. **197**: 104–113.

(80) Sauerbeck, D. 1982. Probleme der Bodenfruchtbarkeit in Ballungsräumen. Mitt. Dtsch. Bodenkundl. Gesellsch. **33**: 179–193.

(81) Sauerbeck, D. 1982. Zur Bedeutung des Cadmiums in Phosphatdüngemitteln. Landbau. Vol. **32**: 192–197.

(82) Sauerbeck, D. 1983. Welche Schwermetallgehalte in Pflanzen dürfen nicht überschritten werden, um Wachstumbeeinträchtigungen zu vermeiden? Landwirt. F. **39**: 108–129.

(83) Sauerbeck, D., and Rietz, E. 1981. Zur Cadmiumbelastung von Mineraldüngern in Abhängigkeit von Rohstoff und Herstellungsverfahren. Landwirt. F. **37**: 685–696.

(84) Sauerbeck, D., and Rietz, E. 1982. Soil-chemical evaluation of different extractants for heavy metals in soils. **In** Environmental Effects of Organic and Inorganic Contaminants in Sewage Sludge, ed. CEC, pp. 147–160. Proceedings of a Workshop held at Stevenage, England, May 25–26, 1982.

(85) Schmitz, F.-J., and Kluge, G. 1982. Das Düngemittelrecht mit fachlichen Erläuterungen. Münster: Landwirtschaftsverlag GmbH.

(86) Schönhard, G. 1979. Vergleich der Wirkung von Kalk und einem Kationenaustauscher bei der Festlegung von Schwermetallen im Boden. Landwirt. F. **32**: 395–404.

(87) Smilde, K.W. 1976. Toxische gehalten aan zware metalen (Zn, Cu, Cr, Ni, Pb en Cd) in grond en gewas. Instituut voor Bodemvruchtbarheid, nota 25, Haren (Gr.).

(88) Smilde, K.W. 1981. Heavy-metal accumulation in crops grown on sewage sludge amended with metal salts. Plant Soil **62**: 3–14.

(89) Smith, M.A. 1980. Redevelopment of contaminated land: tentative guidelines for acceptable levels of selected elements in soils. Interdepartmental Committee on the Redevelopment of Contaminated Land, ICRCL 38/80, Central Directorate on Environmental Pollution, London.

(90) Styperek, P.; Sauerbeck, D.; and Timmermann, F. 1983. Cd-Verfügbarkeit in verschiedenen behandelten Böden in Abhängigkeit von Menge und Bindungsform. Landwirt. F. **39**: 183–195.

(91) Subcommittee of Mineral Toxicity in Animals. 1980. Mineral tolerance of domestic animals. Washington: National Academy of Sciences.

(92) Thormann, A., and Kloke, A. 1979. Hausmüll als Düngemittel. VDI-Umwelt **2**: 94–95.

(93) Tietjen, C. 1975. Principal problems of the use of city wastes for crop production and soil conservation. FAO Soils Bulletin No. 27, Organic Materials As Fertilizers, pp. 211–226. Report of the FAO/SIDA Expert Consultation, Rome, Italy.

(94) Tjell, J.C.; Hovmand, M.F.; and Mosbaek, H. 1979. Atmospheric lead pollution of grass grown in a background area of Denmark. Nature **280**: 425–426.

(95) Tjell, J.C., et al. 1980. Prediction of cadmium concentrations in Danish soils. **In** Characterization, Treatment, and Use of Sewage Sludge, pp. 650–664. Proceedings of the Second European Symposium, Vienna, October 21–23, 1980. London: Reidel.

(96) Umweltbundesamt. 1981. Cadmium-Bericht. Ein Beitrag zum Problem der Umweltbelastung durch nicht- oder schwerabbaubare Stoffe dargestellt am Beispiel Cadmium. Texte 1/81, Umweltbundesamt Berlin.

(97) Underwood, E.J. 1977. Trace elements in human and animal nutrition. New York: Academic Press.

(98) USEPA. 1978. Cadmium additions to agricultural lands via commercial phosphatic fertilizers. EPA Report SW-718. Washington, D.C.: U.S. Environmental Protection Agency.

(99) USEPA; USFDA; and USDA. 1981. Land application of municipal sewage sludge for the production of fruits and vegetables. A statement of federal policy and guidance. Joint Policy Statement SW-905. Washington, D.C.: USEPA.

(100) Verloo, M.; Cottenie, A.; and Landschoot, G. van. 1983. Analytical and biological criteria with regard to soil pollution. Landwirt. F. **39**: 394–403.

(101) Vetter, H. 1983. Schwermetalle in der Nahrungskette — Belastungsgrenzen für Pflanzen. Landwirt. F. **39**: 12–27.

(102) Vetter, H.; Kowalewsky, H.-H.; and Säle, M. 1983. Datenerhebung über die Cadmiumbelastung von Böden und Pflanzen in der

Bundesrepublik Deutschland im Auftrag des Bundesministeriums für Ernährung, Landwirtschaft und Forsten. Darmstadt: VDLUFA.

(103) Vetter, H., and Meyer, H.-H. 1978. Vergleich der Schadstoffbelastung durch Atemluft und Nahrung in der Nachbarschaft einer Blei- und Zinkhütte. Symposium über Carry-over von Schadstoffen in der landwirtschaftlichen und tierischen Produktion, Kulmbach. Bonn: AID.

(104) Wohlrab, B.; Ehlers, M.; and Mollenhauer, K. 1982. Deponien verschiedenster Art, Probleme ihrer Rekultivierung und Integration in die Stadtlandschaft. Mitteilgn. Dtsch. Bodenkundl. Gesellsch. **33**: 109–120.

Changing Metal Cycles and Human Health, ed. J.O. Nriagu, pp. 143–167.
Dahlem Konferenzen 1984. Berlin, Heidelberg, New York, Tokyo: Springer-Verlag.

TRANSPORT OF TRACE METALS IN MARINE FOOD CHAINS

M. Bernhard* and M.O. Andreae**
*Centro Studi Ambiente Marino, ENEA
19100 La Spezia, Italy
**Dept. of Oceanography, Florida State University
Tallahassee, FL 32306, USA

Abstract. For most trace metals the first trophic level (seawater — phytoplankton) shows the highest concentration increase. In the next trophic level (zooplankton) only Cd, Cu, and Zn concentrations increase. Plankton feeding fishes have higher Cu and Zn levels than fishes preying on invertebrates and fish. The Cu, Zn, and probably Cd concentrations are higher in shrimps feeding on invertebrates than in fishes feeding on shrimps. For these three trace metals the concentrations at first increase along the food chain reaching a maximum with crustaceans, and then decrease in fishes. The concentration of mercury increases with size (i.e., age) of the organisms. Its increase along the food chain does not become evident in short-lived organisms. This increase with size and the fact that long-lived organisms are at the top of the food chain are responsible for the mercury increase along the food chain. Mercury is also the only trace metal which poses a risk to heavy consumers of seafood, especially in areas with high natural mercury levels such as the Mediterranean. The significance of the elevated mercury level in hair and blood of these persons also needs to be evaluated in light of the antagonistic action of the selenium present in seafood. The reevaluation of the lead levels in marine foods shows that only very little Pb enters man through the marine food chain and that, therefore, Pb is not a marine pollutant of concern. Very high As concentrations have been observed, especially in benthic organisms. Although most of the As is in a very low toxic or nontoxic organic form in organisms with high total levels, still considerable amounts may be present in the toxic inorganic form. At present not enough data are available to assess possible risk.

Only for mercury and arsenic can speciation and bioavailability be considered: for the other trace metals discussed the analytical tools are not developed sufficiently to allow the determination of the forms which are actually accumulated by the organisms.

INTRODUCTION

The concept of the food chain (or food web) which assigns an organism to a trophic level is very useful because it facilitates considerably the modelling of transport processes. However, it must be realized that marine organisms rarely represent a single trophic level. In fact, the food web of a marine organism in nature includes many prey orgnanisms which may belong to different levels and which themselves feed on prey belonging to various levels. Even autotrophs can take up dissolved and particulate organic matter, and so with regard to these pathways they are not on the first level of the food chain. Therefore, assigning an organism to a single level of the food chain can be valid only as a first approximation, and this level should be selected on the basis of the most prominent food organism. For some trace metals, the first step from seawater to plant results in the highest enrichment of the entire food chain, but for some other metals (e.g., strontium), the levels in the biota can be lower than in seawater. It must be remembered here that it is conceptually problematic to calculate an enrichment factor on the basis of a concentration in water, on one hand, and of a concentration in biomass (as wet weight or even dry weight), on the other. When used in this fashion, the concentration factor integrates the effects of a number of processes which result in the concentration of an element in the organism relative to seawater. In plants the concentration factor can, therefore, give only a rough idea of the relationship between organism and water. For some elements, it may therefore be more illustrative to use the ratio of the trace element to the essential element of similar chemical characteristics with which it is taken up together (e.g., Cs/K, Sr/Ca, Pb/Ca, As/P) when studying enrichment of elements during their passage through the food chain. For heterotrophic animals, the concentration in the body tissues is the result of uptake from food organisms and directly from seawater. During further transport within the food chain, the concentration of the trace metal can increase, decrease, or remain constant. Changes in concentration can occur within a particular step of the food chain if the concentration of a trace metal increases or decreases during the organism's life span. However, the enrichments or depletions at lower trophic levels are not always reflected at higher trophic levels. In fact, the higher levels often have lower trace metal concentrations than the lower trophic levels.

For an evaluation of the relative contribution of toxic trace metals to the human diet, the most important step in the transport of these metals is the last step of the food chain, i.e., from marine foods to man.

TRACE METAL CONCENTRATION AND BODY SIZE

Several authors have investigated the relation between size of marine organisms and concentration of trace metal. The clearest evidence for an increase of trace element concentration with size exists for mercury (e.g. (24)). This increase is more evident in marine organisms which reach a large size and consequently have high mercury levels (43). It is due to efficient uptake, especially of methylmercury, and a very low elimination rate which results in a nearly complete retention of the methylmercury taken up. This could be shown by modelling the Hg concentration increase with size in sardines and tunas (Fig. 1), as will be discussed below.

For other trace metals the picture is less clear. For example, Boyden (10) found that trace element concentrations decrease in some mollusc species while in others they remain constant or increase, e.g., Cd in Patella vulgata (Table 1). A marked increase of Cd in the soft part of the organism with animal size is observed in limpets (P. vulgata) from a Cd polluted site (Fig. 2.). On the other hand, the mollusc Haliotis tuberculata showed only a slight Cd increase with size (13). Similarly, fishes such as the whiting (3) had only slight Cd concentration increases with size. In the Pacific hake Zn and Cd likewise increased with size mainly in small specimens. In fish the few data available show small variations with size (e.g., Fig 3). In tissues such as muscle, brain, etc., at least for the essential elements (Co, Cr, Cu, Fe, Mn, Mo, Ni, Se, Sn, V, and Zn), the trace metal concentrations can be expected to be regulated and hence more or less constant. Varying levels in excretory organs are to be expected since they are involved in regulating processes.

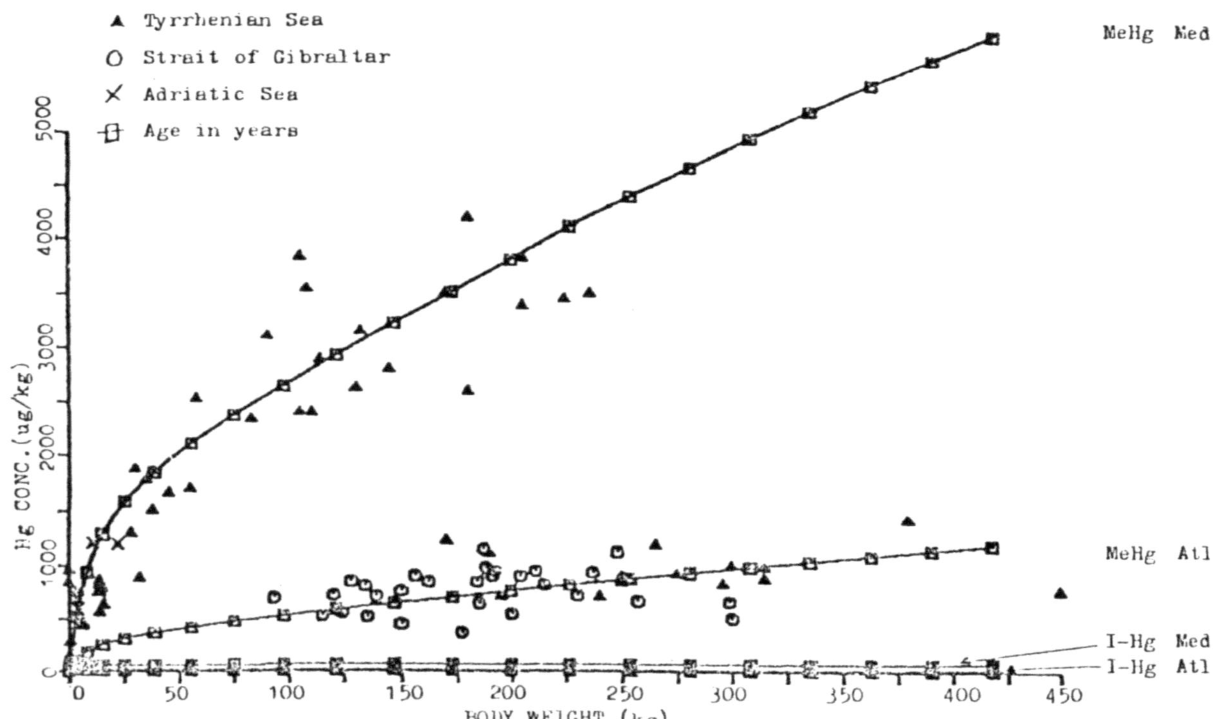

FIG. 1 — Mercury concentration versus body weight in T. thynnus. Continuous line prediction from food chain model; MeHg Med. = methylmercury in "Mediterranean tuna"; MeHg Atl. = methylmercury in "Atlantic tuna"; I-Hg Med. = inorganic mercury in "Mediterranean tuna"; I-Hg Atl. = inorganic mercury in "Atlantic tuna" (6).

	As	Cd	Cu	Fe	Mn	Ni	Pb	Zn	Ref.
Molluscs									
Patella vulgata		+	-	-			-	-	11
Patella vulgata		+	-					-	47
P. intermedia			-	-			-	-	10
Littorina littorea		=	=	-	-		-	-	11
Buccinium undatum		+	=	=	-	-	-	+	11
Scaphander lignaris		=	=		=	=	=	=	11
Crepidula fornicata		=	+		-	=	=	=	11
Cerastoderma edule		-	-	=		=	-	-	10
Mytilus edulis		=	-	-	-	-	-	-	11
Mytilus edulis		-	-	-		-		-	19
Mercenaria mercenaria		-	-	-		=	=	=	11
Ostrea edulis		=	=	-	=	-	-	=	11
Crassostrea gigas		-	-	-	=	-	-	=	11
Venerupis decussata		-	-	=		=	-	=	11
Chlamps opercularis		=	=			-	=	=	11
Pecten maximus		(+)	-		=		-	-	11
Haliotis tuberculata		=		-	-	-	-	-	13
Fish									
Morone saxatilis		-						-	48
juv. estuarian fish				-	-			-	22
Boreogadus saida	+	=		-	-			+	9
Pomatomus saltatrix			=	-	=			=	23
Antimora rostata			=	=	=			=	23

"+" = trace metal concentration increases with size
"=" = trace metal concentration remains constant with size
"-" = trace metal concentration decreases with size

TABLE 1 — Qualitative relation between trace metal concentration and size.

The change in concentration of an element during the organism's life span supplies us with information about the relation between uptake and loss processes. The accumulation of the total amount of an element (Q), the body burden, during the life span of an organism can be described by the following equation:

$$\frac{dQ}{dt} = I\,W - \lambda\,Q, \qquad Q(0) = 0, \qquad (1)$$

where t = the age of the individual; W = the body weight of the individual; I = the amount of element taken up per unit time and per unit body weight; $\lambda = \log_2/T^{1}\!/_{2}$, with $T^{1}\!/_{2}$ the biological half-life.

By introducing the concentration $C(t) = \frac{Q(t)}{W(t)}$ and substituting Q with C W in Eq. 1, one obtains

$$C\,\frac{dW}{dt} + W\,\frac{dC}{dt} = I\,W - \lambda\,C\,W,$$

$$\frac{dC}{dt} = I - \left(\lambda + \frac{1}{W}\frac{dW}{dt}\right)C. \qquad (2)$$

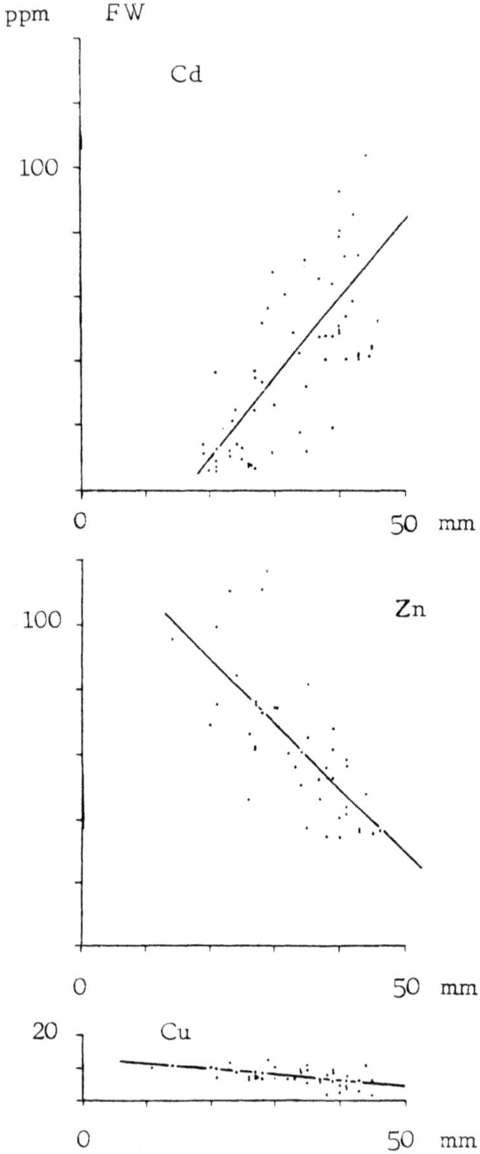

FIG. 2 — Metal concentration vs. shell length in P. vulgata (47).

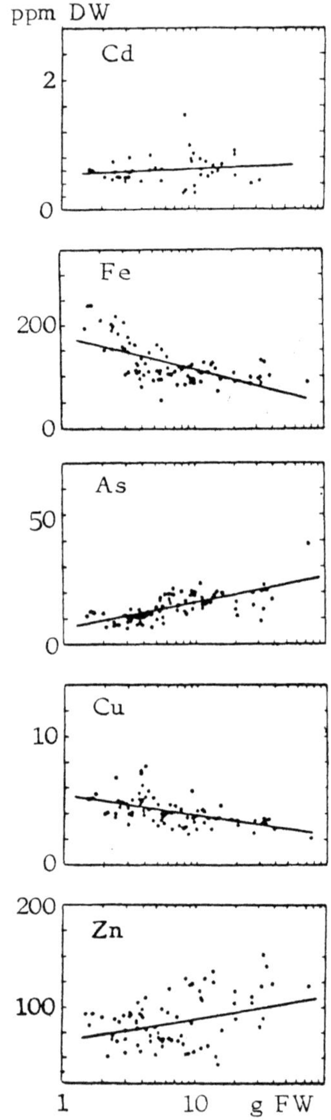

FIG. 3 — Metal concentration vs. body weight in Arctic cod (9).

If the concentration is constant during the life of the organism, i.e., $dC/dt = 0$, then the intake (I) must be equal to the loss (λC) plus the amount of element incorporated into the new body tissue synthesized during growth (second term on the right side of Eq. 3:

$$I = \lambda C + \frac{1}{W}\frac{dW}{dt} C. \qquad (3)$$

Now, if the intake is always greater than the loss plus the amount incorporated into newly grown tissue, then the concentration of an element increases during the life of an organism as, for instance, in the case of mercury accumulation. If the intake is smaller than loss plus "growth incorporation," then the concentrations obviously decrease during the life span. These considerations can be useful in checking laboratory results against field observations.

TRACE METAL CONCENTRATIONS AT DIFFERENT TROPHIC LEVELS

The trace metal concentration at different levels along the food chain can show whether the trace metal concentration increases, decreases, or remains constant. Increase with trophic levels has been called biomagnification.

The uncertainty in the results of trace metal analyses makes it very difficult to select a good data base for the comparison of data from the vast number of publications (e.g., review in (25)). Many publications report trace metal data without reference to quality controls such as intercalibration exercises and analysis of reference standards. A large number of especially older publications report just one value per species. The ICES monitoring exercise and the MED POL projects are examples of how by cooperation the analysts of a region can achieve a certain uniformity of procedures and guarantee a certain accuracy through intercalibration. Nevertheless, there are indications of systematic errors in these data also.

A second consideration may concern the selection of the habitat. Pelagic environments are much less subject to pollution influences and thus will bear out basic relationships more easily than coastal and estuarian environments where migration in and out of polluted areas during the life cycle can change trace metal concentrations in the organisms.

In this paper we will not consider lead in detail. Ever since Patterson and collaborators (58) showed that the great majority of the Pb concentrations published far exceed the true levels because of contamination with Pb during sampling, storage, and analysis, very few reliable Pb data have been reported. As a guidance, the following levels (from Patterson and collaborators (58)) may be considered typical:

sea water (open sea)	0.001	ug/L
sea water (coastal)	0.016-1.3	ug/L
algae (Valonia)	200	ug/kg
Mytilus califorinanus	25	ug/kg
Spiny lobster, muscle	5	ug/kg
Thunnus alalunga	0.3	ug/kg

These "typical levels" indicate that the Pb concentration increases about 100,000 times from seawater to the first trophic level but then decreases with higher trophic levels. These very low Pb levels in marine foods as compared with the high Pb levels in the terrestrial environment indicate that the pathway of Pb from marine foods to man is small enough to be neglected (5).

Seawater presents the starting point of the food chain. Comparing the estimated fresh weight concentrations in plankton with the "open sea" seawater concentration, one notices that for all trace elements listed in Table 2, with the exception of Sr, the trace metal concentration in phytoplankton increase several orders of magnitude over seawater. The data also show that only the concentrations of Cd, Cu, and Zn increase significantly from phytoplankton (first level) to copepods (second level).

Moving up in the food chain from copepods to plankton feeder (anchovy, herring, sardine, and capelin), one notices that the Cd concentration decreases by a factor of 10 (Table 3). Cu and Zn have about the same concentration in herring as in copepods and tend to decrease slightly in hake and whiting. The data in Tables 2 and 3 also show that there may be an increase in element concentration in the lower levels of the food chain and a decrease in the higher. Organisms in higher trophic levels are often more developed and thus are able to control their internal body concentration more efficiently. Hg, on the other hand, tends to increase from copepods to plankton feeders. Shrimps and prawn which feed on small invertebrates and other crustaceans possess higher body levels of Cd, Cu, and Zn than the plankton feeding fishes. The predators (plaice and flounder) have lower Cd, Cu, and Zn concentrations while the Hg levels are about the same. Observations made in the heavily polluted Severn Estuary show that Cd and Zn increase with a pollution gradient in invertebrates while the Zn concentration in flounders decreases with size. Gray mullets caught in polluted areas have "normal" concentrations. The Cd levels of macrophytes (Fucus) and molluscs (Littorina, Patella, Thais, and Nucella) increase towards the pollution sources (29). The concentration in Fucus is

	Sr	Ba	Al	Fe	Mn	Cr	Cu	Ni	Zn	Ag	Cd	Pb	Hg
seawater	80000	2	2	2	0.03	0.3	0.3	0.3	1	0.04	0.04	0.01	0.05
phytoplankton (acid soluble org. fraction, group I, n > 10)													
median	7700	1700	5800	11800	320	ND	170	100	1000	10	80	1	10
range	4800 -13700	260 -8100	390 -33200	2700 -39500	110 -680	ND -70	68 -1000	ND -305	160 -2800	ND -25	21 -250	ND -680	5 -14
n	14	14	12	14	14	10	14	12	14	11	14	14	14
copepods													
median	11600	1600	6500	18300	410	-	975*	185	10500**	<10	380**	310	10
range	7700 -26900	460 -2400	1850 -10500	5100 -26200	560 -950	- -	850 2100	ND -1100	5800 -15750	ND -18	150 -560	ND -830	5.6 -14
n	10	10	10	10	10		10	10	10	10	10	10	10

Plankton data from Martin and Knauer (45); DW have been converted into FW using the authors' FW/DW ratio of 19 for phytoplankton and 10.78 for copepods; n = number of samples analyzed; * = $P < 0.05$; ** = $P < 0.001$; ND = not detected.

TABLE 2 — Comparison between "typical" seawater concentration (ug/l) and levels in "open sea" phytoplankton ("no Ti") and copecods (ug/kg FW) (data from (12,45,63)).

about the same as in Littorina and Patella (herbivores) and in Mytilus (plankton and filter feeder), while the concentrations are higher in Thais and Nucella, which are carnivores and feed on Littorina and Patella (14). Higher trophic levels such as the shrimp (Crangon crangon) which feeds on small benthic crustaceans, annelids, and molluscs show no further increase in Cd concentrations. In fact, it decreases to the level of the macrophytes. A further step in the food chain was investigated in this area. The shrimp Crangon crangon is an important food organism of the flounder (P. flessus). Hardisty and collaborators (32) could show that the shrimp's Cd concentration ranged from 5 to 124 mg Cd/kg DW while the flounders had only levels of 3.5 to 7.3 mg Cd/kg DW. Similar observations have been made by other authors (50, 60).

	Median and range of means				
	Cd	Cu	Hg	Zn	loc. ref.
plankton feeder					
herring	<30	1500	40	8000	N. Sea
	30-700	600-3600	20-240	3000-22000	(34)
herring	20	1200	20	5800	N. Atl.
	20-20	760-1800	10-35	3300-7500	(35)
herring	20	1180	40	9670	Irish c.
	ND-80	280-2500	10-23	3450-18500	(38)
"typical"	20	1200	40	8000	
sardine	60	1500	60	9500	N. Atl.
	9-60	1100-6800	6-80	8900-10000	(35)
sardine			250		Medit.
			150-390		(65)
sardinella	730	730	-	9350	Medit. (65)
sprat	75	790	65	15550	Irish c.
	60-90	180-1050	60-140	12100-21350	(38)
capelin	<70	1100	10	5600	N. Atl.
	10-30	700-1800	10-30	4800-7200	(35)
anchovy	18	1050	160	18500	Medit.
	5-20	885-1200	145-180	17100-20400	(65)
feed on invertebrates					
brown shrimp	640	19000	110?	27500	N. Sea
	<500-1000	7200-22000	50-230	6300-40000	(34)
brown shrimp	100	14500	140	34000	N. Sea
	20-280	1600-23000	70-390	24000-49000	(36)
brown shrimp	60	12000	80	35500	N. Sea
	<20-230	1500-31000	30-300	9000-49000	(37)
"typical"	100	14500	110	34000	
deep sea prawn	300	16500	25	13800	W. Greenl.
	<200-950	12550-18000	<20-30	12500-17000	(35)
Norway lobster	10	6125	960	15100	Medit.
	2.5-200	5850-6300	290-970	14650-15700	(65)

	Median and range of means				loc.
	Cd	Cu	Hg	Zn	ref.
feed on invertebrates					
cod	<30 20-500	430 260-2000	100 30-480	3750 2400-6000	N. Sea (34)
cod	10 <1-30	490 210-810	100 60-300	4200 3700-6000	N. Sea (35)
cod	- 100-900	360? 20-150	40 40-50	3750 2900-6300	N. Atl. (35)
cod	-	400	260	2700	Irish c. (35)
cod	25 2-40	500 100-900	140 70-370	7700 3400-13200	Irish c. (38)
cod	40 27-53	650 200-2100	70 50-140	4600 1900-7300	NW-Atl. (35)
cod	6 6-7	450 400-500	80 70-90	3950 3900-3950	NW-Atl. (38)
"typical"	20	450	100	3950	
feed on crustaceans and fish					
hake	40 20-60	500 300-3700	90 30-130	2700 1800-6400	N. Atl. (35)
hake			30-850	1000-3750	Medit. (65)
haddock	40 ND-130	650 220-1400	50 20-60	9650 5190-15800	Irish c. (38)
haddock	4	400	50	3290	NW-Atl. (38)
whiting	10 ND-150	240 ND-550	80 30-90	4650 1570-9810	Irish c. (38)
tuna	125 55-195	-	715 300-1485	-	Medit. (65)
mullet	28 4-70	455 155-800	190 55-1300	4250 3050-5050	Medit. (65)
Greenland halibut	<200	300 160-300	40 30-50	1700 1200-4000	N. Atl. (35)
plaice	350 <20-600	500 220-800	90 20-260	4600 3100-8000	N. Sea (34)
plaice	4 3-4	300 140-800	120 20-500	5000 3600-6100	N. Atl. (35)
plaice	25 ND-80	480 270-850	25 10-80	8530 4600-13300	Irish c. (38)
"typical"	25	480	90	5000	
sole	20 <10-50	340 220-750	150 50-320	5100 3600-6500	N. Atl. (35)

? = values selected from source data; + = value estimated from range; ND = not detected.

TABLE 3 — Selected trace metals (ug/kg FW) in some fish (muscle) and shellfish species (whole body).

Arsenic is significantly enriched in marine algae (typically 5 mg As/kg FW) relative to seawater (1.5 ug/L) (1,42,57). There is a tendency for the arsenic concentration to decrease in the sequence: algae-crustaceans-molluscs-fish, with concentrations of circa 5 mg As/kg FW typical for crustaceans and circa 2 mg/kg FW typical for marine fish (20,41,42,61). Exceptions, however, are the very high concentrations in some molluscs and benthic fishes in which concentrations in some individuals may reach 100 mg As/kg FW. All these observations show that in higher levels of the food chain As, Cd, Cu, Pb, and Zn do not increase but rather tend to decrease.

On the other hand, mercury increases along the food chain and levels reached seem to depend on the Hg seawater concentration, as has been illustrated in two pelagic food chain of tuna (6). Tunas (T. thynnus) caught in the Mediterranean can be divided into two populations on the basis of the mercury concentrations: a "low-mercury population" and a "high-mercury population" (Fig. 1). In both food chains Hg increases with size of the organism and in higher trophic levels. Fisheries biology data suggest that the "low-mercury population" is of Atlantic origin and the "high-mercury population" is indigenous to the Mediterranean. The "Atlantic tunas" enter the Mediterranean only for spawning. In fact, all tunas caught in the Strait of Gibraltar entering or leaving the Mediterranean belong to the "low-mercury populations." Similarly, anchovies, mackerels, and sardines (food organisms for tuna) caught in the Strait of Gibraltar have lower Hg concentrations than specimens caught in the Tyrrhenian Sea. A simple mathematical model simulating the food chain of tuna (seawater-plankton-sardine-tuna) show that very small differences in the Hg concentration between seawater from the Atlantic and the Mediterranean are sufficient to explain the large differences in Hg concentration/size relationships observed first in the tuna's prey and then in tuna (Fig. 1). The model also illustrates that the food chain is by far the most important route of uptake and that increase in concentration in higher trophic levels depends on the fact that organisms with a longer life are in the higher positions of the food chain. Phytoplankton needs a few days between cell divisions, zooplankton (copepods) 20 to 40 days, sardines reach 5 years, but the great majority of adult fish has 1 to 3 years, while tunas can reach 15 years and more. Only 10% of inorganic mercury ingested is taken up compared with 90% of methylmercury. Inorganic mercury is eliminated with a biological half-life of about 60 days against 7 years for methylmercury. Due to the physiological differences in the uptake and loss processes of the two Hg forms simultaneously with an increase of total mercury the proportion of the methylmercury increases with size at each trophic level. Since predators in higher trophic positions feed on prey which has already "preconcentrated" methylmercury, each higher level feeds on prey with a higher relative methylmercury content.

It may be worthwhile mentioning that the pelagic habitat of these animals is remote from anthropogenic activities and, therefore, the high Hg concentrations found in tuna and other large (and old) marine organisms (swordfish, porpoises, whales, etc.) are not due to anthropogenic sources but are caused by the very efficient uptake of naturally occurring methylmercury from food organisms and its very low elimination rate. In the pelagic biota of the Western Mediterranean much higher Hg levels have been observed than in specimens of the same species from the adjacent Atlantic. These are attributed to high environmental levels resulting from natural geochemical anomalies (7), since the high Hg concentrations in riverine and estuarine sediments and the concentration gradient in marine sediments adjacent to rivers in this region cannot be explained on the basis of anthropogenic inputs.

BIOAVAILABILITY

There are numerous data available from the radioecological literature which demonstrates that different forms of trace metals are accumulated to different extent by marine organisms (e.g., (33,55)).

Usually the ionic forms are more readily accumulated than the particulate, complexed, or chelated forms. For example, Bernhard and Zattera (8) could show that ionic 65-zinc was taken up practically completely by a growing Phaeodactylum tricornutum culture while the zinc bound to EDTA or other complexing agents was not taken up at all in the presence of ionic and particulate zinc. Only a few percent were accumulated when all zinc present was chelated with EDTA. Indirect evidence came from Sunda and Guillard (62) who showed that the copper toxicity to Thalassiosira pseudonana is related to the activity of the cupric ion and not to the total Cu concentration. Many others have observed that the Cu toxicity is reduced by natural and synthetic chelators such as amino acids, humic acids, soil extract, EDTA, detergents, etc. (4,46,59). Several authors could demonstrate that certain phytoplankton species can produce organic substances which complex Cu (e.g., (31)). Arsenate and arsenite are both toxic to the marine diatom Skeletonema costatum at concentrations as low as 5 ug As/L, while dimethylarsenic acid showed no toxic effect at 25000 ug As/L (56). Arsenite is rapidly chemically oxidized into arsenate, and arsenate in seawater is partially converted to dimethylarsenic acid by phytoplankton (2).

Similar to the results obtained for autotrophs, the reduction of trace metal uptake by marine animals as a result of changing the physicochemical form has been observed in many instances. Reduced uptake in the presence of chelators has been demonstrated for many radioisotopes (e.g., (33)). Trace elements contained in particles are much more highly enriched than metals contained in solution by organisms such as filter feeders. For example, particulate Cr(III) was enhanced with respect to soluble Cr(VI) in the filter-feeding clam while the worm Hermione hystrix accumulated preferentially Cr(VI) to Cr(III) (16,17).

Hg-alkyl compounds are taken up faster and can penetrate cell membranes more easily because they are more lipid-soluble than the much less toxic inorganic Hg salts (18,52,53). The accumulation of both inorganic and organic mercury is reduced by glutathione (18). The availability of metals and hence their absorption from the gut depends on the chemical form of the metals, as has been shown by supplying metals to marine organisms in different matrixes (starch, gelatine, food organism Neris diversicolor) (51).

Arsenic undergoes a number of chemical transformations in marine food chains which regulates its bioavailability. It is present in seawater mostly as arsenate, but also is converted by algae to arsenite, mono- and dimethylarsinate, and a number of other organoarsenic compounds (2,40). In macro-algae 50 to 80% is organic arsenic (57). Inorganic arsenic taken up from the seawater was retained twice as effectively as the mostly organic arsenic taken up from the food. Organoarsenic compounds are taken up by marine organisms with the food or synthesized from inorganic arsenic taken up from seawater (39). At higher trophic levels, organoarsenic compounds predominate (typically 90%) (42). Although in seafood most arsenic is in the less (or non-) toxic organic form (see (69) and Zielhuis and Wibowo, this volume), nevertheless, at high body levels the concentration of the much more toxic inorganic form still could be considerable.

Like methylmercury, ingested "seafood arsenic" is rapidly taken up in the gastrointestinal tract of higher animals and humans, but unlike methylmercury, "seafood arsenic" is rapidly excreted without being broken down to toxicologically active arsenic species (15,21,64). Contrary to mercury, where the microbial methylation results in an increase in toxicity, the formation of organoarsenic compounds in the marine food chain results in the protection of the entire food chain, including man, from the toxicity of inorganic arsenics.

It is obvious from these examples that the total concentration of a trace metal will be of little use in predicting its rate of accumulation and its environmental and public health impact.

TRANSFER OF TRACE METALS FROM MARINE FOODS TO MAN

Marine foods of man include many trophic levels. Marine plants (seaweeds) are consumed in many nations, especially in the Far East but also in Europe, e.g., layerbread from the Irish Sea or agar. Phytoplankton and zooplankton are not eaten, but the plankton eaters such as oysters, mussels, clams, scallops, herrings, and sardines are consumed. Krill is used for animal fodder.

Higher trophic levels include the carnivorous gastropods, clams, cephalopods, crabs and shrimps, and foremost fishes. The world (population 4150 million in 1977) average intake of fish and shellfish protein is 3.8 g per day and per person out of a total of 68.8 g total protein/day and person (27). These 3.8 g protein correspond to about 20 g edible fish and shellfish/day or 140 g edible tissue, or about one meal of fish and shellfish/week. As can be seen from Table 4, aquatic plants and seaweeds make up only 4% of the world's harvest of marine and freshwater foods. Of the total catches of aquatic animals, 10% are caught in freshwater and the great majority (90%) are marine. Seventy-seven percent of the world's landings are marine fish, followed by molluscs (7%) and crustaceans (4%). Other marine animals contribute 1% or less. The NW Pacific and the NE Atlantic are the most productive areas (Table 5). Among fishes, pelagic plankton feeders such as herrings, sardines, and anchovy make up 28%

	in thousand tons living weight (FW)	
World, grand total	74370	
Aquatic plants and seaweeds	3080	4 % (*)
Total aquatic animals	71290	96 % (*)
Freshwater organisms		10 %
Crustaceans	110	0.2%
Molluscs	260	0.4%
Freshwater fish	6070	9 %
Diadromous fish	1030	1 %
Miscellaneous aquatic animals + products	4	
Marine organisms		90 %
Crustaceans	2960	4 %
Molluscs	4720	7 %
Diadromous fish	820	1 %
Marine fish	55130	77 %
Miscellaneous aquatic animals + products	190	0.3%

(*) percentage of grand total; all other percentages are based on total aquatic animals.

TABLE 4 — Nominal catches in 1979 (metric tons) (data from FAO (28), rounded off).

	in thousand tons FW	% of subtotal
Atlantic, NW	2840	4
NE	11710	18
W-central	1780	3
E-central	2820	4
Mediterranean + Black Sea	1320	2
Atlantic, SW	1470	2
SE	2520	4
Antarctic	450	1
Indian Ocean, W	2220	3
E	1350	2
Antarctic	70	
Pacific, NW	18320	29
NE	1970	3
W-central	5690	9
E-central	2020	3
SW	360	1
SE	6900	11
subtotal fishing areas	63810	

TABLE 5 — Nominal catches in 1979 of marine animals according to major fishing areas (data from FAO (28), rounded off).

of the world's landings, followed by bentho-pelagic fishes (cod, hake, and haddock) which feed on invertebrates, fish larvae, clupoids, and anchovy.

Looking at seafood consumption per nation (Table 6), one notices that small populations on islands can consume relatively large amounts of marine foods. According to FAO food balance sheets (27), the population (40000 inhabitants) of the Faeroe Islands consume on the average 38.6 g protein (equal to 193 g FW of edible tissue) originating from marine foods. Assuming 150 g of edible tissue constitutes one meal, this corresponds to more than one meal per day of seafood. The Japanese average is the highest of larger populations. By comparison the average consumption of marine foods in the USSR, USA, and Australia is small.

However, individuals are reported to consume much greater amounts. Fishermen at sea, especially in less affluent regions, will consume exclusively marine foods. About 800 g per day were consumed on fishing boats in southern Italy (7), Canadian Indians reach 1300 g per day during the fishing season, and 800–1500 g per day were eaten by fishermen and their families from Minamata (54). From these data and the concentrations listed in Table 3 for some metals, average transports for populations and individual transports can be estimated (Table 7). These estimations can only be considered approximate because for

	Population in mill.	Tot. prot.	FW	Fresh-water	Seawater fish	Seawater crust.	Seawater molluscs
Faeroe Isl.	0.04	38.6	193	0.1	31.4 [82]	6.0 [16]	1.1 [3]
Maldives	0.14	37.1	186	-	37.1 [100]	-	-
Vanuatu	0.1	23.2	116	-	23.0 [99]	0.2 [2]	-
Japan	113.9	22.5	113	1.5 [7]	17.4 [77]	1.0 [4]	2.3 [10]
Iceland	0.1	19.2	96	0.6 [3]	15.9 [83]	2.4 [13]	-
Bermuda	0.06	15.8	79	-	13.8 [87]	2.0 [13]	-
Yemen Rep.	1.9	13.0	65	-	13.0 [100]	-	-
Portugal	8.8	10.2	51	-	13.8 [87]	0.2 [2]	0.3 [3]
USSR	258.9	9.4	47	1.6 [17]	7.6 [31]	-	0.1 [1]
Australia	13.8	3.4	17	0.3 [9]	2.3 [68]	0.8 [24]	-
USA	216.8	3.2	16	0.2 [6]	2.0 [63]	0.7 [22]	0.2 [6]

percentage is given in brackets [ ].

TABLE 6 — Consumption of fish and seafood in g FW (living weight) and g protein per day by populations of selected countries (27).

Average intake

World population (1977) 4150 million

20 g FW seafoods/day
7.3 kg FW seafoods/year

	Typical conc. ug/kg	Mean intake ug/week	Total world intake kg/year
Pb	1	0.14	30
Cd	25	3.5	750
Hg	100	14	3000
Cu	1000	140	30000
As	2000	280	60000
Zn	5000	700	150000

TABLE 7 — Total and average intake of trace metals from seafoods.

many groups of species trace metal data did not seem to be available, and because it is difficult to judge the accuracy of the available data.

These values can be used to estimate the trace metal flux from the sea to man. The average fluxes for populations and individual or critical groups of high seafood consumers can be compared with the tolerable intakes recommended by various agencies (e.g., (26, 66, 67, 69)). At present provisional intakes are available only for very few trace metals (Table 8).

The Hg limits have been reconfirmed in recent years (68), with the additional restriction for pregnant and lactating women to half the PTWI. The values for lead have certainly to be revised in the light of recent evidence. Also the PTWI's for As need further reevaluation since insufficient information is

	per kg body weight	per 70 kg man
(seafood - As?	350	24500)*
Cd	5	300
Cu	3500	245000
methyl Hg	3	200
(Pb	40	2800)*

* These values probably have to be revised.

TABLE 8 — Provisional tolerable weekly intakes (PTWI) in ug.

available on the occurrence of the chemical forms of very different toxicity ((69), and Zielhuis and Wibowo, this volume). Based on the PTWI and the average trace metal content in seafood, it is possible to estimate the number of meals (150 g FW each) of seafood per week necessary to reach the PTWI (Table 9). From Table 9 one can conclude that only mercury presents a problem, since even high consumers of marine food cannot eat a sufficient number of meals to reach the PTWI's for the other trace metals. However, if the concentrations are much higher than the typical concentration, e.g., in polluted areas, then the PTWI may also be reached for elements other than Hg.

Mercury presents a potential risk to persons eating seafood twice a day or tuna once a week because they can easily reach the PTWI, especially in regions with high Hg environmental levels such as the Mediterranean. These high consumption rates are, however, characteristic only for critical groups such as fishermen, fish vendors and their families, and persons who have a preference or easy access to species with high mercury levels such as tuna, swordfish, halibut, etc. Up to 14 meals per week of seafoods may be consumed by persons in these critical groups. Especially tuna fishermen from Japan and Korea, who are confined on tuna fishing vessels for long periods, and tuna fish retailers consume large amounts of high mercury tuna meat (review in (24)). Much

	Typical conc. ug/kg FW	kg consumed to reach PTWI	Number of meals per week to reach PTWI
(seafood - As?	2000	12	80)
Cd	25	12	80
Cu	1000	245	1600
methyl Hg	100	2	13
" in tuna	1000	0.1	0.7

TABLE 9 — Number of meals of seafood necessary to reach PTWI's.

higher than average mercury levels in hair and blood have been observed in these heavy fish eaters. In several cases the levels observed were so high that effects were expected, but none could be identified.

Recent findings that selenium acts antagonistically to mercury and that in both man and marine mammals high mercury concentrations are associated with high selenium concentrations (called co-accumulation by Kosta et al. (40)) may point to a natural protection mechanism against mercury poisoning (see review in (44)). Ever since Parizek and Ostadalova (49) observed that small amounts of selenite could clearly counteract the toxicity of mercury(II)chloride, many other authors have investigated the antagonistic effects of selenium on inorganic and organic mercury compounds. Of special interest are investigations which have shown that methylmercury toxicity is considerably reduced in experimental animals when tuna (2000–3000 ug Se/kg FW) was supplied to the diet (30). There can be no doubt that the high mercury levels in pelagic fishes and mammals (tuna, swordfish, whales, and porpoises) are of natural origin since their habitat is far away from potential mercury pollution sources or possible anthropogenic contamination. Similarly, naturally occurring inorganic arsenics may present hazards to high consumers of seafoods. Although most of the arsenic is in the much less toxic (or non-toxic) organic form at the very high body levels observed in certain benthic fishes and molluscs, even a few percent of a total of 100 mg as/kg FW could still result in several milligrams of inorganic arsenic per kg FW of seafood.

REFERENCES

(1) Andreae, M.O. 1978. Distribution and speciation of arsenic in natural waters and marine algae. Deep-Sea A. **25**: 391–402.

(2) Andreae, M.O., and Klumpp, D.W. 1979. Arsenic uptake and metabolism by marine phytoplankton. Env. Sci. Tec. **13**: 738–741.

(3) Badscha, K.S., and Sainsburg, M. 1977. Uptake of zinc, lead, and cadmium by young whiting in the Severn Estuary. Mar. Poll. Bull. **8**: 164–166.

(4) Bernhard, M. 1955. Die Kultur von Seeigellarven (Arbacia lixula L.) in kuenstlichem und natuerlichem Meerwasser mit Hilfe von Ionenaustauschsubstanzen und Komplexbildner. Publ. Staz. Zool. Napoli **29**: 80–95.

(5) Bernhard, M. 1980. The relative importance of lead as a marine pollutant. **In** Lead in the Marine Environment, eds. M. Branica and Z. Konrad, pp. 345–352. Oxford: Pergamon Press.

(6) Bernhard, M.; Buffoni, G.; and Renzoni, A. 1982. Mercury in Mediterranean tuna. Why is their level higher than in Atlantic tuna? A model. Thalassia Jugoslavia. **18**: 231–243.

(7) Bernhard, M., and Renzoni, A. 1977. Mercury concentration in Mediterranean marine organisms and their environment: natural or anthropogenic origin. Thalassia Jugoslavica **13**: 265–300.

(8) Bernhard, M., and Zattera, A. 1969. A comparison between the uptake of radioactive and stable zinc by a marine unicellular alga. **In** II National Symposium on Radioecology, eds. J. Nelson and D. Evans, pp. 389–398. Washington: USAEC.

(9) Bohn, A., and McElroy, R.O. 1976. Trace metal (As, Cd, Fe and Zn) in Arctic cod, Boreogadus saida and selected zooplankton from Strathcona Sound, Northern Baffin Island. J. Fish Res. Bd. Can. **33**: 2836–2840.

(10) Boyden, C.R. 1974. Trace element content and body size in molluscs. Nature **251**: 311–314.

(11) Boyden, C.R. 1977. Effect of size upon metal content of shellfish. J. Marine Biol. Ass. U. K. **57**: 675–714.

(12) Brewer, P.G. 1975. Minor elements in seawater. **In** Chemical Oceanography, eds. J.P. Riley and G. Sharrow, pp. 415–496. London: Academic Press.

(13) Bryan, G.W.; Potts, G.W.; and Forster, G.R. 1977. Heavy metals in the gastropod mollusc Haliotis tuberculata (L.). J. Marine Biol. Ass. U. K. **57**: 379–390.

(14) Butterworth, J.; Lester, P.; and Nickless, G. 1972. Distribution of heavy metal in the Severn Estuary. Mar. Poll. Bull. **3**: 72–73.

(15) Charbonneau, S.M.; Spencer, K.; Bryce, F.; and Sandi, E. 1978. Arsenic excretion by monkeys dosed with arsenic-containing fish or with inorganic arsenic. Bull. Env. Contam. Toxicol. **20**: 470–477.

(16) Chipmann, W.A. 1966. Uptake and accumulation of chromium-51 by the clam Tapes decussatus in relation to physical and chemical form. **In**

Disposal of Radioactive Wastes into Seas, Oceans and Surface Waters, IAEA Symposium, pp. 571–582. Vienna: IAEA.

(17) Chipmann, W.A. 1967. Some aspects of the accumulation of 51-Cr by marine organisms. **In** Radioecological Concentration Processes, eds. A. Aberg and F.P. Hungate, pp. 931–941. Oxford: Pergamon Press.

(18) Corner, E.D.S., and Rigler, F.H. 1958. The modes of action of toxic agents. III. Mercury chloride and N-amylmercuric chloride on crustaceans. J. Marine Biol. Ass. U. K. **37**: 85–96.

(19) Cossa, D.; Bourget, E.; and Piuze, J. 1979. Sexual maturation as a source of variation in the relationship between cadmium concentration and body weight of Mytilus edulis L. Mar. Poll. Bull. **10**: 174–176.

(20) Costa, M.R.M., and Da Fonseca, M.I.C. 1967. The amount of natural arsenic in shellfish. Rev. Port. Farm. **17**: 1–19 (in Portuguese).

(21) Crecelius, E.A. 1977. Changes in the chemical speciation of arsenic following ingestion by man. Env. Health Prosp. **19**: 147–150.

(22) Cross, F.A., and Brooks, J.H. 1973. concentrations of Mn, Fe, and Zn in juveniles of five estuarine-dependent fishes. **In** Third National Symposium on Radioecology, ed. D.J. Nelson, pp. 254–265. Washington: AEC.

(23) Cross, F.A.; Hardy, L.H.; Jones, N.Y.; and Barber, R.T. 1973. Relation between total body weight and concentrations of manganese, iron, copper, zinc and mercury in white muscle of bluefish (Pomatomus saltatrix) and a bathyl-demersal fish Antimora rostrata. J. Fish. Res. Bd. Can. **30**: 1287–1291.

(24) Doi, R., and Ui, J. 1975. The distribution of mercury in fish and its form of occurrence. **In** Heavy Metals in the Aquatic Environment, ed. P.A. Krenkel, pp. 197–222. Oxford: Pergamon Press.

(25) Eisler, R. 1981. Trace Metal Concentrations in Marine Organisms. New York: Pergamon Press.

(26) FAO. 1973. Evaluation of mercury, lead, cadmium and the food additives amoranth, diethylpyrocarbonate and octylgallate. FAO Nutrition Meetings Report Series No. 51A. Rome: FAO.

(27) FAO. 1980. Food Balance Sheets and per Caput Food Supplies. Rome: FAO.

(28) FAO. 1980. Yearbook of Fishery Statistics, Catches and Landings, vol. 48. Rome: FAO.

(29) Fage, R., and James, K.H. 1974. Trace metal concentration in Fucus from the Bristol Channel. Mar. Poll. Bull. **5**: 9–12.

(30) Ganther, H.E., and Sunde, M.L. 1974. Effect of tuna fish and selenium on the toxicity of methylmercury. J. Food Sci. **39**: 1–5.

(31) Gnassia-Barelli, M.; Romeo, M.; Laumond, F.; and Pesano, D. 1978. Experimental studies on the relationship between natural copper complexes and their toxicity to phytoplankton. Mar. Biol. **47**: 15–19.

(32) Hardisty, M.W.; Huggins, R.J.; Kartar, S.; and Sainsbury, M. 1974. Ecological implication of heavy metal in fish from the Severn Estuary. Mar. Poll. Bull. **5**: 12–15.

(33) Hiyama, Y., and Shimizu, M. 1964. On the concentration factor of radioactive Cs, Sr, Cd, Zn and Ce in marine organisms. Records Oceanog. Works Japan **7**: 43–77.

(34) ICES. 1974. Report of Working Group for the International Study of the Pollution of the North Sea and its Effects on Living Resources and their Exploitation. ICES Cooperative Research Report No. 39. DK-Charlottenlund: ICES.

(35) ICES. 1977. A Baseline Study of the Level of Contaminating Substances in Living Resources of the North Atlantic, ICES Cooperative Research Report No. 69. DK-Charlottenlund: ICES.

(36) ICES. 1977. The ICES Coordinated Monitoring Programme in the North Sea, 1974 ICES Cooperative Research Report No. 58. DK-Charlottenlund: ICES.

(37) ICES. 1977. The ICES Coordinated Monitoring Programme, 1975 and 1976, ICES Cooperative Research Report No. 72. DK-Charlottenlund: ICES.

(38) ICES. 1980. Extension to the Baseline Study of Contaminant Levels in Living Resources of the North Atlantic, ICES Cooperative Research Report No. 95. DK-Charlottenlund: ICES.

(39) Klumpp, D.W., and Peterson, P.J. 1981. Chemical characteristics of arsenic in a marine food chain. Marine Biol. **62**: 297–305.

(40) Kosta, L.; Byrne, A.R.; and Zelenko, V. 1975. Correlation between selenium and mercury in man following exposure to inorganic mercury. Nature **254**: 238–239.

(41) Leatherland, T.M., and Burton, J.D. 1974. The occurrence of some trace metals in coastal organisms. J. Marine Biol. Ass. U. K. **54**: 457–468.

(42) Lunde, G. 1977. Occurrence and transformation of arsenic in the marine environment. Env. Health Prosp. **19**: 47–52.

(43) MacKay, N.J.; Kazacos, M.N.; Williams, R.J.; and Leedow, M.I. 1975. Selenium and heavy metals in black marlin. Mar. Poll. Bull. **6**: 57–61.

(44) Magos, L., and Webb, M. 1979. Synergism and antagonism in the toxicology of mercury. **In** The Biogeochemistry of Mercury in the Environment, ed. J.O. Nriagu, pp. 581–599. Amsterdam: Elsevier-North Holland.

(45) Martin, J.H., and Knauer, G.A. 1973. The elemental composition of plankton. Geochim. Cosmochim. Acta **37**: 1639–1653.

(46) Morris, O.P., and Russell, G. 1973. Effect of chelation on toxicity of copper. Mar. Poll. Bull. **4**: 159–160.

(47) Noel-Lambot, F.; Bouquegneau, J.M.; Frankenne, F.; and Disteche, A. 1980. Cadmium, zinc and copper accumulation in limpets (Patella vulgata) from the Bristol Channel with special reference to metallothioneins. Mar. Ecol. Prog. Ser. **2**: 81–89.

(48) O'Rear, C.W. 1971. Some environmental influences on the zinc and copper content of striped bass, Morone saxatilis (Walbaum). Ph.D. Thesis, Virginia Polytechnic and State University, Bracksburg, VA.

(49) Parizek, J., and Ostadalova, I. 1967. The protective effect of small amounts of selenium in sublimate intoxication. Experienta **23**: 142–145.

(50) Peden, J.D.; Crothers, J.H.; Waterfall, C.E.; and Beasley, J. 1973. Heavy metals in Somerset marine organisms. Mar. Poll. Bull. **4**: 7–9.

(51) Pentreath, R.J. 1973. The role of food and water in the accumulation of radionuclides by marine teleost and elasmobranch fish. **In** Radioactive Contamination of the Marine Environment, pp. 421–436. Vienna: IAEA.

(52) Pentreath, R.J. 1976a. The accumulation or inorganic mercury from sea water by the plaice, Pleuronectes platessa L. J. Exp. Mar. B. **25**: 103–120.

(53) Pentreath, R.J. 1976b. The accumulation of organic mercury from sea water by the plaice, Pleuronectes platessa L. J. Exp. Mar. B. **25**: 121–132.

(54) Piotrowski, J.K., and Inskip, M.J. 1981. Health effects of methylmercury, Marc Report No. 24. London: University of London.

(55) Polikarpov, G.G. 1966. Radioecology of Aquatic Organisms. New York: Reinholt.

(56) Sanders, J.G. 1979. Effects of arsenic speciation and phosphate concentration on arsenic inhibition of Skeletonema costatum (Bacillariophyceae). J. Phycol. **15**: 424–428.

(57) Sanders, J.G. 1979. The concentration and speciation of arsenic in marine macro-algae. Est. Coas. M. **9**: 95–99.

(58) Settle, D.M., and Patterson, C.C. 1980. Lead in albacore: guide to lead pollution in Americans. Science **207**: 1167–1176.

(59) Steeman-Nielsen, E., and Wium-Anderson, S. 1970. Copper ions as poison in the sea and fresh-water. Marine Biol. **6**: 93–97.

(60) Stickney, R.R.; Windom, H.L.; White, D.B.; and Taylor, F.E. 1975. Heavy-metal concentrations in selected Georgia estuarine organisms with comparative food-habit data. **In** Proceedings of the Symposium on Mineral Cycling in Southeastern Ecosystems, eds. F.G. Howell et al., pp. 257–267. Washington: Technology Information Center, USERDA.

(61) Stoeppler, M.; Mohl, C.; and Nuernberg, H.W. 1980. Total arsenic in seawater and marine organisms: a comparative study from the Mediterannean Sea and selected regions of the oceans. V. Journees Etud. Pollutions, pp. 281–284. Principaute' di Monaco: CIESM.

(62) Sunda, W., and Guillard, R.R.L. 1976. The relationship between cupric ion activity and the toxicity of copper to phytoplankton. J. Mar. Res. **34**: 511–529.

(63) Topping, G. 1980. A review of the past and present measurements of selected trace metals in sea water in the Oslo commission and ICNAF:

NAFO areas. ICES Cooperative Research Report No. 97. DK-Charlottenlund: ICES.

(64) Tam, G.K.H.; Charbonneau, S.M.; Bryce, F.; and Sandi, E. 1982. Excretion of single oral dose of fish-arsenic in man. Bull. Env. Contam. Toxicol. **28**: 669–673.

(65) UNEP. 1980. Co-ordinated Mediterranean pollution monitoring and research programme (MED POL). Part I: Summary, Scientific Report February 1975–June 1980. Geneva: UNEP.

(66) WHO. 1976. Environmental Health Criteria 1. Mercury. Geneva: WHO.

(67) WHO. 1977. Environmental Health Criteria 3. Lead. Geneva: WHO.

(68) WHO. 1980. Report on consultation to re-examine the WHO Environmental Health Criteria for Mercury (21–25 April 1980 Geneva), EHE/EHC/80.22. Geneva: WHO.

(69) WHO. 1981. Environmental Health Criteria 18. Arsenic. Geneva: WHO.

(70) Zook, E.G.; Powell, J.J.; Hackley, B.M.; Emerson, J.A.; Brooker, J.R.; and Knobl, Jr., G.M. 1976. National Marine Fisheres Service preliminary survey of selected seafoods for mercury, lead, cadmium, chromium and arsenic content. J. Agr. Food **24**: 47–53.

Changing Metal Cycles and Human Health, ed. J.O. Nriagu, pp. 169–186.
Dahlem Konferenzen 1984. Berlin, Heidelberg, New York, Tokyo: Springer-Verlag.

ADDITIONAL EXPOSURE ROUTES

P.J. Landrigan and K. Bridbord
National Institute for Occupational Safety and Health
Cincinnati, OH 45226, and Rockville, MD 20857, USA

Abstract. Toxic trace metals are widely dispersed in the environment. Adults are primarily exposed to these metals in the workplace. Children may be exposed from numerous sources, including contaminated air, dust, soil, lead-based paint, pottery, sewage sludge, and battery burning. New sources of exposure will undoubtedly be discovered as additional industrial and commercial uses are found for the toxic metals.

INTRODUCTION

Trace metals have an enormous range of uses in modern industrialized societies. Consequently man may be exposed to these elements through a wide variety of environmental pathways (33). In addition to exposures through food chains and beverages, human absorption may result from inhalation or ingestion, from exposure at work, from contact with industrial waste products, from the use of commercial products, or merely from residence in a contaminated city or near a point-source of environmental pollution.

In this paper we shall catalogue the major routes of exposure to the trace metals (excluding foods and beverages), and we shall describe several episodes of excessive exposure to the metallic elements. Although these examples will focus principally on lead, cadmium, arsenic, and mercury, they are intended to serve as models for the discussion of exposures to other metals.

OCCUPATIONAL EXPOSURES

Adults have their most intense exposure to trace metals in the workplace (Table 1). Not surprisingly, industrial workers are the subset of the population who have been shown to have heaviest body burdens of metals (13). Although some occupational absorption may occur as the result of ingestion from fingertips, cigarettes, or food, by far the major route of industrial absorption has been shown to be by inhalation (37). In regard to lead, Sir Thomas Legge, the first Medical Inspector of Factories in the United Kingdom, noted axiomatically that: "Practically all industrial lead poisoning is due to the inhalation of dust and fume; and if you stop their inhalation, you will stop the poisoning (37)."

Occupational absorption of heavy metals may produce severe acute intoxications or chronic diseases of a nature and severity seldom encountered in other exposure settings. Among the acute and subacute intoxications recorded

Metal	Industries and Processes
Lead	Lead smelting, refining Storage battery manufacture Spray painting Bronze manufacture Pouring leaded iron Steel burning (ship burning) Manufacture of plastics with lead stearate Lead pigment manufacture Firing range operations Welding
Cadmium	Lead, cadmium, and zinc smelting Battery manufacture Jewelry manufacture Welding
Arsenic	Copper and gold smelting Brass and bronze manufacture Insecticide and rodenticide formulation Tanning Semiconductor manufacture Pharmaceutical manufacture
Mercury (inorganic)	Amalgam manufacture Battery manufacture Jewelry manufacture Taxidermy

TABLE 1 - Industrial processes associated with trace metal exposures.

among workers exposed occupationally to metals are anemia, peripheral neuropathy, and encephalopathy in lead workers (7); pneumonitis in jewelry workers exposed to cadmium fume (8); "metal fume fever" among workers in zinc plants; memory impairment and pathologic shyness progressing to paranoid psychosis, termed "erethism" in workers exposed to inorganic mercury (27); bronchitis and green discoloration of the tongue in workers inhaling vanadium; and acute pneumonia among workers exposed to beryllium fume.

Disabling or fatal chronic illnesses include nephritis and renal failure in lead and cadmium workers (39); nephrosis in workers exposed to mercury (29); skin cancer, lung cancer, angiosarcoma of the liver, and possibly lymphatic cancer in smelter and pesticide workers exposed to arsenic (32); berylliosis and probably lung cancer in beryllium workers (57); cancer of the nasal sinuses in nickel refiners (22); and erosion of the nasal septum and lung cancer in workers exposed to chromium (4).

Prevention of occupational exposures to trace metals is accomplished most effectively by substitution of less hazardous materials, by the complete enclosure of hazardous processes, or by the installation of exhaust ventilation. Worker education and the use of proper work practices are important ancillary measures. Personal respirators are the least effective means of reducing occupational exposures to metals and should be used only as a temporary measure pending the installation of proper engineering controls.

PARA-OCCUPATIONAL EXPOSURES

Members of workers' families are at risk of exposure to metals transported home from the workplace on workers' clothing, shoes, skin, hair, and automobiles (18). Early in this century Oliver (46) described the occurrence of lead poisoning with peripheral neuropathy (wrist drop) in the wives of lead workers who washed their husbands' lead-laden work clothing at home by hand.

More recently, in an evaluation of 91 lead smelter workers' children in Memphis, Tennessee, we found elevated blood lead levels (30 ug/dl or above) in 38 (41.8%) children; 10 had blood lead levels of 80 ug/dl and above and /or erythrocyte protoporphyrin (EP) concentrations above 190 ug/dl (5). The childrens' blood lead levels were correlated positively with household dust lead concentrations, with fathers' blood lead levels, and with duration (in months) of fathers' employment at the smelter (Tables 2 and 3). Under regulations established in 1978 by the U.S. Occupational Safety and Health Administration, industries using lead are now required to provide changing and showering facilities for their workers (55).

Dust Lead Concentration (PPM)	No. of Children	Children's Blood Lead Level (ug/dl)
0 - 1000	18	21.8 ± 7.8*
1000 - 2000	5	41.8** ± 11.1
2000 - 3000	5	46.2** ± 21.1
5000 - 6000	4	65.0** ± 10.4
7000 - 80,000	6	73.3** ± 34.0

* Mean ± SD.
** Significantly > mean level of children exposed to 0 - 1000 ppm level of lead in dust ($P < 0.001$).
From Baker et al. (3).

TABLE 2 – Comparison of lead concentration in household dust with children's blood lead levels, Memphis, November 1975.

Months of Employment Completed	Children's Blood Lead Levels (ug/dl) Number	Mean (+ S.D.)
0 - 2	5	26.4 (± 5.0)
3 - 4	8	43.7 (± 21.1)
5 - 12	4	59.5 (± 13.4)

From Baker et al. (3).

TABLE 3 – Comparison of duration of father's employment at smelter with children's blood lead levels, Memphis, November 1975.

EXPOSURES TO AIRBORNE METALS

Lead

Automotive and industrial emissions are the major contributors to the lead content of air. Of the approximately 100,000 tons of lead particulates released into the air each year in the United States, over 95% derives from the combustion of leaded gasoline in motor vehicles (43).

Concentrations of airborne lead are generally higher in cities than in rural areas and correlate in most locations with the density of motor vehicle traffic. In an international comparative study, a high airborne lead content was noted in Mexico City which has dense traffic and uses heavily leaded gasoline; by contrast, low air lead levels were noted in Tokyo, which also has dense traffic, but uses mainly unleaded gasoline (56). Airborne concentrations of lead have been shown to be elevated near busy expressways and to decrease logarithmically with distance from those roadways (28). Studies of children living near freeways in New Jersey (11) and in California (28) have shown somewhat higher blood lead levels in children living within 30 meters of major roadways than in those living at greater distances (Table 4). Also, the children's blood lead levels were correlated positively with the average daily traffic volume on the roadways by which they lived (11). In the past decade, coincident with major reductions in the amount of lead added to gasoline in the United States, significant reductions in urban air lead concentrations and in children's blood levels have been noted (9).

Although industrial releases account for only a small fraction of airborne lead, these stationary sources can produce concentrated zones of high-dose exposure, especially where climatic conditions such as aridity, low wind velocity, and frequent thermal inversions minimize opportunities for dispersal of airborne lead. The worst such situation in the United States has existed in the vicinity of a large lead ore smelter in Kellogg, Idaho (34). Elevated concentrations of lead have been documented in air in Kellogg, and air lead levels were found to be significantly higher within one kilometer of the smelter than at greater distances (Fig. 1). A blood lead survey of children in the area of the Kellogg smelter conducted in 1974 showed that 385 (41.9%) of 919 1-9-year-olds had blood lead levels of 40-79 ug/dl, and that an additional 41 (4.5%) had levels of 80 ug/dl or greater (35). Blood lead levels were highest within one kilometer of the smelter, where 99% of 172 children examined had blood lead levels above 40 ug/dl, and 21% had levels above 80 ug/dl (Fig. 2).

Distance of Residence From Roadway (feet)	Number of Children	Distribution (%) of Blood Lead Levels		
		40 ug/dl	40-59 ug/dl	60 + ug/dl
< 100	758	42.6	49.3	8.1
100 - 200	507	72.4	24.2	3.4
> 200	3,961	68.4	26.9	4.7

* From Caprio et al. (9).

TABLE 4 – Blood lead levels in one- to nine-year-old children in relation to distance of residence from a major roadway in Newark, New Jersey, 1971.*

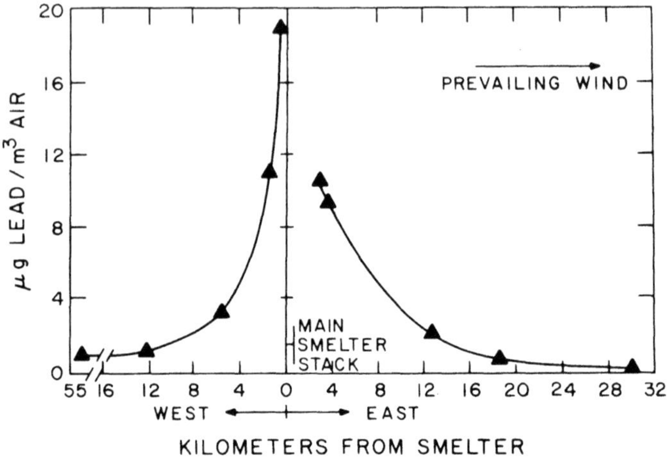

FIG. 1 – Air lead levels by distance from a lead ore smelter, Kellogg, Idaho – 1974.

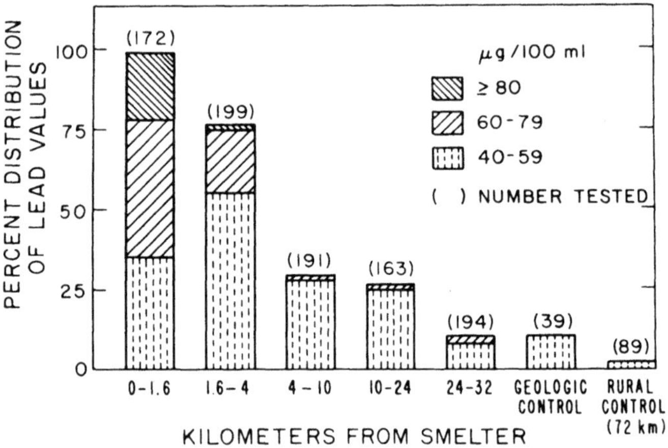

FIG. 2 – Blood lead levels in one- to nine-year-old children by distance from a lead ore smelter, Kellogg, Idaho – 1974.

Cadmium

Cadmium is also dispersed into the air by industrial sources, and populations may be exposed to cadmium from the same sources as lead. Cadmium is released into the air by smelting plants, particularly by zinc and lead smelters.

Near such plants cadmium levels have been found to be high in air, dust, and surface soil, and children living near zinc smelters have been shown to have significantly elevated body burdens of cadmium (6). Cadmium may also be released into the air at scrap recycling plants (42) from the burning of cadmium in pigment or plastics stabilizer. Cigarettes and cigarette smoke contain measurable amounts of cadmium, and children living in the same households as smokers have significantly higher blood cadmium levels than do the children of nonsmokers (59).

Arsenic

Arsenic is also dispersed into the atmosphere by metallurgical plants, by gold and copper smelters in particular (10,40). Highly elevated arsenic exposures, as measured by elevated urine total arsenic concentrations, have been recorded in children living within 0.5 km of a large copper smelter in Tacoma, Washington; arsenic exposures were found to decrease logarithmically with distance from that plant (40).

Kinetics of Respiratory Absorption

Respiratory absorption of trace metals depends principally upon the aerodynamic diameter of metal-bearing airborne particles; that diameter, in turn, is a function of particle density, size, and shape (43). For most common lead aerosols of mixed particle sizes, it has been estimated that between 30 and 37% of the inhaled lead mass will be retained in the pulmonary alveoli and subsequently be absorbed (43,54). The fraction deposited in the alveoli ranges, however, from above 60% for those particles with a diameter less than 0.1 uM to virtually zero for those with a diameter above 10 uM. The larger particles deposit in the nose, throat, and upper airways where they are cleared by ciliary action and are then either expectorated or swallowed (41).

The dose-response relationship between air and blood lead levels has been evaluated extensively in both adults and children (54). Four types of study have figured in that evaluation: community-based epidemiologic studies in relation to mobile emission sources; community-based epidemiologic studies in relation to stationary sources; occupational studies; and clinical studies of human volunteers.

The mathematical form of the air lead-blood lead relationship is complex and varies among the different age and sex groups in the population. It is complicated further by the fact that air lead levels have a statistically normal distribution, while blood lead levels are distributed log-normally (54). Thus arithmetic increases in air lead concentration do not give constant blood lead increases;

instead the incremental rise in blood lead becomes smaller with increasing air lead levels. Also the form of the curve is probably influenced by variations in size and chemical composition of airborne lead particles. Most published studies have tended to describe the air lead-blood lead relationship in terms of a linear regression model (55). Although this model is an oversimplication, it appears to provide an adequate description of the data over the range of population exposures usually encountered and in persons without massive skeletal lead stores (54).

In the linear model, the quantitative dose-response relationship between air lead and blood lead is reported usually as the slope of the regression of blood lead upon air lead in the equation:

$$\text{blood lead} = a + b\,(\text{air lead}),$$

where a is the intercept of air lead on blood lead, or the background blood lead concentration in persons with no air lead exposure, and b is the slope.

The slope is then defined as the ratio:

$$\frac{\text{blood lead change (ug/dl)}}{\text{unit air lead change (ug/M}^3\text{)}}.$$

The U.S. Environmental Protection Agency (54) has tabulated the estimated slopes of air lead upon blood lead in a number of epidemiological and clinical studies (Table 5). That review found that in all of the published studies there existed a positive dose-response relationship between air lead and blood lead concentrations. The slopes ranged from 0.11 to 6.8, with most of the values falling between 1.0 and 2.0. It appears that males tend to have higher slopes than females. Also children tend to have higher slopes than adults; that finding may reflect the higher respiratory intake per unit weight of children as compared to adults. Also children exposed to airborne lead are in double jeopardy of absorption, in that they may not only inhale suspended particulates but may also, as a result of their oral behavior, ingest previously airborne particles which have settled into dust and surface soil (16). Finally, the data suggest that slopes may be higher in the presence of declining air lead levels than in a steady state or with increasing exposures (54).

It is important to recognize that each reported slope of the air lead-blood lead relationship is not an absolutely precise measure, but is rather a calculated value which can (when appropriate data are provided) be set within confidence limits. The precision of the calculation depends upon the precision of both the air lead and blood lead determinations. The 95% confidence intervals

Study	Population	- Sample - Size	Ratio
Epidemiological			
Angle et al. (1)**	Children	831	1.40
Azar et al. (3)**	Adult Males	149	1.75
Baker et al. (6)	Children	187	2.30
Tepper et al. (52)**	Adult Females	1935	1.09
Nordman (44)	Adult Males	536	0.42
Nordman (44)	Adult Females	478	0.11
Fugas et al. (23)	Adults	330	2.64
Johnson et al. (28)**	Adult Males	106	1.30
	Adult Females	139	0.80
	Boys (under 16 years)	87	2.20
Tsuchiya et al. (53)**	Adult Males	591	3.09
Goldsmith (24)	Children Males	202	2.30
Goldsmith (24)	Children Females	203	1.70
Yankel et al. (60)	Children	879	1.27
Williams et al. (58)**	Adults	482	0.80
Daines et al. (20)	Black Females	(unknown)	2.30
Clinical			
Griffin et al. (25)**	Adult Males	11 at exposure of 10.9 ug/M^3	1.45
Griffin et al. (25)**	Adult Males	14 at exposure of 3.2 ug/M^3	1.80
Rabinowitz et al. (48)	Adult Males	2	1.7, 2.5
Chamberlain et al. (15)	Adults	7	1.20
Kehoe (30)**	Adults	9	0.73
Hammond et al. (26)***	Adults	- at baseline blood lead of 5 ug/dl	6.8
	Adults	- at baseline blood lead of 10 ug/dl	2.9
	Adults	- at baseline blood lead of 20 ug/dl	0.8

* From U.S. Environmental Protection Agency (54).
** Data reanalyzed by Chamberlain (14).
*** Data reanalyzed by Nutrition Foundation (45).

TABLE 5 – Estimated blood lead to air lead ratios (ug/dl per ug/M^3).*

for the air lead-blood lead slopes reported in published studies typically ranged from ± 0.5 to ± 0.7, although in some cases they were as wide as ± 1.0 (50).

Overall, it appears that in adults without prolonged or intense prior exposure to lead, each 1 ug/M^3 increase in air lead exposure will increase the population mean blood lead level by approximately 1 ug/dl. In children, this ratio is higher, and each 1 ug/M^3 increase in air lead exposure brings about a mean increase of nearly 2 ug/dl in the blood lead level (54).

EXPOSURES FROM DUST AND SOIL

Concentrations of lead in dust and soil range from near zero to several hundred thousand parts per million. Highest concentrations have been found near smelters (34, 49) and roadways (28, 35). Dustfall studies indicate that near smelters and roadways precipitation of suspended airborne particulates is the major source of lead in dust and soil (28). An additional source of dust and soil lead in urban areas is the powdering and flaking of lead-based paint from the walls of buildings (51) and from exterior steel structures such as bridges and elevated freeways (35).

Children appear to obtain lead from dust and soil as a result of their normal oral exploratory behavior (16). Several studies have demonstrated elevated levels of lead on the hands of children living near sources of environmental lead pollution (49) or in households with elevated dust lead content (16). Positive correlations between blood lead levels and concentrations of lead in house dust and soil have been observed among both preschool and school-age (11-year-old) children living near smelters (34,49), near roadways (28), in urban areas (16), and in the homes of smelter workers (5).

Increases in blood lead levels begin to occur in children exposed to lead in dust and soil at concentrations above 500-1,000 ppm; a 3 to 6% increase in blood lead content results from every doubling of dust or soil lead content (54).

Concentrations of cadmium (36), zinc (36), and arsenic (40) have been shown to be elevated in dirt and soil samples obtained near smelters. Although the kinetics of absorption of these metals have not been studied, the mechanisms are presumably similar to those described for lead.

OTHER EXPOSURES

Lead-based Paint

Lead-based paint was widely used to coat the interior surfaces of houses in the United States prior to World War II. Ingestion of this paint has become the major source of high-dose lead exposure and of clinically evident lead poisoning among children in the United States (17). Although paint produced for household use today must, by regulation, contain no more than 0.06% lead by dry weight, interior paint manufactured prior to the 1940s in some instances contained more than 50% lead.

Lead paint poisoning classically occurs in children under the age of five years who live in deteriorated prewar housing (19). A triad of causative factors has been described: "a dilapidated old house, a young child with pica, and parents with inadequate resources to cope with the family's needs and thus to keep the child away from the lead"(17). Although cases of childhood lead paint poisoning have been reported from all regions of the United States, and from both urban and rural settings, most cases occur in the "lead belts" of the older cities of the northeastern and north central states.

Over the past decade, the incidence of pediatric lead paint poisoning has declined greatly in the United States (12). The major cause of this decline appears to have been the massive federally supported program of screening, paint removal, and education which has been conducted in lead paint "target areas" under the provisions of the Lead-Based Paint Poisoning Prevention Act of 1971. Since passage of that Act, more than 3.7 million children and 300,000 housing units have been screened for lead, 236,000 children have been found to have elevated blood lead levels and have been entered into follow-up programs, and 109,000 housing units have been deleaded (12).

Lead-glazed Pottery

Although pottery is not a widespread source of increased lead absorption, pottery glazed with lead can release large amounts of the metal into food and drink and has been responsible for outbreaks of serious poisoning (31,47). Such episodes have typically been associated with the storage of acidic foods, such as orange juice in an improperly glazed vessel. A large outbreak of lead poisoning among members of a rural commune in Oregon resulted from the fermentation of home-made plum wine in a lead-glazed bathtub (47).

Burning of Lead Batteries

Lead fume, produced by the burning of lead batteries in home stoves to generate heat during the winter, is a recently rediscovered source of childhood lead poisoning (21). Lead exposure from battery burning was a serious cause of lead intoxication during the Great Depression (17) and may increase again in frequency if fuel prices continue to rise.

Sewage Sludge

Populations may be exposed to cadmium through ingestion of vegetables grown in soils enriched with sewage sludge; sewage sludge used as fertilizer has been shown to have a high cadmium content (38).

DISCUSSION

Toxic metals are ubiquitous environmental contaminants in an industrialized society. Although all of these metals are natural constituents of the earth's crust, their distribution has been radically altered by human activity, and they are now dispersed widely in air, food, soil, and water.

Absorption of these metals has become the norm rather than the exception among residents of the developed countries, and the margin of safety between present-day background levels and the levels associated with toxicity has become vanishingly small.

The extent of the problem of toxic metal exposure among residents of the United States is well illustrated by recently developed data on pediatric blood lead levels derived from the national Health and Nutrition Evaluation Survey (HANES-II); HANES-II is based on a probability sample of the entire noninstitutionalized population of the United States. These data indicate that 3.9% of all United States children between the ages of six months and five years examined from 1976 to 1980 had blood lead levels of 30 ug per dl or more (2). With projection of that finding to the entire population of the United States, it may be estimated that 675,000 young children have elevated blood levels. There was, in addition, a pronounced racial discrepancy in the HANES-II data. While only 2.0% of white children had elevated blood lead levels, the prevalence in black children was 12.2% (Table 6), and among black children residing in the central core areas of large cities in families with an annual income of less than $6,000, the prevalence was 18.6%. Given recent evidence that blood lead levels above 30 ug per dl in young children are associated with the development of irreversible intellectual and behavioral deficits, the HANES-II

	Race		
	White	Black	All Races
Both sexes	2.0	12.2	3.9
Boys	2.1	13.5	4.3
Girls	1.8	10.9	3.5

* From Annest & Roberts (1).

TABLE 6 — Percent of children ages six months to five years with blood lead levels of 30 ug/dl or more* — United States — 1976–1980.

data have profoundly disturbing implications not only for the future health of the affected children, but also for the integrity of the whole fabric of urban society.

REFERENCES

(1) Angle, C.R., and McIntire, M.S. 1979. Environmental lead and children: the Omaha study. J. Tox. Env. H. **5**: 855–870.

(2) Annest, J.L., and Roberts, J. 1981. Blood Lead Levels for Persons 6 months to 74 years of Age. Hyattsville, MD: National Center for Health Statistics.

(3) Azar, A.; Snee, R.D.; and Habibi, K. 1975. An epidemiologic approach to community air lead exposure using personal air samples. Envir. Qual. Safety **(Supplement II)**: 254–288.

(4) Baetjer, A.M. 1950. Pulmonary carcinoma in chromate workers. Arch. Indust. Hyg. **2**: 487–504.

(5) Baker, E.L.; Folland, D.S.; Taylor, T.A.; Frank, M.; Peterson, W.; Lovejoy, G.; Cox, D.; Housworth, J.; and Landrigan, P.J. 1977. Lead poisoning in children of lead workers — home contamination with industrial dust. New Eng. J. Med. **296**: 260–261.

(6) Baker, E.L., Jr.; Hayes, C.G.; Landrigan, P.J.; Handke, J.L.; Leger, R.T.; Housworth, W.J.; and Harrington, J.M. 1977. A nationwide survey of heavy metal absorption in children living near primary copper, lead, and zinc smelters. Am. J. Epidem. **106**: 261–273.

(7) Baker, E.L., Jr.; Landrigan, P.J.; Barbour, A.G.; Cox, D.H.; Folland, D.S.; Ligo, R.N.; and Throckmorton, J. 1979. Occupational lead poisoning in the United States: clinical and biochemical findings related to blood lead levels. Brit. J. Indust. Med. **36**: 314–322.

(8) Baker, E.L., Jr.; Peterson, W.A.; Holtz, J.; Mann, J.M.; Coleman, C.; and Landrigan, P.J. 1979. Subacute cadmium intoxication in jewelry workers; an evaluation of diagnostic procedures. Arch. Env. Health **34**: 173–177.

(9) Billick, I.H., and Gray, V.E. 1978. Lead Based Paint Poisoning Research, Review and Evaluation 1977. Washington, D.C.: U.S. Department of Housing and Urban Development (Publication No. HUD–POR–347).

(10) Birmingham, D.J.; Key, M.M.; Holaday, D.A.; and Perone, V.B. 1965. An outbreak of arsenical dermatoses in a mining community. Arch. Dermatol. **91**: 457–464.

(11) Caprio, R.J.; Margulis, H.L.; and Joselow, M.M. 1974. Lead absorption in children and its relation to urban traffic densities. Arch. Env. Health **28**: 195–197.

(12) Centers for Disease Control. 1981. Surveillance of childhood lead poisoning — United States. Morbid. Mortal. Weekly Rep. **30**: 556–567.

(13) Centers for Disease Control. 1983. Lead burdens and occupational lead exposure — United States. Morbid. Mortal. Weekly Rep., in press.

(14) Chamberlain, A.C. 1983. Effect of airborne lead on blood lead. Atmos. Env., in press.

(15) Chamberlain, A.C.; Clough, W.S.; Heard, M.J.; Newton, D.; Stoth, A.N.B.; and Wells, A.C. 1975. Uptake of lead by inhalation of motor exhaust. Proc. Roy. Soc. Lond. **192**: 77–110.

(16) Charney, E.; Sayre, J.; and Coulter, M. 1980. Increased lead in inner city children: where does the lead come from? Ped. **65**: 226–231.

(17) Chisolm, J.J., Jr. 1971. Lead poisoning. Sci. Am. **224**:15–23.

(18) Chisolm, J.J., Jr. 1978. Fouling one's own nest. Ped. **62**: 614–617.

(19) Chisolm, J.J., Jr., and Harrison, H.E. 1956. The exposure of children to lead. Ped. **18**: 943–957.

(20) Daines, R.H.; Smith, D.W.; Feliciano, A.; and Trout, J.R. 1972. Air levels of lead inside and outside of homes. Ind. Med. Surg. **41**: 26–28.

(21) Dolcourt, J.L.; Finch, C.; and Coleman, G.D.; et al. 1981. Hazard of lead exposure in the home from recycled automobile storage batteries. Ped. **68** : 225–230.

(22) Doll, R.; Matthews, J.D.; and Morgan, L.G. 1977. Cancers of the lung and nasal sinuses in nickel workers: reassessment of the period of risk. Brit. J. Indust. Med. **34**: 102–110.

(23) Fugas, M.D.; Majic, D.J.; Paukovic, R.; Wilder, B.; and Skuric, Z. 1974. Biological significance of some metals as air pollutants. Research Triangle Park, NC (U.S. Environmental Protection Agency Publication No. 02-302-3).

(24) Goldsmith, J.R. 1974. Food Chain and Health Implications of Airborne Lead. Washington, D.C. (U.S. Department of Commerce NTIS No. PB-248745).

(25) Griffin, T.B.; Coulston, R.; Wills, H.; Russel, J.C.; and Knelson, J.H. 1975. Clinical studies on men continuously exposed to airborne particulate lead. Envir. Qual. Safety **(Supplement II)**: 221–240.

(26) Hammond, P.B.; O'Flaherty, E.; and Gartside, P.S. 1981. The impact of air lead on blood lead in man — a critique of the recent literature. Food Cosmet. **19**: 631–638.

(27) Hunter, Sir D. 1969. The Diseases of Occupations, 4th ed. Boston: Little, Brown, and Company.

(28) Johnson, D.E.; Tillery, J.B.; and Prevost, R.J. 1975. Levels of platinum, palladium, and lead in populations of Southern California. Env. Health Perspec. **12**: 27–33.

(29) Kazantsis, G. 1965. Chronic mercury poisoning — clinical aspects. Ann. Occup. Hyg. **8**: 65–71.

(30) Kehoe, R.A. 1961. The metabolism of lead in man in health and disease. The Harben Lectures. J. Roy. Inst. Pub. Health Hyg. **24**: 1–81, 101–120, 129–143, 177–203.

(31) Klein, M.; Namer, R.; and Harpur, E. 1970. Earthenware containers as a source of fatal lead poisoning: case study and public health considerations. New Eng. J. Med. **283**: 669–672.

(32) Landrigan, P.J. 1981. Arsenic — state of the art. Am. J. Indust. Med. **2**: 5–14.

(33) Landrigan, P.J. 1982. Occupational and community exposures to toxic metals: lead, cadmium, mercury and arsenic. West. J. Med. **137**:531–538.

(34) Landrigan, P.J.; Baker, E.L., Jr.; Feldman, R.G.; Cox, D.H.; Eden, K.V.; Orenstein, W.A.; Mather, J.A.;Yankel, A.J.; and von Lindern, I.H. 1976. Increased lead absorption with anemia and slowed nerve conduction in children near a lead smelter. J. Ped. **89**: 904–910.

(35) Landrigan, P.J.; Baker, E.L., Jr.; Himmelstein, J.S.; Stein, G.F.; Weddig, J.P.; and Straub, W.E. 1982. Exposure to lead from the Mystic River bridge: the dilemma of deleading. New Eng. J. Med. **306**: 673–676.

(36) Landrigan, P.J.; Gehlbach, S.H.; Rosenblum, B.F.; Shoults, J.M.; Candelaria, R.M.; Barthel, W.F.; Liddle, J.A.; Smrek, A.J.; Staehling, N.W.; and Sanders, J.F. 1975. Epidemic lead absorption near an ore smelter: the role of particulate lead. New Eng. J. Med. **292**: 123–129.

(37) Legge, Sir T. 1934. Industrial Maladies. London: Oxford University Press.

(38) Linnman, L.; Anderson, A.; Nelsson, K.O.; and Lind, B. 1973. Cadmium intake by wheat from sewage sludge used as a plant nutrient source. Arch. Env. Health **27**: 45–47.

(39) Malcolm, D., and Barnett, H.A.R. 1982. A mortality study of lead workers 1925–76. Brit. J. Indust. Med. **39**: 404–410.

(40) Milham, S., Jr., and Strong, T. 1974. Human arsenic exposure in relation to a copper smelter. Env. Res. **7**: 176–182.

(41) Muir, D.C.F. 1972. Deposition and Clearance of Inhaled Particles. **In** Clinical Aspects of Inhaled Particles, ed. D.C.F. Muir. London: Wm. Heineman Medical Books, Ltd.

(42) Muskett, C.J.; Roberts, L.H.; and Page, B.J. 1979. Cadmium and lead pollution from secondary metal refinery operations. Sci. Total Env. **11**: 73–87.

(43) National Academy of Sciences. 1972. Airborne Lead in Perspective. Division of Medical Sciences. Washington, D.C.: National Research Council, Committee on Biological Effects of Atmospheric Pollutants.

(44) Nordman, C.H. 1975. Environmental Lead Exposure in Finland. A Study on Selected Population Groups. Ph.D. Dissertation, University of Helsinki.

(45) Nutrition Foundation. 1982. Assessment of the Safety of Lead and Lead Salts in Food. A Report of the Nutrition Foundation, Expert Advisory Committee, Nutrition Foundation Inc., Washington, D.C.

(46) Oliver, Sir T. 1914. Lead Poisoning: From the Industrial, Medical, and Social Points of View, Lectures delivered at the Royal Institute of Public Health. New York: P.B. Hoeber.

(47) Osterud, H.T.; Tufts, E.; and Holmes, M.A. 1973. Plumbism at the Green Parrot Goat Farm. Clin. Tox. **6**: 1–7.

(48) Rabinowitz, M.B.; Wetherill, G.W.; and Kopple, J.D. 1976. Kinetic analysis of lead metabolism in healthy humans. J. Clin. Inv. **58**: 260–270.

(49) Roels, H.; Buchet, J.P.; Lauwerys, R.R.; Bruaux, P.; Claeys-Thoreau, F.; LaFontaine, A.; and Verduyn, G. 1980. Exposure to lead by the oral and the pulmonary routes of children living in the vicinity of a primary lead smelter. Env. Res. **22**: 81–94.

(50) Snee, R.D. 1981. Evaluation of studies of the relationship between blood lead and air lead. Int. Arch. Occup. Env. Health **48**: 219–242.

(51) Spittler, T.M., and Feder, W.A. 1979. A study of soil contamination and plant lead uptake in Boston urban gardens. Commun. Soil Sci. Plant Analy. **10**: 1195–1210.

(52) Tepper, L.B., and Levin, L.S. 1975. A survey of air and population lead levels in selected American communities. Envir. Qual. Safety **(Supplement II)**: 152–195.

(53) Tsuchiya, R.; Sugita, M.; Seki, Y.; Koboryashi, Y; Hori, M.; and Park, C.B. 1975. Study of lead concentrations in atmosphere and population in Japan. Envir. Qual. Safety **(Supplement II)**: 95–147.

(54) U.S. Environmental Protection Agency. 1977. Air Quality Criteria for Lead. Washington, D.C. (Publication No. EPA–600/ 8–77–017).

(55) U.S. Occupational Safety and Health Administration. 1978. Occupational exposure to lead — final standard. Federal Register **43**: 52952–53014.

(56) Vahter, M., ed. 1982. Assessment of Human Exposures to Lead and Cadmium Through Biological Monitoring. Stockholm: National Swedish Institute of Environmental Medicine.

(57) Wagoner, J.K.; Infante, P.F.; and Bayliss, D.L. 1980. Beryllium: an etiologic agent in the induction of lung cancer, non-neoplastic respiratory disease and heart disease among industrially exposed workers. Env. Res. **21**: 15–32.

(58) Williams, M.K.; King, E.; and Walford, J. 1969. An investigation of lead absorption in an electric accumulator factory with the use of personal samplers. Br. J. Ind. Me. **26**: 202–216.

(59) Wysowski, D.K.; Landrigan, P.J.; and Ferguson, S.W. 1978. Cadmium exposure in a community near a smelter. Am. J. Epidem. **107**: 27–35.

(60) Yankel, A.; Lindern, I. von; and Walter, S.D. 1977. The Silver Valley lead study. The relationship between childhood blood lead levels and environmental exposure. J. Air Pollu. **27**: 763–767.

Changing Metal Cycles and Human Health, ed. J.O. Nriagu, pp. 187–198.
Dahlem Konferenzen 1984. Berlin, Heidelberg, New York, Tokyo: Springer-Verlag.

BIOAVAILABILITY OF TRACE ELEMENTS IN FOODSTUFFS AND BEVERAGES

J.M. McKenzie
Dept. of Nutrition, University of Otago
Dunedin, New Zealand

Abstract. Many aspects of the bioavailability of trace elements remain to be elucidated. Host variables affecting bioavailability include species, physiological state, nutritional status, stress, disease, homeostatic mechanisms, and individual variation. Dietary variables include both extrinsic and intrinsic factors; among these interfering substances, interactions with other dietary components and chemical speciation are important. Metabolic variables include absorptive and transport mechanisms, biotransformations, accumulation, and excretory mechanisms. Concurrent exposure to several metals alters their biological effects.

INTRODUCTION

Investigations into the bioavailability to man of toxic and essential elements have been a relatively recent research activity, although it has been known for many years that some forms of metals had a greater toxicity (and thus in some instances a higher bioavailability) than did other forms. For example, "metal fume fever" resulted from exposure to the fumes rather than the solid form of metals. The bioavailability of metals is of concern to the toxicologist in terms of what might minimize the bioavailability and hence the toxic effects of metals. On the other hand, a nutritionist is concerned with maximizing the bioavailability of the essential elements in order to fulfil the nutritional requirements of a species. The possibility of hazardous exposure to metals is increasing because the use of these nonbiodegradable materials for both industrial and nonindustrial purposes is increasing. Rarely are people exposed to only one metal, and therefore some of the most recent bioavailability studies have attempted to study several metals at one time. Such studies are complex and have brought attention to how little is yet known about many aspects of the bioavailability of metals.

This discussion will be restricted to the bioavailability of trace elements; that is, as well as metallic elements some nonmetallic elements such as arsenic, selenium, and fluorine will be included because of their toxic effects.

DEFINITIONS

The definition of bioavailability used here is adapted from that proposed by Spivey Fox et al. (8): "Bioavailability is a quantitative measure of the utilization of an element under specified conditions to affect the organism's normal structure and physiological processes." Although this definition has found most application with the essential elements, it can also be used for the toxic elements. Departure from the normal structure and physiological processes will be displayed more often as signs of toxicity of that element rather than as signs of deficiency. Utilization will be taken to include absorption, transport to a site of metabolic/toxic activity, biotransformation to a metabolically active/toxic form, retention/accumulation, and excretion. Bioavailability therefore includes many more aspects of the metabolism of a trace element than just its absorption. Since we are considering only foodstuffs and beverages, neither cutaneous nor pulmonary absorption will be considered (although either route can allow the entry of hazardous amounts of trace elements). We are particularly unprotected from high concentrations of many metals in air, whereas the intestinal wall provides one of our most efficient mechanisms for the rejection of environmental metals. It is worth noting also that translocation of inhaled particles to the gastrointestinal tract occurs, e.g., much of the inhaled arsenic may be swallowed and absorbed from the gastrointestinal tract. Knowledge is sparse in the areas of transport mechanisms or of biotransformation mechanisms of many trace elements; the same is true for excretory mechanisms. The elements having the greatest disease potential are those that tend to accumulate in the body. The target organ (critical organ) is that which displays the first identifiable toxic effects and is often one of the principal sites of trace element accumulation.

Trace elements can be defined as those elements present in concentrations of less than 0.005%, i.e., less than 50 ppm in the tissues. The trace elements with known or possibly essential functions also display toxic effects when present in higher concentrations. This discussion therefore will include both those trace elements known principally for their potentially toxic biological affects as well as those known more usually for their nutritional properties.

Many factors can affect the bioavailability of trace elements. These will be discussed under three headings: host variables, dietary variables, metabolic variables.

HOST VARIABLES AFFECTING BIOAVAILABILTY

These are listed in Table 1. Many of these factors must be borne in mind when animal models are used in bioavailability and metal toxicity studies, that is, weanling male rats from an inbred strain probably would not respond in a similar manner to a human population group comprised of men and women of mixed ages and of various states of health. How different the response would be is often as much a matter of guesswork as it is of scientific extrapolation in the current state of knowledge.

In the newborn, a greater bioavailability is often observed. For example, there is a markedly greater retention of $^{54}MnCl_2$ in newborn than in mature animals, presumably because the intestinal barrier to manganese has not completely matured. For some trace elements this greater bioavailability continues into infancy and childhood, the best known example being the possibility that infants and children might absorb lead up to five times more efficiently than do adults. Although the absorptive ability of the gastrointestinal tract is thought to decline in the elderly, no studies on the bioavailability of trace elements in them appear to have been conducted. During pregnancy, especially the latter stages, there is an increase in the gastrointestinal absorption of essential trace elements such as iron, zinc, copper, and manganese; the decrease to normal occurs during lactation. Whether there is a general increase in the absorption

Species	Stress
Sex	Presence of disease
Stage of maturity	Homeostatic mechanisms
Physiological state	Individual variation
Nutritional status	

TABLE 1 – Host variables affecting bioavailability of trace elements.

of other (including nonessential) trace elements during pregnancy and/or lactation is not known. Such information is important since the placenta seems to be an ineffective barrier to lead or mercury, although it does seem to provide some barrier to cadmium.

The influence of small alterations in general nutritional status on trace element bioavailability is unknown. However, malnutrition, especially protein-energy malnutrition, can affect several aspects of bioavailability. The gut has a diminished absorptive ability and there might be a decreased concentration of serum transport proteins, but detoxification mechanisms are usually impaired. The overall effect could be a reduced bioavailability of those trace elements which could help to protect against the toxic effects of other trace elements against which the body has reduced capacity to defend itself. In thiamin-deficient animals, the uptake of calcium and zinc was decreased, but the uptake of cadmium was increased; in a chronic pyridoxine deficiency there appeared to be a nonspecific increase in the uptake of calcium, zinc, and cadmium (7). Stress could affect trace element bioavailability in many ways, such as inducing gastric acid secretion. The low pH of the gastric juice, for example, can enhance the gastrointestinal absorption of arsenic trioxide which is only slightly soluble in water but is dependent on the pH of the gastric juice (2). The effect of diseases on metal bioavailability will be most noticeable when the condition affects the normal functioning of the gastrointestinal tract. Such conditions usually reduce the absorptive capacity of the gastrointestinal tract and include the inborn errors of zinc and copper metabolism in which the infant is unable to absorb sufficient trace elements to meet its nutritional requirements. The practice of pica (i.e., the ingestion of nonfood items) among children has resulted in trace element toxicities exemplified by lead; in adults it has contributed to a trace element deficiency (zinc).

Homeostatic mechanisms, including the regulation of their bioavailability, have been proposed for many of the essential trace elements. Although many of the details of such mechanisms are yet only hypotheses, circumstantial evidence for the essential trace elements includes the smaller variability in their tissue concentrations as compared with the nonessential trace elements. For cadmium, the observation that a single high dose is much less toxic in an animal already exposed to high levels than in an animal not previously exposed is taken to indicate some homeostatic regulation.

Individual variation is displayed even within an inbred strain of experimental animals; much greater individual biological variation is to be expected within a genetically heterogenous group of people.

DIETARY VARIABLES AFFECTING BIOAVAILABILITY

These have been subdivided into extrinsic variables (i.e., other components of the diet that might inhibit or enhance utilization) or intrinsic variables (i.e., related to the trace element itself) and are listed in Table 2.

Extrinsic Factors

Compositional changes occur as food crops mature; changes occur in both the content and the chemical nature of components such as trace elements and other components which could influence trace element bioavailability. Processing of foods could include canning of acidic foods such as fruits and juices in unlacquered tin containers and subsequent enhancement of the tin content of such foods.

The most important extrinsic factors, however, are the latter two listed in Table 2. The principle behind the treatment of some acute metal poisonings depends on a reduction of metal bioavailability by interfering substances. For example, a reduction in gastrointestinal absorption can be achieved by the use of milk or egg white which denature and precipitate with heavy metals in the gut while an enhancement of excretion is possible by the use of chelating drugs such as calcium disodium edatate or dimercaprol (BAL) (7). For the essential metals, interfering substances such as goitrogens (for iodine), phytate, oxalate, or dietary fiber have received much attention because of the need to avoid deficiency states. Many such interfering substances reduce the availability of trace elements in the diet by reducing the absorption from the lumen of the gut. They can also decrease bioavailability by affecting the enterohepatic and enteropancreatic circulation of trace elements such that their subsequent loss from fecal excretion can exceed the total dietary intake, for example, phytate can have this effect on zinc. How much of a role phytate might play in protecting against metal toxicities seems to be unexplored. Since zinc bioavailability can be inhibited to such an extent as to produce a nutritional deficiency,

A. Extrinsic Factors	Age or maturity of foods
	Processing
	Interfering/enhancing substances
	Interactions with other dietary components
B. Intrinsic Factors	Chemical speciation
	Dietary concentration

TABLE 2 – Dietary variables affecting the bioavailability of trace elements.

presumably the bioavailability of a chemically similar trace element such as cadmium can also be influenced. In the rat, vitamin C in pharmacological doses reduces the absorption of copper from the gut, but it enhances the absorption of iron and zinc. In man there appears to be no effect of pharmacological doses of vitamin C on zinc absorption, and whether self-medication with vitamin C for the prevention or treatment of the common cold also has a side-benefit in reducing the bioavailability of potentially toxic elements such as cadmium is unknown.

Interactions with other dietary components include interactions among the trace elements themselves. Some such interactions were rationalized by the hypothesis (3) that those elements whose physical and chemical properties are similar will act antagonistically to each other biologically. Thus a high zinc intake will protect against cadmium toxicity, and it can also produce a copper deficiency. However, many element interactions do not readily fit this hypothesis nor can such interactions always be predicted from a knowledge of the physicochemical properties of elements — for example, selenium is the best known protection against cadmium toxicity, but these two trace elements do not have similar physical or chemical properties. Furthermore, an element need not interact directly with another to affect its bioavailability. Beryllium is an element which is being used increasingly in missiles and nuclear reactor components. As a strong inhibitor of alkaline phosphatase, it has been suggested that beryllium ions will inhibit cereal phytase and thus the hydrolysis of phytate (in the stomach of pigs). A reduced phosphorus absorption results (6), but what effect the higher phytate content would then have on the bioavailability of other elements in the gut is a matter for speculation. It would not seem unreasonable that beryllium might, by this means, indirectly interact with other trace elements. Improved utilization of iron and/or zinc is thought to be the most likely explanation for the ability of added dietary ascorbic acid to alleviate or prevent almost all aspects of cadmium toxicity (9). Element-element interactions are not limited to just two or even three elements. For example, a complex interaction involving nickel, iron, zinc, and copper has been described. Trace elements interact with other nutrients as well as among themselves. Some trace elements appear to have a higher bioavailability when animal protein or when high concentrations of protein are present concurrently in the diet. Amino acids such as histidine or cysteine enhance the absorption of trace elements such as zinc; amino-sugar complexes promote the urinary excretion of trace elements; simple and, in particular, complex carbohydrates can influence trace element bioavailability; lipids also can have an effect.

Intrinsic Factors

Chemical speciation is by far the more important of the two intrinsic factors listed in Table 2 as affecting trace element bioavailability, even though it is one of the aspects of trace element bioavailability about which knowledge is most lacking. Some differences in bioavailability among chemical species can be explained on the basis of aqueous or lipid solubility, for example, the toxicity of barium compounds depends on their solubility. The use of insoluble barium sulfate as an X-ray contrast material for gastrointestinal examinations depends on its lack of absorption from the alimentary tract, in contrast to the poisoning which has resulted from accidental or suicidal ingestion of soluble barium compounds (2). Inorganic forms of trace elements are often not as well absorbed as are organic complexes, but there are many exceptions to this generalization. Thus, some organic selenium compounds, including selenodiacetic and selenopro-prionic acids, are less toxic to rats per unit of selenium than selenite, presumably as a consequence of lower absorption. At high enough dietary concentrations, inorganic complexes are sufficiently available to produce severe toxic effects, for example, among the hazards of imbibing illicit whisky is poisoning by the high concentration of metals such as lead. Storage of acidic fruit juices in galvanized containers is likewise hazardous. Among inorganic compounds, the valence state can have a major influence on the bioavailability — for example, no adverse effects are observed following exposure to trivalent chromium, but hexavalent chromium is very toxic to man. The urinary excretion of pentavalent antimony is greater than for trivalent antimony compounds. Among the organic complexes of trace elements, great differences in bioavailability also occur, again affecting all aspects of trace element utilization. For example, distribution, biotransformation, and excretion differ among the organic tin compounds; short-chain alkyl tin compounds have a relatively high gastro-intestinal absorption, whereas long-chain alkyl compounds are absorbed to a very small degree.

METABOLIC VARIABLES AFFECTING BIOAVAILABILITY

Information is lacking on many aspects of the metabolism of trace elements as outlined in Table 3 (including those trace elements with known essential functions).

Absorption can occur by passive diffusion or by active transport or by a combination of the two. In addition, microscopic solid particles of silica (SiO_2) are absorbed as such from the alimentary tract. Trace elements are often broadly classified as anions or as cations; some of the multivalent elements may be present as colloidal species and therefore would not behave as might be

Absorptive mechanisms
Transport mechanisms
Biotransformation
Storage/accumulation
Excretory mechanisms

TABLE 3 – Metabolic variables affecting the bioavailability of trace elements.

expected (1). In general anions are well absorbed from the gut and thus, for example, fluorine and iodine are well absorbed; vanadates and perhaps arsenates and antimonates appear to be exceptions to this generalization. Univalent cations also seem to diffuse readily across the gut wall, but divalent cations are less readily absorbed, especially if they have large atomic weights (1). A general transport mechanism has been proposed for the absorption of divalent cations from the lumen of the gut — for example, cadmium appears to share a common (or similar) absorptive mechanism with several other metals including zinc, copper, iron, manganese, and mercury. The absorption and retention of ingested lead, however, is greatly affected by the dietary levels of calcium, phosphorus, iron, copper, and zinc. Specific transport mechanisms for some divalent cations have also been postulated, e.g., for iron and for zinc. Some sharing of such specific mechanisms seems to occur — for example, cobalt shares an intestinal mucosal transport pathway with iron. Multivalent cations are almost all poorly absorbed across the gut wall, possibly because they hydrolyze to form colloidal suspensions at neutral pH (1). Apart from these general principles, very little is known about the mechanism of absorption of many of the trace elements.

Knowledge is also lacking concerning transport mechanisms for many of the trace elements. For some there are specific transport proteins in blood plasma — for example, ceruloplasmin (ferroxidase I) transports copper. The manganese in human serum is selectively and almost totally bound in the trivalent form by a $\beta_1$globulin, whereas nickeloplasmin, a nickel-containing α-macroglobulin carries nickel (although albumin is the main transport protein for nickel). Plasma albumin presumably acts as a carrier for many cations; transferrin carries several cations including iron, zinc, indium, and trivalent chromium; an α_2-macroglobulin carries zinc and some manganese; amino acids might also carry some trace elements such as zinc and copper. Plasma fluorine exists in both ionic and bound forms, but for most trace elements only very low concentrations of their ionic forms can be tolerated in the plasma.

Beryllium appears to be transported in the form of a colloidal phosphate, probably adsorbed on plasma α-globulin. The red blood cells contain higher concentrations of many trace elements than in the plasma; this has been observed for zinc, cadmium, hexavalent chromium, selenium, alkyl mercury, and lead.

Studies of the biotransformation of trace elements into metabolically active or toxic forms in the tissues present great methodological difficulties. For an element such as selenium, the problem is made easier since there is only one known metabolically active form of selenium, namely the selenoenzyme glutathione peroxidase. It is also known that selenite selenium has to undergo a chemical transformation by the erythrocytes in order to be bound by the plasma proteins. There is some knowledge of the biotransformation of arsenic — for example, trivalent arsenic is oxidized in vivo to pentavalent arsenic prior to excretion in the urine (10). The in vivo methylation of inorganic arsenic has been demonstrated in both animals and humans; a few data indicate that arsenic from seafood is not biotransformed in the body but is eliminated in the urine as such (Zielhuis and Wibowo, this volume). Mercury vapor, which can be physically dissolved in blood, is rapidly oxidized to mercuric mercury in tissues. Methylmercury undergoes biotransformation to inorganic mercury by demethylation in the tissues, although this biotransformation is unlikely to occur in the brain. The phenyl and methoxyethyl mercury compounds are decomposed, mainly in the liver, also to inorganic mercury; the rate of biotransformation of methoxyethyl mercury is more rapid than that of phenyl mercury (2).

Many metal ions form coordination complexes with organic molecules, often with constituents in the body of great metabolic significance. Some metals have an especially high affinity for sulfur — for example, the metallothioneins, which in various species bind strongly to elements including cadmium, copper, zinc, mercury, silver, and nickel, have an uncharacteristically high content (up to 30%) of the sulfur-containing amino acid cysteine. Such proteins can also contain extraordinary concentrations of metals — for example, horse kidney metallothionein contains as much as 5–6% cadmium and 2% zinc. Some amino acids such as histidine and cysteine form complexes which have high stability constants with trace elements such as zinc and nickel. For many trace elements, little is known of their biotransformation consequent upon normal or toxic exposure — for example, the form in which many of them appear in the urine is unknown.

For the trace elements known primarily for their toxic effects, the amount stored in the tissues is often referred to as the "body burden." There are many exceptions to generalizations regarding tissue accumulation. Among the halides, an extremely high proportion of the total body burden of fluorine is in the skeleton, whereas the iodine pool is mainly in the thyroid gland as well as in the extracellular space, the red blood cells, and the salivary and gastric glands (9). Among the Group IIB elements, zinc is not accumulated in any particular storage organ, copper is stored primarily in the liver and secondly in central nervous system tissue, whereas the kidneys accumulate about one-third of the body burden of cadmium; furthermore, the biological half-life of cadmium is up to 33 times longer than that for zinc. The skeleton is the site of storage/accumulation for several trace elements including fluorine, lead, barium, strontium, and uranium. Among the soft tissues, the liver is the main or one of the main sites of accumulation of iron, antimony, silver, copper, cobalt, selenium, and cadmium; the kidneys are an important site of accumulation of cadmium, bismuth, and thallium; the lungs accumulate titanium, aluminum, tin, and cadmium. There is very little accumulation in any particular organ for molybdenum, nickel, manganese, zinc, chromium, vanadium, bromine, rubidium, and germanium.

For a few trace elements, specific storage proteins occur — for example, ferritin is the water soluble and hemosiderin is the water insoluble storage form of iron; metallothionein synthesis is induced specifically by metals such as cadmium, zinc, or copper. The protective effect of prior zinc administration against cadmium toxicity has been ascribed to the more rapid accumulation of hepatic and renal cadmium as metallothionein (the production of which had been induced by zinc), especially as cadmium so bound is not mobilized during subsequent pregnancy. The protective effect of selenium against cadmium toxicity appears to be due, at least in part, to a diversion in the binding of the cadmium in the soluble fraction to higher molecular weight proteins in the testis, liver, or kidney (11).

Excretory mechanisms generally operate via the kidneys or the gut. Anions and univalent cations are generally readily excreted from the kidneys, whereas many divalent cations are excreted from the alimentary tract, that is, in bile and/or pancreatic juice and/or intestinal secretions. The homeostasis of some trace elements is regulated by excretory mechanisms or by a combination of absorption and excretory mechanisms. Thus cobalt homeostasis is regulated largely through its urinary excretion. Because of its low absorption and its relatively rapid excretion, vanadium is less bioavailable and hence less toxic than other metals that do not share these characteristics (10). Other modes of excretion such as sweating out or transfer to hair and nails are usually of minor quantitative significance. Some nonmetals such as arsenic, antimony, and selenium (and perhaps also noble metals such as gold) are characteristically excreted into hair and nails (1).

There have been few studies to determine the effects of concurrent exposure to toxic metals; even fewer studies have investigated concurrent exposures to both essential and toxic trace elements. Concomitant exposure to well-tolerated dietary doses of lead, cadmium, and arsenic appeared to alter the cellular uptake and biological effects of each element. Furthermore, exposure to more than one toxic metal appeared to alter the impact of a single metal on tissue levels of essential elements (4). The small effect of high dietary cadmium intakes on blood cadmium concentrations could have been explained, at least in part, by concurrent higher than usual dietary intakes of selenium and zinc (5).

This review of the bioavailability of trace elements in foodstuffs and beverages has tried to cover what is known and to highlight those aspects about which there is little information. As man is exposed to unaccustomed forms and combinations of trace elements in his food supply because of changing biogeochemical cycles of metals, research must continue in order to determine the consequences on human health.

REFERENCES

(1) Bowen, H.J.M. 1979. Environmental Chemistry of the Elements. London: Academic Press Inc.

(2) Friberg, L.; Nordberg, G.F.; and Vouk, V.B. 1979. Handbook on the Toxicology of Metals. Amsterdam: Elsevier/North Holland Biomedical Press.

(3) Hill, C.H., and Matrone, G. 1970. Chemical parameters in the study of in vivo and in vitro interactions of transition elements. Fed. Proc. Am. Soc. Exp. Biol. **29**: 1474–1481.

(4) Mahaffey, K.R.; Capar, S.G.; Gladem, B.C.; and Fowler, B.A. 1981. Concurrent exposure to lead, cadmium and arsenic. J. Lab. Clin. Med. **98**: 463–481.

(5) McKenzie, J.M.; Kjellström, T.E.; and Sharma, R.P. 1982. Cadmium Intake, Metabolism and Effects in People with a High Intake of Oysters in New Zealand. U.S. Environmental Protection Agency, Report No. R807058–01–0.

(6) Moore, J.H., and Tyler, C. 1955. Studies on the intestinal absorption and excretion of calcium and phosphorus in the pig. 3. The effect of beryllium carbonate on the absorption of phosphorus. Brit. J. Nutr. **9**: 389–397.

(7) Prasad, R.; Lyall, V.; and Nath, R. 1982. Effect of vitamin B_6 and B_1 deficiencies on the intestinal uptake of calcium, zinc and cadmium. Ann. Nutr. Metab. **26**: 324–330.

(8) Spivey Fox, M.R.; Jacobs, R.M.; Jones, A.O.L.; Fry, B.E.; Rakowska, M.; Hamilton, R.P.; Harland, B.F.; Stone, C.L.; and Tao, S.-H. 1981. Animal models for assessing bioavailability of essential and toxic elements. Cereal Chem. **58**: 6–11.

(9) Underwood, E.J. 1977. Trace Elements in Human and Animal Nutrition, 4th ed. New York: Academic Press Inc.

(10) Waldron, H.A. 1980. Metals in the Environment. London: Academic Press Inc.

(11) Whanger, P.D.; Ridlington, J.W.; and Holcomb, C.L. 1980. Interactions of zinc and selenium on the binding of cadmium to rat tissue proteins. **In** Micronutrient Interactions. Ann. N.Y. Acad. Sci. **355**: 333–346.

Changing Metal Cycles and Human Health, ed. J.O. Nriagu, pp. 199–223.
Dahlem Konferenzen 1984. Berlin, Heidelberg, New York, Tokyo: Springer-Verlag.

BACTERIAL TRANSFORMATIONS OF AND RESISTANCE TO HEAVY METALS

S. Silver
Biology Dept., Washington University
St. Louis, MO 63130, USA

Abstract. Bacteria carry out chemical transformations of heavy metals. These transformations (including oxidation, reduction, methylation, and demethylation) are sometimes by-products of normal metabolism and confer no known advantage upon the organism responsible. Sometimes, however, the transformations constitute a mechanism of resistance. Many species of bacteria have genes that control resistances to specific toxic heavy metals. These resistances often are determined by extrachromosomal DNA molecules (plasmids). The same mechanisms of resistance occur in bacteria from soil, water, industrial waste, and clinical sources. The mechanism of mercury and organomercurial resistance is the enzymatic detoxification of the mercurials into volatile species (methane, ethane, metallic Hg^0) which are rapidly lost from the environment. Cadmium and arsenate resistances are due to reduced net accumulation of these toxic materials. Efficient efflux pumps cause the rapid excretion of Cd^{2+} and AsO_4^{3-}. The mechanisms of arsenite and of antimony resistance, usually found associated with arsenate resistance, are not known. Silver resistance is -due to lowered affinity of the cells for Ag^+, which can be complexed with extracellular halides, thiols, or organic compounds. Sensitivity is due to binding of Ag^+ more effectively to cells than to Cl^-.

INTRODUCTION

Bacterial cells divide the Periodic Table into three classes. Some elements are necessary for intracellular metabolism (67); some elements are not used generally within the cell but abound in natural environments and can be coupled to extracellular structural or regulatory functions; and finally, some elements have no useful biological function (70,79). Potassium and phosphorus are examples of the first class; calcium and chlorine are not needed at all for most bacteria; and arsenic, mercury, and cadmium are examples of toxic elements without biological utility. This report will deal with toxic elements and their compounds. In some cases, such as with mercury and arsenic, microbes can transform elements from relatively less toxic inorganic ions into relatively more toxic methylated forms. The same or other microbes can degrade organometallic compounds (15,79); oxidation and reduction by microbial enzymes also affect the bioavailabiity and toxicity of heavy metals to higher organisms.

Free-living bacterial cells have evolved resistance mechanisms to cope with heavy metal pollution. These resistance mechanisms are highly specific. The genes determining the resistance mechanisms occur on small nonchromosomal DNA molecules called plasmids. These resistance plasmids also have genes controlling resistances to most known antibiotics. This subject has already been reviewed elsewhere (68-70,79). I shall only try to provide a general picture and newer information.

MERCURY AND MERCURIAL TRANSFORMATIONS AND RESISTANCES

The mercury cycle (Fig. 1) is the best known case of microbial metabolism affecting the chemical form of a heavy metal. Microbial activity is associated with mercury methylation, demethylation (15), and the oxidation and reduction of inorganic mercury. I will deal first with the transformations from highly toxic methylmercury (found in fish) to less toxic ionic Hg^{2+} (the predominant form in seawater) to least toxic Hg^0. Both of these transformations are carried out by enzymes governed by bacterial resistance plasmids and transposons (moveable DNA sequences) and not by chromosomal genes. Without the plasmid genes, the cells remain mercury sensitive. Finally, I will address microbial oxidation and methylation of mercury.

$$CH_3Hg^+ \rightleftarrows Hg^{2+} \rightleftarrows Hg^0$$

FIG. 1 — Environmental transformations of mercury.

The earliest studies of enzymatic detoxification of Hg^{2+} were with a multiply drug-resistant E. coli (37) and with a soil pseudomonad (21,84). The frequency of Hg^{2+} resistance among clinical isolates can be over 50% (44,45,92). Recently, mercuric- and organomercurial-resistant strains with very similar properties have been found in a wide variety of bacterial species from soil, water, and marine environments (20,34,52,53,61,83). Mercuric resistant microorganisms are found in much higher frequencies in polluted waters than in nearby cleaner waters (48).

Bacteria with plasmids showed a small number of patterns of resistance to organomercurials (65,89): (a) In E. coli, about 90% of the mercuric resistance plasmids also controlled resistance to the organomercurials merbromin and flourescein mercuric acetate, but to no other tested organomercurial. The other 4% determined additional resistances to phenylmercuric acetate and thimerosal (65) (Fig. 2). (b) The plasmids in Pseudomonas aeruginosa also fell into classes with regard to resistance to organomercurials (12). However, about 50% fell into each class. All Pseudomonas plasmids controlled resistance to p-hydroxymercuribenzoate and some Pseudomonas plasmids showed additional resistances to methylmercuric and ethylmercuric compounds (12,89). (c) Initially, a single pattern was reported with S. aureus plasmids (88,89). This pattern is different from those with the gram-negative bacteria because S. aureus plasmids control resistances to phenylmercuric acetate, p-hydroxymercuribenzoate, and fluorescein mercuric acetate, but not to thimerosal or to merbromin. Recently, the first thimerosal-resistant S. aureus were found (60).

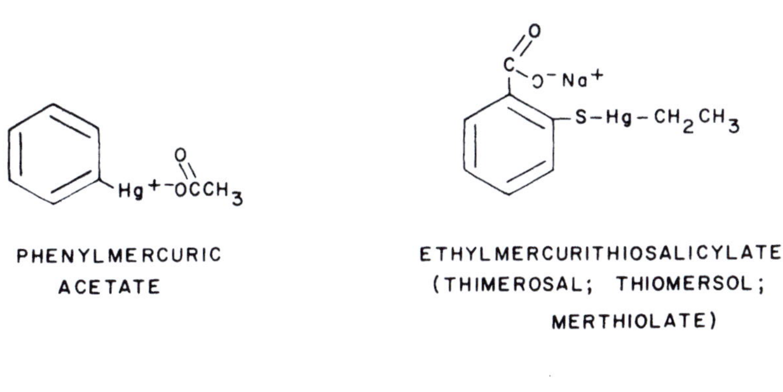

FIG. 2 — Structures of phenylmercuric acetate and thimerosal. These organomercurials are degraded by plasmid-mediated enzymes.

These new clinical isolates volatilize mercury more rapidly from thimerosal (Fig. 3) than do the previous S. aureus strains. Of course, the sensitive strain without a plasmid cannot volatilize mercury from thimerosal at all (Fig. 3). The new strains also showed activity for thimerosal as an inducer of synthesis of this mercuric detoxification system, which is only made after exposure to low levels of Hg^{2+} or organomercurials (60,88). The limited number of patterns of resistance can be understood in terms of the biochemistry of the enzymes involved.

Since this was a workshop on biogeochemical cycles, it is worth emphasizing that mercurial-detoxifying and mercurial-resistant strains have been found in every type of bacteria tested. These include soil and marine bacteria (Bacillus (34,61), Pseudomonas (12), and Mycobacterium (Falkinham, personal communication)), mineral-leaching Thiobacillus (53,54), and even antibiotic-producing Streptomyces (Silver, unpublished).

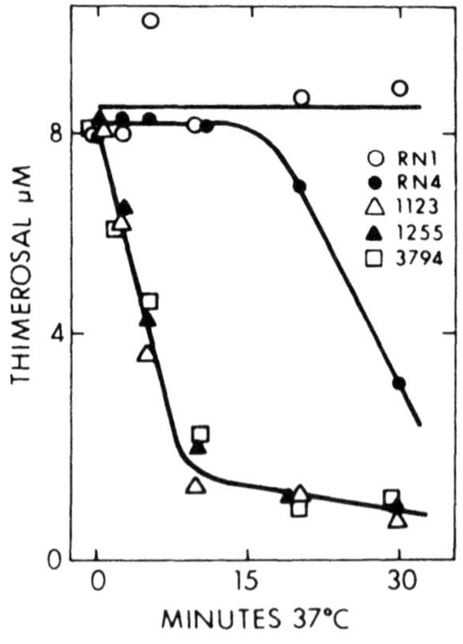

FIG. 3 — Detoxification of (loss of mercury from) thimerosal by a plasmidless sensitive strain (RN1), by a thimerosal-sensitive plasmid-containing strain (RN4), and by new thimerosal-resistant isolates (1123,1255, and 3794) (from (60)).

Enzymatic Mechanism of Mercurial Detoxification

Hg^{2+} resistance in bacteria results from enzymatic detoxification of mercury compounds. The mercury is volatilized (37,78,84) as metallic Hg^0 and the enzyme responsible is called mercuric reductase (Fig. 4).

Several organomercurials are also enzymatically detoxified to volatile compounds (Figs. 3 and 4). Benzene is produced from phenylmercury, methane from methylmercury, and ethane from ethylmercury. The enzymes responsible for cleaving the Hg-C bond are organomercurial lyases. Tezuka and Tonomura (81,82) separated two small soluble lyase enzymes from a Pseudomonas strain. Both have molecular weights of about 19,000. One enzyme cleaved phenylmercuric acetate, p-hydroxymercuribenzoate, and methylmercury, while the other enzyme cleaved only phenylmercuric acetate and p-hydroxymercuribenzoate (81,82). With an organomercurial resistant E. coli, there was no evidence for cleavage of p-hydroxymercuribenzoate (89), and Schottel (64) was unable to separate the two lyases. Nevertheless, kinetic analysis indicated that there were two enzymes active toward phenylmercuric acetate and only one active toward methyl- and ethylmercury. The general properties of the enzymes from the soil pseudomonad and E. coli were similar, except that the E. coli organomercurial lyases had a somewhat greater molecular weight (64).

Mercuric reductase has been studied in greater detail (17,23,35,64). We once thought the enzyme was strictly NADPH-dependent, but it now appears that some mercuric reductases can utilize either NADPH or NADH (e.g., the Streptomyces strain in Fig. 5C).

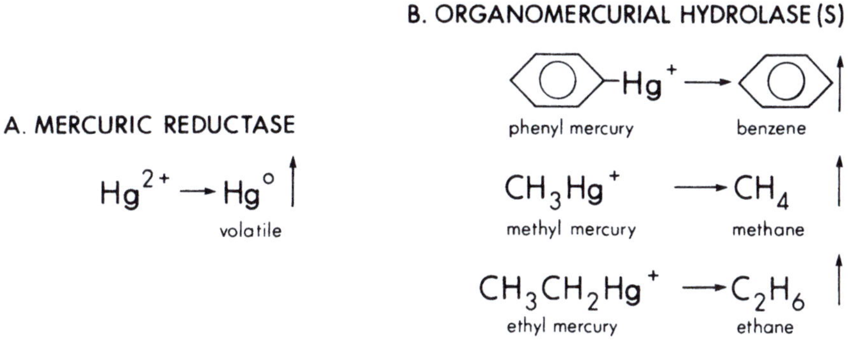

FIG. 4 — Enzymatic detoxification of Hg^{2+} and organomercurials.

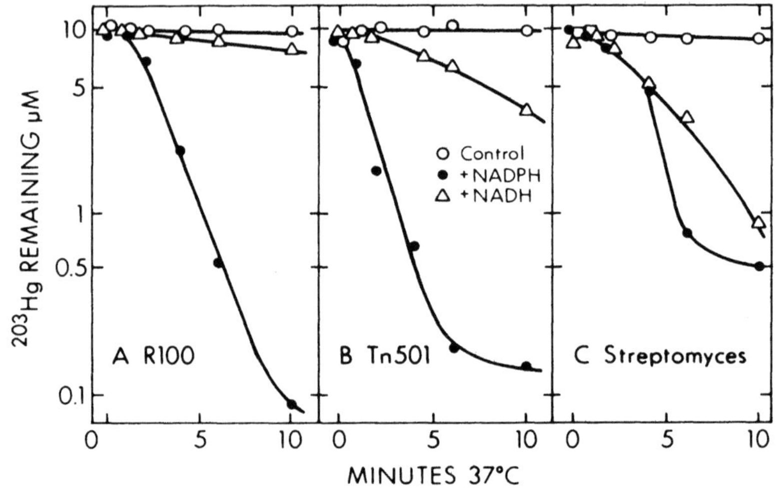

FIG. 5 — Volatilization of mercury from Hg^{2+} by enzyme preparations from cells bearing (A) plasmid R100, (B) transposon Tn501, or (C) a new Streptomyces strain. Conditions were as given by Schottel (64), except 25 µM NAD(P)H was added with 10 µM $^{203}Hg^{2+}$ (Silver, unpublished experiment).

Characterization of the mercuric reductase enzyme (17) has shown it to be similar to a class of NAD(P)H-dependent FAD-containing flavoproteins including glutathione reductase and lipoamide dehydrogenase (91). These enzymes have the following reaction sequence: (a) NAD(P)H-reducing enzyme-bound FAD to $FADH_2$. (b) $FADH_2$ next reduces the disulfide bond between two neighboring cysteines in the enzyme active site. (c) Hg^{2+} is bound to the two cysteine SH groups and (d) finally Hg^{2+} is reduced to Hg^0 (17). The structure and partial amino acid sequences of E. coli, yeast and human glutathione reductase, and of E. coli and pig lipoamide dehydrogenase are known (38,91). One can thus identify the location and sequence of the mercuric reductase gene from the DNA sequence analysis (11). Mercuric reductase contains a fifteen amino acid sequence:

THR ILE GLY GLY THR CYS VAL ASN VAL GLY CYS VAL PRO SER LYS

which is essentially identical to sequences at the active sites of glutathione reductase and lipoamide dehydrogenase (38,91) and which contains the active cystine. This is the first application of DNA sequencing technology to studies of heavy metal resistances. It provides a suggestion of the source from where enzymes able to detoxify heavy metals might have arisen, namely, enzymes of normal cellular metabolism.

The subunit structure of the mercuric reductase enzyme from various sources is currently under reevaluation. Furukawa and Tonomura (22) reported that the enzyme from a soil pseudomonad had a molecular weight of about 65,000. Schottel (64) reported that the enzyme from plasmid R831 that originated in Serratia was a trimer of 170,000 daltons containing identical monomer subunits, each approximately the size of the Pseudomonas enzyme. Each subunit contained a single bound FAD, for a total of three FAD's per 170,000 molecular weight. Fox and Walsh (17) reported that the molecular weight of another mercuric reductase (from transposon Tn501, which was found in P. aeruginosa) was about 125,000, and the enzyme appeared to be dimeric. We have confirmed the dimeric structure of the mercuric reductase of Tn501.

Antibodies were prepared against purified mercuric reductases coded by two plasmids in E. coli (73). Reductases obtained from all but one different gram-negative source reacted with these antibodies, as shown by inhibition of enzyme activity and by formation of precipitin bands on double-diffusion gels. The enzymes fell into two major subclasses, based on only partial immunological identity. The prototype of the first enzyme class is coded by transposon Tn501 (4). This enzyme class also includes mercuric reductases governed by a variety of plasmids found in clinical, soil, and marine bacteria. One strong conclusion from studies of plasmid-determined mercuric resistance is that the same system appears widely in bacterial isolates from diverse environments. Newer Hg^{2+}- (and in one case phenylmercuric-) resistant transposons from soil microbes showed different patterns of digestion by DNA restriction endonucleases (61).

The second immunological subgroup of gram-negative mercuric reductases has as its prototype the enzyme coded for by plasmid R100. It is with R100 that the genetic structure of the mercuric resistance operon is being studied in detail (16,47,49). The enzyme from R100 may be trimeric (73), as was the enzyme from plasmid R831 (64). This subgroup includes enzymes from a wide variety of sources including the enzyme determined by a second Pseudomonas mercury transposon Tn502 ((73), and V. Stanisich, personal communication). All of the mercuric reductases from gram-negative bacteria were immunologically related with the single exception of the enzyme from Thiobacillus ferrooxidans

(54). The antibodies prepared against the two classes of gram-negative enzymes did not interact with mercuric reductases from S. aureus stains and marine and soil bacilli. These enzymes from gram-positive sources showed otherwise similar functional requirements to those from the gram-negative bacteria (73,88) and yet they are immunologically distinct.

To summarize briefly the current understanding of plasmid-determined mercuric and organomercurial resistances: (a) They occur widely in all bacterial types tested and are the best understood of all plasmid-coded heavy metal resistances. (b) Resistance is due to enzymatic detoxification of the mercurials to volatile compounds of lesser toxicity. (c) The enzymes responsible (mercuric reductases and organomercurial lyases) have been purified and studied from the viewpoint of enzymology. (d) The newer tools of DNA cloning and sequencing are being applied to these systems.

Oxidation and Methylation of Mercury

Much less is known about the oxidation of Hg^0 to Hg^{2+} and the subsequent methylation of Hg^{2+} to CH_3Hg^+ and $(CH_3)_2Hg$ than is known about the reverse processes. However, microbial activities have been implicated in both reactions. Since most atmospheric mercury is Hg^0 and most aqueous mercury (in the rivers, lakes, and seas) is Hg^{2+} (see review (79) for original references), how is the geochemical mercury cycle maintained? A wide range of bacteria can oxidize Hg^0 to Hg^{2+} (32). The bacteria that oxidize Hg^0 do not (in general) carry out the further conversion of Hg^{2+} to methylmercury (32). The ubiquitous enzyme catalase (found in animal tissues and bacteria) can oxidize Hg^0 to Hg^{2+} and is probably responsible for the rapid conversion of inhaled mercury vapor into ionic mercury in the blood (26).

There appears to be general agreement that Hg^{2+} can be methylated abiotically and extracellularly by methylcobalamin (vitamin B12) (10,40,93). Cobalamin is synthesized only by microbial cells and is either excreted or released on cell lysis. Other bacteria (such as E. coli) and cells of higher organisms (plants and animals, all of which require this vitamin for growth) can accumulate cobalamin and methylate it enzymatically (93) to provide for methyl transfer reactions. Many common bacteria carry out mercury methylation (87) presumably by this means. Mercury can be methylated by microbial activities in such environments as river sludge (29) and human feces (15). Both bacteria and fungi can methylate mercury (87). Although only the methylcobalamin-dependent process has been studied in detail (93), there are reasons (such as lack of dependency upon added cobalamin with some but not with other bacteria (87)) to question whether a second biologically important methyl donation process may exist. This is a topic where further work is needed (10). It is also unclear as to whether all of the methylmercury found in fish and in man arose initially by methylation by microbes growing in intestines and on gills (as suggested in (79)), or whether mammalian enzymes also function to produce methylmercury.

ARSENIC AND ANTIMONY RESISTANCES

Arsenic and antimony resistances are governed by plasmids that also code for antibiotic and other heavy metal resistances (30,51,76). Arsenate, arsenite, and antimony(III) resistances are coded for by an inducible operon-like system in both S. aureus and E. coli (71). Each of the three ions induces all three resistances. In E. coli, Bi(III) is a gratuitous inducer of arsenate resistance, even though the plasmid system does not confer Bi(III) resistance. S. aureus has a genetically separate plasmid-mediated Bi(III) resistance determinant (50, 51) of unknown mechanism.

The mechanism of arsenate resistance is a reduced accumulation of arsenate by induced resistant cells. Arsenate is normally accumulated via the cellular phosphate transport systems, of which many bacteria appear to have two (Fig. 6). The distinction between arsenate and arsenite resistance was shown initially by finding that phosphate did not protect against arsenite inhibition of growth (71). Genetic studies have also demonstrated that the arsenate resistance gene is different from, but closely linked to, the arsenite resistance gene. Arsenite and antimony resistances may be determined by separate genes (50). The presence of the resistance plasmid does not alter the kinetic parameters of the cellular phosphate transport systems; even the K_i for arsenate as a competitive inhibitor of phosphate transport (Fig. 6) is unchanged. This finding, along with direct evidence for plasmid-governed energy-dependent efflux of arsenate, indicated that the block on net uptake of arsenate resulted from rapid efflux. The energy dependence of the efflux process was shown by its sensitivity to "uncouplers" and ionophore antibiotics such as nigericin and monensin (72). Hg^{2+} also inhibited the efflux system, and the Hg^{2+} inhibition was readily reversed by mercaptoethanol (72). Mobley and Rosen (43) demonstrated that glucose but not succinate could energize arsenate efflux in an E. coli strain that could not synthesize ATP from respiratory substrates. Although all of these data are indirect, AsO_4^{3-} efflux appears to involve directly an ATPase-linked transport system (Fig. 6). An interesting question about the arsenate efflux system concerns its specificity. Arsenate generally functions as a phosphate analogue and is accumulated by bacteria via phosphate transport

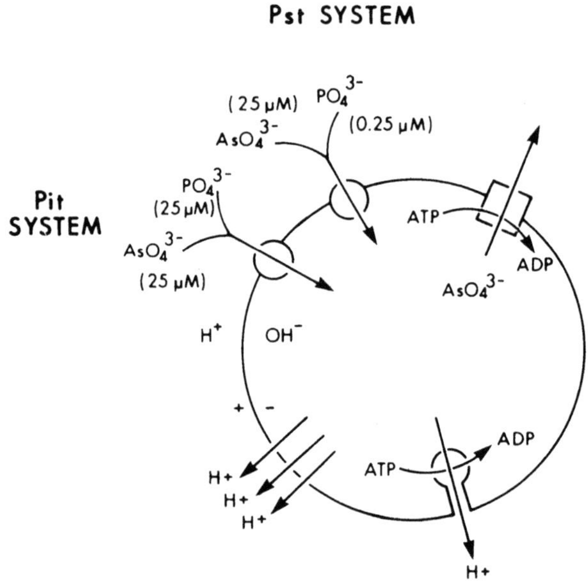

FIG. 6 — The phosphate (arsenate) transport systems and the arsenate efflux system of E. coli. 0.25 µM or 25 µM, where indicated, is the K_m (for PO_4^{3-}) or the K_i (for AsO_4^{3-} as a competitive inhibitor of PO_4^{3-} transport) for the Pit and Pst phosphate transport systems.

systems (67). The arsenate-resistance efflux system should not excrete phosphate, since the cells would then become phosphate-starved, a situation no more advantageous than being arsenate-inhibited! A basic conclusion from our work on arsenate (and also on Cd^{2+} resistance; see below) is that toxic heavy metals often get into cells by means of transport systems for normally-required nutrients (67,79). Energy-dependent efflux systems functioning as resistance mechanisms should be highly specific for the toxic anion or cation to prevent loss of the required nutrient.

The mechanism(s) of plasmid-determined arsenite and antimony resistances are not known. Arsenite is not oxidized to the less toxic arsenate by plasmid-bearing E. coli or S. aureus (71), although Alcaligenes strains have been isolated that do have an inducible arsenite oxidizing system (55,58). Arsenicals and antimonials are toxic by virtue of inhibiting thiol-containing enzymes (1). Plasmid-containing resistant cells did not excrete soluble thiol compounds into the medium (71). The absence of "detoxification" (measured by experiments inoculating sensitive cells into medium "preconditioned" by growth of resistant cells) eliminated all other mechanisms involving changes in extracellular chemical states. It is possible that arsenite resistance (like arsenate resistance; above) results from a change in cell membrane transport. Arima and Beppu (3,5) isolated an arsenite resistant Pseudomonas pseudomallei which had an inducible decrease in the energy-dependent accumulation of arsenite. No further work has been done on this system.

Oxidation and Methylation of Arsenic

Oxidation and methylation of arsenic compounds by microbes was known well before the modern era of molecular genetics. Arsenite-resistant bacteria capable of oxidizing the more toxic arsenite ion to somewhat less toxic arsenate were first isolated from cattle-dipping tanks in South Africa in 1918 and later in Australia (reviewed in (79)). Similar strains have appeared over the years and these arsenite-oxidizing strains contain an inducible soluble arsenite dehydrogenase enzyme (55,58,79). There is no evidence of involvement of plasmids in the oxidation of arsenite by these bacteria.

Fungal growth producing a garlic-like smell of methylarsine was first found in moldy wallpaper in the 1940s (reviewed in (40,79)). The process of methyl donation required CH_3-cobalamin, as does the methylation of Hg^{2+} (40). An enzymatic reaction sequence of (a) reduction of AsO_4^{3-} to AsO_3^{3-}, (b) methylation by methylcobalamin of arsenite to $CH_3AsO_3^{2-}$, followed by (c) methylation to $(CH_3)_2AsO_2^-$, and then followed by (d) reduction to $(CH_3)_2AsH$ (dimethylarsine) was postulated (40) and later directly demonstrated (14, 59). The volatile arsenic compound found was entirely dimethylarsine. Clearly this sequence is a more complex one than the abiotic methylation of Hg^{2+} by methylcobalamin. However, under some conditions, the entire volatile methylation product of Hg^{2+} was $(CH_3)_2Hg$ (36), and it is not possible at the moment to tell how different or how similar the methylation of different heavy metals and metalloids is at the biochemical level (10,40,93).

CADMIUM AND ZINC RESISTANCE

Plasmid-determined cadmium resistance has been found only in S. aureus (51). In some collections, Cd^{2+} resistance is the most common S. aureus plasmid resistance, more common than mercury or penicillin resistance (44). Gram-negative cells without plasmids are just as resistant to Cd^{2+} as are staphylococci with plasmids (45), probably because of relatively reduced Cd^{2+} uptake. However, there are occasional gram-negative bacteria that are sensitive or even "hypersensitive" to Cd^{2+} ((52), and Barkay, unpublished data). The basis of Cd^{2+} sensitivity and resistance in other bacterial species is not known. There is no evidence, however, for chemical transformations associated with resistance.

In staphylococci, Cd^{2+} is accumulated by a membrane transport system utilizing the cross-membrane electrical potential (57,86,90) (Fig. 7). This uptake system is highly specific for Cd^{2+} and Mn^{2+} with respective K_m's pf 10 μM and 16μM in whole cells (90) and 0.2 μM and 0.95 μM in membrane vesicles (57).

Two separate plasmid genes are responsible for the Cd^{2+} resistance of S. aureus strains (50, 51) (Fig. 8). The cadA and cadB genes confer, respectively, a large and a small increase in Cd^{2+} resistance (Fig. 8). When both genes are present, the effect of cadA masks the cadB gene effect; cadA$^+$ cadB$^+$ strains are no more resistant that are cadA$^+$ cadB$^-$ strains. Both genes confer increased Zn^{2+} resistance (Fig. 8). This and the genetic linkage of Cd^{2+} and Zn^{2+} resistance (57) indicate that the cadA and cadB genes are also responsible for Zn^{2+} resistance.

Cd^{2+} resistance is due to lessened accumulation of Cd^{2+} by resistant cells (86, 90). The cadA gene product causes this lessened Cd^{2+} accumulation (57, 90) (Fig. 9). A plasmid-encoded efflux system rapidly excretes Cd^{2+} from resistant cells (Fig. 10). Although not directly demonstrated, it seems plausible that the

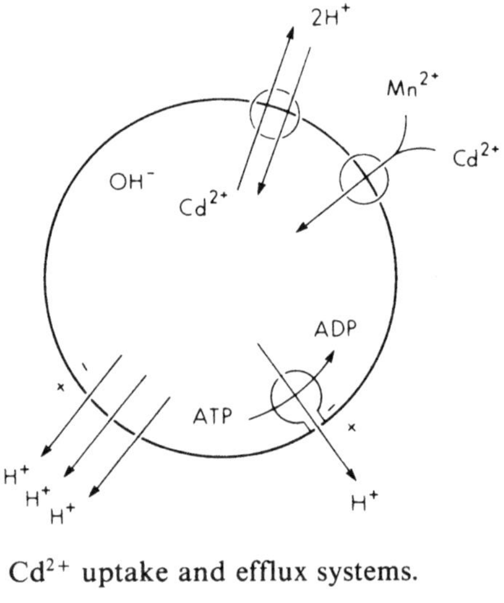

FIG. 7 — Model for Cd^{2+} uptake and efflux systems.

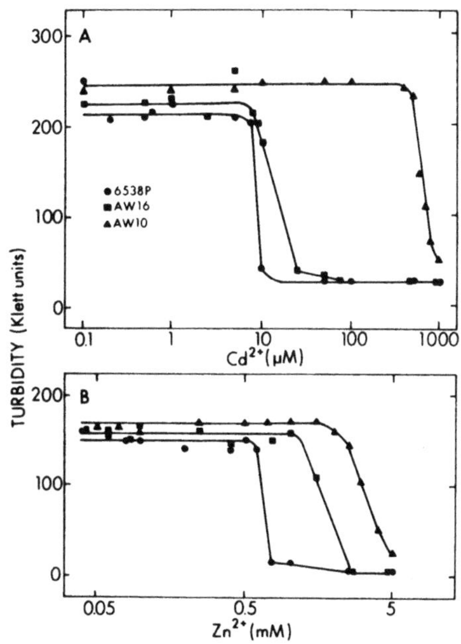

FIG. 8 — Cd^{2+} and Zn^{2+} resistance of strains with and without plasmids. Overnight cultures were diluted 1:200 into broth containing varying concentrations of (A)$CdCl_2$ or (B)$ZnSO_4$. Culture turbidities were measured after 7 h of growth at 37°C. Strains 6538 P (sensitive), AWIO (*cadA*$^+$-resistant), and AW16 (*cadB*$^+$-resistant) (from (57)).

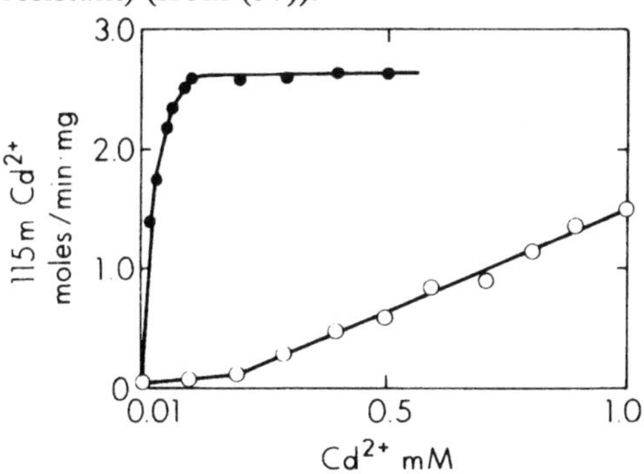

FIG. 9 — Cd^{2+} uptake by sensitive and by resistant S. aureus. Initial rate of accumulation of $^{115m}Cd^{2+}$ by a resistant plasmid-containing strain (O) and its sensitive plasmidless variant (●) (80).

cadA gene product might also cause Zn^{2+} efflux. However, the presence of the $cadA^+$ gene neither reduces Mn^{2+} uptake nor causes rapid efflux of accumulated Mn^{2+} (57). $CadA^+$ resistant strains possess this efflux system, but sensitive cells and $cadB^+$ resistant strains do not (74).

The cadA-encoded efflux system is energy dependent. Cd^{2+} efflux was abolished by dinitrophenol and at low temperature (Fig. 10). Inhibitor studies indicate that the efflux system is a $Cd^{2+}/2H^+$ exchange system (Fig. 7), analogous to the tetracycline efflux system (42) responsible for plasmid-mediated resistance to that antibiotic.

We do not know the mechanism of cadB gene function. Our clues at the moment include the significant resistance to Zn^{2+} conferred by cadB (Fig. 8) and an inducible Cd^{2+} binding-activity governed by this gene (74). This binding activity is not energy-dependent, and we hypothesize that it might be due to a membrane component analogous to metallothionein, the small Cd^{2+}-binding protein made by both mammals and microbes (Kägi and Hapke, this volume). This is no more than a working hypothesis at this time. When seeking Cd^{2+}-resistant mutants of mammalian cells, Corrigan and Huang (13) found what appear to be transport mutants that accumulate less Cd^{2+} and more Zn^{2+} than wild-type cells. Whether the metallothionein studies or mammalian transport studies will provide a basis for understanding cadB resistance is an open question.

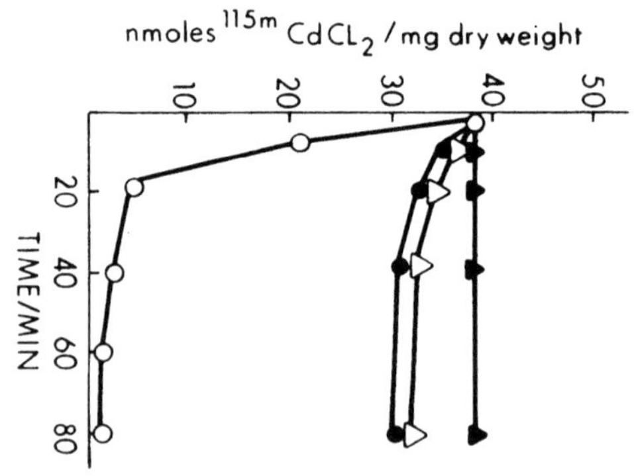

FIG. 10 — Retention of $^{115m}Cd^{2+}$ by cadmium-resistant S. aureus. Cell suspensions were pre-incubated with lmM $^{115m}Cd^{2+}$ for 10 min at 37°C. Cells were washed free of Cd^{2+}, and then $^{115}Cd^{2+}$ efflux was assayed. 0, control cells at 37°C; ▲, cells at 4°C; △, cells with 10 mM dinitrophenol; and ●, cells with 100 μM dicyclohexylcarbodiimide (85).

SILVER RESISTANCE

Microbial silver toxicity is found in situations of industrial pollution, resulting especially from the use of photographic film. In hospitals, silver salts are preferred antimicrobial agents for burns (18,19,63). It is thus not surprising that silver-resistant bacteria have been found in urban and industrial polluted sites (77) or that silver-resistant bacteria (2,9,31) and silver-resistant plasmids (41) have been described. The plasmid-determined resistance is very great, and the ratio of minimum-inhibitory Ag^+ concentrations (resistant:sensitive) can be greater than 100:1 (69,74). The level of resistance is strongly dependent upon available halide ions; without Cl^-, there was relatively little difference between cells with or without resistance plasmids (74). Br^- and I^- protected both sensitive and resistant cells at concentrations far below those used for Cl^-. Both sensitive and resistant cells bind Ag^+ tightly and are killed by effects on cell respiration and other cell surface functions (8,66). Once bound extracellularly, Ag^+ enters the cell and is found in high speed centrifugal supernatant fluids (Silver, unpublished data). Our current hypothesis is that the sensitive cells bind Ag^+ so tightly that they extract it from AgCl and other bound forms, whereas cells with resistance plasmids do not compete successfully with Ag^+-halide precipitates for Ag^+.

OTHER HEAVY METAL RESISTANCES

There are several other plasmid-mediated heavy metal resistances (75,77,79), but nothing today is known about the mechanisms of resistance to bismuth, boron, cobalt, lead, nickel, or tellurium ions. Chromate resistance in a pseudomonad isolated from river sediment is due to reduction of toxic Cr(VI) to less toxic Cr(III), and this resistance appears to be plasmid determined (7). Cupric ion resistance has been found both with plasmids in E. coli (33, 80) and in S. aureus (25), but no studies have been reported on the mechanism of this resistance.

Methylation of Other Heavy Metals

In addition to arsenic and mercury, chemical or biogenic methylation reactions have been reported for palladium, thallium, selenium, lead, platinum, gold, tin, chromium, and sulfur (10,62,93). Tin-resistant bacteria were isolated from polluted areas of Chesapeake Bay (27). These bacteria were also resistant to dimethyltin. Bacteriy from Chesapeake Bay were able to synthesize dimethyl $\frac{1}{n}$ and trimethlytin (28) as was previously proposed during a general discussion of methylcobalamin-dependent methylation (93). It is not known whether the bacteria that are resistant to methyltin synthesize or degrade methyltin. Resistance may (as in the case of mercury) be unrelated to the ability to synthesize these organometallic compounds. Once methylated by biological processes, abiotic transmethylation reactions occur (10), and heavy metal compounds can be destroyed abiotically by oxidation (56).

Over the last thirteen years, our laboratory has been studying the mechanisms and genetics of plasmid-mediated heavy metal resistances. In the absence of any obvious source of direct selection, one may ask why these resistances occur with such high frequencies. Selective agents in hospitals and "normal human" environments are only beginning to be examined. It has been suggested that heavy metal resistances may have been selected in earlier times, and that they are merely carried along today "for a free ride" with selection for antibiotic resistances. I doubt that there is such a thing as a "free ride" as far as these determinants are concerned. For example, in Tokyo in the late 1970s, both heavy metal resistances and antibiotic resistances were found with high frequencies in E. coli isolated from hospital patients, whereas heavy metal resistance plasmids without antibiotic resistance determinants were found in E. coli from an industrially polluted suburban river (46). Selection occurs for resistances to both types of agents in the hospital, but only that for toxic heavy metals in the river environment. Radford et al. (61) found Hg^{2+}-resistant microbes in agricultural soil with no known human mercurial input. In such settings, the prevalence of resistant microbes may be very low, but it may come into much greater quantitative prominance after industrial or agricultural pollution (21,52,77,83). This situation may be closely analogous to that with antibiotic-resistance plasmids which are found in low frequencies in antibiotic-virgin populations (24,39) but which become dominant with extensive use of antibiotics.

The "take home message" for this report is that transformations of toxic heavy metals that occur on a global scale are generally brought about by microorganisms. Considerations of human health problems must take into account the chemical forms of heavy metals available. Microbes might be utilized to convert more toxic forms of heavy metals into less toxic forms (perhaps the mercury cycle), or other microbes might be fooled into accumulating heavy metals in order to clean up sources of concern. In both cases, understanding of the basic biochemical and genetic bases for microbial resistances to heavy metals and microbial transformations of these materials is essential for rational planning.

Acknowledgements. The work in our laboratory on these topics over the last decade has been contributed to by many talented people whose names often appear as co-authors in the References section. Support has been provided most recently by National Science Foundation Grant PCM 82–05592 and by U.S. Public Health Service Grant AI15672.

REFERENCES

(1) Albert, A. 1973. Arsenicals, antimonials and mercurials. **In** Selective Toxicity, 5th ed., pp. 392–397. London: Chapman and Hall.

(2) Annear, D.I.; Mee, B.J.; and Bailey, M. 1976. Instability and linkage of silver resistance, lactose fermentation and colony structure in Enterobacter cloacae from burn wounds. J. Clin. Path. **29**: 441–443.

(3) Arima, K., and Beppu, M. 1964. Induction and mechanisms of arsenite resistance in Pseudomonas pseudomallei. J. Bact. **88**: 143–150.

(4) Bennett, P.M.; Grinsted, J.; Choi, C.L.; and Richmond, M.H. 1978. Characterization of Tn*501*, a transposon determining resistance to mercuric ions. Mol. Gen. Genet. **159**: 101–106.

(5) Beppu, M., and Arima, K. 1964. Decreased permeability as the mechanism of arsenite resistance in Pseudomonas pseudomallei. J. Bact. **88**: 151–157.

(6) Bisogni, Jr., J.J. 1979. Kinetics of methylmercury formation and decomposition in aquatic environments. **In** The Biogeochemistry of Mercury in the Environment, ed. J.O. Nriagu, pp. 211–230. Amsterdam: Elsevier/North Holland Press.

(7) Bopp, L.H.; Chakrabarty, A.M.; and Ehrlich, H.L. 1983. Plasmid-determined resistance to Cr(VI) and reduction of Cr(VI) to Cr(III). J. Bact. **155**: 1105–1109.

(8) Bragg, P.D., and Rainnie, D.J. 1974. The effect of silver ions on the respiratory chain of Escherichia coli. Can. J. Microbiol. **20**: 883–889.

(9) Bridges, K.; Kidson, A.; Lowbury, E.J.L.; and Wilkins, M.D. 1979. Gentamicin- and silver-resistant Pseudomonas in a burns unit. Brit. Med. J. **1**: 446–449.

(10) Brinckman, F.E.; Olson, G.J.; and Iverson, W.P. 1982. The production and fate of volatile molecular species in the environment: metals and metalloids. **In** Atmospheric Chemistry, ed. E.D. Goldberg, pp. 231–249. Dahlem Konferenzen. Berlin: Springer-Verlag.

(11) Brown, N.L.; Ford, S.J.; Pridmore, R.D.; and Fritzinger, D.C. 1983. Nucleotide sequence of a gene from the Pseudomonas transposon Tn*501* encoding mercuric reductase. Biochem. **22**: 4089–4095.

(12) Clark, D.L.; Weiss, A.A.; and Silver, S. 1977. Mercury and organomercurial resistance determined by plasmids in Pseudomonas. J. Bact. **132**: 186–196.

(13) Corrigan, A.J., and Huang, P.C. 1981. Cellular uptake of cadmium and zinc. Biol. Trace Element Res. **3**: 197–216.

(14) Cullen, W.R.; McBride, B.C.; and Pickett, A.W. 1979. The transformation of arsenicals by Candida humicola. Can. J. Microbiol. **25**: 1201–1205.

(15) Edwards, T., and McBride, B.C. 1975. Biosynthesis and degradation of methylmercury in human faeces. Nature **253**: 462-463.

(16) Foster, T.J.; Nakahara, H.; Weiss, A.A.; and Silver, S. 1979. Transposon *A*-generated mutations in the mercuric resistance genes of plasmid R100-1. J. Bact. **140**: 167–181.

(17) Fox, B., and Walsh, C.T. 1982. Mercuric reductase: Purification and characterization of a transposon-encoded flavoprotein containing an oxidation-reduction active sulfide. J. Biol. Chem. **257**: 2498–2503.

(18) Fox, Jr., C.L. 1968. Silver sulfadiazine. A new topical for Pseudomonas in burns. Arch. Surg. **96**: 184–188.

(19) Fox, Jr., C.L., and Modak, S.M. 1974. Mechanism of silver sulfadiazine action on burn wound infections. Antimicrob. Agents Chemother. **5**: 582–588.

(20) Friello, D.A., and Chakrabarty, A.M. 1980. Transposable mercury resistance in Pseudomonas putida. **In** Plasmids and Transposons: Environmental Effects and Maintenance Mechanisms, eds. C. Suttard and K.R. Rozee, pp. 249–260. New York: Academic Press.

(21) Furukawa, K.; Suzuki, T.; and Tonomura, K. 1969. Decomposition of organic mercurial compounds by mercury-resistant bacteria. Agric. Biol. Chem. **33**: 128–130.

(22) Furukawa, K., and Tonomura, K. 1971. Enzyme system involved in the decomposition of phenyl mercuric acetate by mercury-resistant Pseudomonas. Agric. Biol. Chem. **35**: 604–610.

(23) Furukawa, K., and Tonomura, K. 1972. Metallic mercury-releasing enzyme in mercury-resistant Pseudomonas. Agric. Biol. Chem. **36**: 217–226.

(24) Gardner, P.; Smith, D.H.; Beer, H.; and Moellering, Jr., R.C. 1969. Recovery of resistance (R) factors from a drug-free community. Lancet **2**: 774–776.

(25) Groves, D.J., and Young, F.E. 1975. Epidemiology of antibiotic and heavy metal resistance in bacteria: resistance patterns in Staphylococci isolated from populations not known to be exposed to heavy metals. Antimicrob. Agents Chemother. **7**: 614–621.

(26) Halbach, S., and Clarkson, T.W. 1978. Enzymatic oxidation of mercury vapor by erythrocytes. Biochim. Biophys. Acta **523**: 522–531.

(27) Hallas, L.E., and Cooney, J.J. 1981. Tin and tin-resistant microorganisms in Chesapeake Bay. Appl. Env. Microbiol. **41**: 446–471.

(28) Hallas, L.E.; Means, J.C.; and Cooney, J.J. 1982. Methylation of tin by estuarine microorganisms. Science **215**: 1505–1507.

(29) Hamdy, M.K., and Noyes, O.R. 1975. Formation of methylmercury by bacteria. Appl. Microbiol. **30**: 424–432.

(30) Hedges, R.W., and Baumberg, S. 1973. Resistance to arsenic compounds conferred by a plasmid transmissible between strains of Escherichia coli. J. Bact. **115**: 459–460.

(31) Hendry, A.T., and Stewart, I.O. 1979. Silver-resistant enterobacteriaceae from hospital patients. Can. J. Microbiol. **25**: 915–921.

(32) Holm, H.W., and Cox, M.F. 1975. Transformation of elemental mercury by bacteria. Appl. Microbiol. **29**: 491–494.

(33) Ishihara, M.; Kamio, Y.; and Terawaki, Y. 1978. Cupric ion resistance as a new marker of a temperature sensitive R plasmid, Rtsl in Escherichia coli. Biochem. Biophys. Res. Comm. **82**: 74–80.

(34) Izaki, K. 1981. Enzymatic reduction of mercurous and mercuric ions in Bacillus cereus. Can. J. Microbiol. **27**: 192–197.

(35) Izaki, K.; Tashiro, Y.; and Funaba, T. 1974. Mechanism of mercuric chloride resistance in microorganisms. III. Purification and properties of a mercuric ion reducing enzyme from Escherichia coli bearing R factor. J. Biochem. **75**: 591–599.

(36) Jensen, S.; and Jernelöv, A. 1969. Biological methylation of mercury in aquatic organisms. Nature **223**: 753–754.

(37) Komura, I.; Funaba, T.; and Izaki, K. 1971. Mechanism of mercuric chloride resistance in microorganisms. II. NADPH-dependent reduction of mercuric chloride and vaporization of mercury from mercuric chloride by a multiple drug resistant strain of Escherichia coli. J. Biochem. **70**: 895–901.

(38) Krohne-Ehrlich, G.; Schirmer, R.H.; and Untucht-Grau, R. 1977. Glutathione reductase from human erythrocytes. Isolation of the enzyme and sequence analysis of the redox-active peptide. Eur. J. Biochem. **80**: 65–71.

(39) Maré, I.J. 1968. Incidence of R factors among Gram negative bacteria in drug-free human and animal communities. Nature **220**: 1046–1047.

(40) McBride, B.C., and Edwards, T.L. 1977. Role of the methanogenic bacteria in the alkylation of arsenic and mercury. **In** Biological Implications of Metals in the Environment, pp. 1–19. **ERDA** Symposium Series nr. 42. Richland, WA: ERDA.

(41) McHugh, G.L.; Moellering, R.C.; Hopkins, C.C.; and Swartz, M.N. 1975. Salmonella typhimurium resistant to silver nitrate, chloramphenicol, and ampicillin. Lancet **1**: 235–240.

(42) McMurry, L.; Petrucci, Jr., R.E.; and Levy, S.B. 1980. Active efflux of tetracycline encoded by four genetically different tetracycline resistance determinants in Escherichia coli. Proc. Natl. Acad. Sci. USA **77**: 3974–3977.

(43) Mobley, H.L.T., and Rosen, B.P. 1982. Energetics of plasmid-mediated arsenate resistance in Escherichia coli. Proc. Natl. Acad. Sci. USA **79**: 6119–6122.

(44) Nakahara, H.; Ishikawa, T.; Sarai, Y.; and Kondo, I. 1977. Distribution of resistances to metals and antibiotics of Staphylococcal strains in Japan. Zentralb. Bakteriol. Parasitenkd. Infektionskr. Hyg. 1 Abt. Orig. A **237**: 470–476.

(45) Nakahara, H.; Ishikawa, T.; Sarai, Y.; Kondo, I.; Kozukue, H.; and Silver, S. 1977. Linkage of mercury, cadmium, and arsenate and drug

resistance in clinical isolates of Pseudomonas aeruginosa. Appl. Env. Microbiol. **33**: 975–976.

(46) Nakahara, H., and Kozukue, H. 1982. Volatilization of mercury determined by plasmids in E. coli isolated from an aquatic environment. **In** Drug Resistance in Bacteria: Genetics, Biochemistry, and Molecular Biology, ed., S. Mitsuhashi, pp. 337–340. Tokyo: Japanese Scientific Societies Press.

(47) Nakahara, H.; Silver, S.; Miki, T.; and Rownd, R.H. 1979. Hypersensitivity to Hg^{2+} and hyperbinding activity associated with cloned fragments of the mercurial resistance operon of plasmid NRl. J. Bact. **140**: 161–166.

(48) Nelson, Jr., J.D., and Colwell, R.R. 1975. The ecology of mercury-resistant bacteria in Chesapeake Bay. Microb. Ecol. **1**: 191–218.

(49) Ni'Bhriain, N.; Silver, S.; and Foster, T.J. 1983. Tn*5* insertion mutations in the mercuric ion resistance genes derived from plasmid R100. J. Bact. **155**: 690–703.

(50) Novick, R.P.; Murphy, E.; Gryczan, T.J.; Baron, E.; and Edelman, I. 1979. Penicillinase plasmids of Staphylococcus aureus: restriction-deletion maps. Plasmid **2**: 109–129.

(51) Novick, R.P., and Roth, C. 1968. Plasmid-linked resistance to inorganic salts in Staphylococcus aureus. J. Bact. **95**: 1335–1342.

(52) Olson, B.H.; Barkay, T.; and Colwell, R.R. 1979. Role of plasmids in mercury transformation by bacteria isolated from aquatic environment. Appl. Env. Microbiol. **38**: 478–485.

(53) Olson, G.J.; Iverson, W.P.; and Brinckman, F.E. 1981. Volatilization of mercury by Thiobacillus ferrooxidans. Current Microbiol. **5**: 115–118.

(54) Olson, G.J.; Porter, F.D.; Rubinstein, J.; and Silver, S. 1982. Mercuric reductase enzyme from a mercury-volatilizing strain of Thiobacillus ferrooxidans. J. Bact. **151**: 1230–1236.

(55) Osborne, F.H., and Ehrlich, H.L. 1976. Oxidation of arsenite by a soil isolate of Alcaligenes. J. Appl. Bact. **41**: 295–305.

(56) Parris, G.E., and Brinckman, F.E. 1976. Reactions which relate to environmental mobility of arsenic and antimony. II. Oxidation of trimethylarsine and trimethylstibine. Env. Sci. Technol. **10**: 1128–1134.

(57) Perry, R.D., and Silver, S. 1982. Cadmium and manganese transport in Staphylococcus aureus membrane vesicles. J. Bact. **150**: 973–976.

(58) Phillips, S.E., and Taylor, M.L. 1976. Oxidation of arsenite to arsenate by Alcaligenes faecalis. Appl. Env. Microbiol. **32**: 392–399.

(59) Pickett, A.W.; McBride, B.C.; Cullen W.R.; and Manji, H. 1981. The reduction of trimethylarsine by Candida humicola. Can. J. Microbiol. **27**: 773–778.

(60) Porter, F.D.; Ong, C.; Silver, S.; and Nakahara, H. 1982. Selection for mercurial resistance in hospital settings. Antimicrob. Agents Chemother. **22**: 852–858.

(61) Radford, A.J.; Oliver, J.; Kelly, W.J.; and Reanney, D.C. 1981. Translocatable resistance to mercuric and phenylmercuric ions in soil bacteria. J. Bact. **147**: 1110–1112.

(62) Ridley, W.P.; Dizikes, L.J.; and Wood, J.M. 1977. Biomethylation of toxic elements in the environment. Science **197**: 329–332.

(63) Rosenkranz, H.S., and Carr, H.S. 1972. Silver sulfadiazine: effect on the growth and metabolism of bacteria. Antimicrob. Agents Chemother. **2**: 367–372.

(64) Schottel, J.L. 1978. The mercuric and organomercurial detoxifying enzymes from a plasmid-bearing strain of Escherichia coli. J. Biol. Chem. **253**: 4341–4349.

(65) Schottel, J.; Mandal, A.; Clark, D.; Silver, S.; and Hedges, R.W. 1974. Volatilisation of mercury and organomercurials determined by inducible R-factor systems in enteric bacteria. Nature **251**: 335–337.

(66) Schreurs, W.J.A., and Rosenberg, H. 1982. Effect of silver ions on transport and retention of phosphate by Escherichia coli. J. Bact. **152**: 7–13.

(67) Silver, S. 1978. Transport of cations and anions. **In** Bacterial Transport, ed. B.P. Rosen, pp. 221–324. New York: Marcel Dekker Inc.

(68) Silver, S. 1981. Mechanisms of bacterial resistances to toxic heavy metals: arsenic, antimony, silver, cadmium and mercury. **In** Environ-

mental Speciation and Monitoring Needs for Trace Metal-Containing Substances from Energy-Related Processes, eds. F.E. Brinckman and R.H. Fish, pp. 301–314, Spec. Publ. 618. Washington, D.C.: National Bureau of Standards.

(69) Silver, S. 1981. Mechanisms of plasmid-determined heavy metal resistances. **In** Molecular Biology, Pathogenicity and Ecology of Bacterial Plasmids, eds. S.B. Levy, R.C. Clowes, and E.L. Koenig, pp. 179–189. New York: Plenum Press.

(70) Silver, S. 1981. Mechanisms of plasmid-determined resistances to heavy metals in Staphylococcus aureus. **In** Staphylococci and Staphylococcal Infections, ed. J. Jeljaszewicz, pp. 589–599. Stuttgart: Gustav Fischer Verlag.

(71) Silver, S.; Budd, K. Leahy, K.M.; Shaw, W.V.; Hammond, D.; Novick, R.P.; Willsky, G.R.; Malamy, M.H.; and Rosenberg, H. 1981. Inducible plasmid-determined resistance to arsenate, arsenite and antimony (III) in Escherichia coli and Staphylococcus aureus. J. Bact. **146**: 983–996.

(72) Silver, S., and Keach, D. 1982. Energy-dependent arsenate efflux: The mechanism of plasmid-mediated resistance. Proc. Natl. Acad. Sci. USA **79**: 6114–6118.

(73) Silver, S., and Kinscherf, T.G. 1982. Genetic and biochemical bases for microbial transformations and detoxification of mercury and mercurial compounds. **In** Biodegradation and Detoxification of Environmental Pollutants, ed. A.M. Chakrabarty, pp. 85–103. Boca Raton, FL: CRC Press.

(74) Silver, S.; Perry, R.D.; Tynecka, Z.; and Kinscherf, T.G. 1982. Mechanisms of bacterial resistances to the toxic heavy metals antimony, arsenic, mercury and silver. **In** Drug Resistance in Bacteria: Genetics, Biochemistry, and Molecular Biology, ed. S. Mitsuhashi, pp. 347–361. Tokyo: Japan Scientific Societies Press.

(75) Smith, D.H. 1967. R factors mediate resistances to mercury, nickel, and cobalt. Science **156**: 1114–1116.

(76) Smith, H.W. 1978. Arsenic resistance in Enterobacteria: its transmission by conjugation and by phage. J. Gen. Microbiol. **109**: 49–56.

(77) Summers, A.O.; Jacoby, G.A.; Swartz, M.N.; McHugh, G.; and Sutton, L. 1978. Metal cation and oxyanion resistances in plasmids of gram-negative bacteria. **In** Microbiology 1978, ed. D. Schlessinger, pp. 128–131. Washington, D.C.: American Society for Microbiology.

(78) Summers, A.O., and Silver, S. 1972. Mercury resistance in a plasmid-bearing strain of Escherichia coli. J. Bact. **112**: 1128–1136.

(79) Summers, A.O., and Silver, S. 1978. Microbial transformations of metals. Ann. Rev. Microbiol. **32**: 637–672.

(80) Tetaz, T.J., and Luke, R.K.J. 1983. Plasmid-controlled resistance to copper in Escherichia coli. J. Bact. **154**: 1263–1268.

(81) Tezuka, T., and Tonomura, K. 1976. Purification and properties of an enzyme catalyzing the splitting of carbon-mercury linkages from mercury-resistant Pseudomonas K-62 strain. I. Splitting enzyme 1. J. Biochem. **80**: 79–87.

(82) Tezuka, T., and Tonomura, K. 1978. Purification and properties of a second enzyme catalyzing the splitting of carbon-mercury linkages from mercury-resistant Pseudomonas K-62. J. Bact. **135**: 138–143.

(83) Timoney, J.F.; Port, J.; Giles, J; and Spanier, J. 1978. Heavy-metal and antibiotic resistance in the bacterial flora of sediments of New York Bight. Appl. Env. Microbiol. **36**: 465–472.

(84) Tonomura, K., and Kanzaki, F. 1969. The reductive decomposition of organic mercurials by cell-free extract of a mercury-resistant pseudomonad. Biochim. Biophys. Acta **184**: 227–229.

(85) Tynecka, Z.; Gos, Z.; and Zajac, J. 1981. Energy-dependent efflux of cadmium coded by a plasmid resistance determinant in Staphylococcus aureus. J. Bact. **147**: 313–319.

(86) Tynecka, Z.; Gos, Z.; and Zajac, J. 1981. Reduced cadmium transport determined by a resistance plasmid in Staphylococcus aureus. J. Bact. **147**: 305–312.

(87) Vonk, J.W., and Sijpesteijn, A.K. 1973. Studies on the methylation of mercuric chloride by pure cultures of bacteria and fungi. Anton. van Leeuwenhoek J. Microbiol. Serol. **39**: 505–513.

(88) Weiss, A.A.; Murphy, S.D.; and Silver, S. 1977. Mercury and organomercurial resistances determined by plasmids in Staphylococcus aureus. J. Bact. **132**: 197–208.

(89) Weiss, A.A.; Schottel, J.L.; Clark, D.L.; Beller, R.G.; and Silver, S. 1978. Mercury and organomercurial resistance with enteric, staphylococcal, and pseudomonad plasmids. **In** Microbiology 1978, ed. D. Schlessinger, pp. 121–124. Washington, D.C.: American Society for Microbiology.

(90) Weiss, A.A.; Silver, S.; and Kinscherf, T.G. 1978. Cation transport alteration associated with plasmid-determined resistance to cadmium in Staphylococcus aureus. Antimicrob. Agents Chemother. **14**: 856–865.

(91) Williams, Jr., C.H.; Arscott, L.D.; and Schulz, G.E. 1982. Amino acid sequence homology between pig heart lipoamide dehydrogenase and human erythrocyte glutathione reductase. Proc. Natl. Acad. Sci. USA **79**: 2199–2201.

(92) Witte, W.; Van Dip, N.; and Hummel, R. 1980. Resistenz gegen Quecksilber und Cadmium bei Staphylococcus aureus unterschiedlicher ökologischer Herkunft. Z. Allg. Mikrobiol. **20**: 517–521.

(93) Wood, J.M.; Cheh, A.; Dizikes, L.J.; Ridley, W.P.; Rakow, S.; and Lakowicz, J.R. 1978. Mechanisms for the biomethylation of metals and metalloids. Fed. Proc. **37**: 16–21.

Changing Metal Cycles and Human Health, ed. J.O. Nriagu, pp. 225–236.
Dahlem Konferenzen 1984. Berlin, Heidelberg, New York, Tokyo: Springer-Verlag.

NEW APPROACHES TO STUDY EFFECTS OF METALS AND THEIR COMPOUNDS IN VITRO AND IN VIVO

A.J.M. Schoot Uiterkamp
Bureau of Environmental Projects TNO
2600 AE Delft, The Netherlands

Abstract. In vitro and in vivo test systems and some non-invasive instrumental techniques to study effects of metals and their compounds are briefly reviewed. Emphasis is on short-term tests for mutagenicity/carcinogenicity, immunotoxicity, pulmonary toxicity, neurotoxicity, and teratogenicity. The non-invasive instrumental techniques are nuclear magnetic resonance, positron emission tomography, photoacoustic spectroscopy, in vivo neutron activation analysis, and X-ray fluorescence.

INTRODUCTION

Carcinogenicity is but one of the potential health effects of metals and their compounds. Other adverse effects can be manifested in the immunologic, neurologic, behavioral, pulmonary, cardiovascular, renal, skin, intestinal, and teratogenic systems. However, the literature on in vitro and in vivo assay systems is dominated by the end points mutagenicity and carcinogenicity. Only a small range of other effects can be explored in presently available assay systems. This paper focuses on short-term in vitro and in vivo testing and on some non-invasive instrumental techniques.

IN VITRO AND IN VIVO SYSTEMS

Table 1, adapted from Archer (2), summarizes the basic ways in which most in vitro and in vivo effect studies are carried out.

Toxicant exposure	Test environment
A. In vitro	In vitro
B. In vitro	In vivo
C. In vivo	In vitro
D. In vivo	In vivo

TABLE 1 – General scheme for effect studies.

Approach A (e.g., Ames test) is, generally speaking, most economical, fastest, and most artificial. Approach D (e.g., exposure in humans) is, generally speaking, most expensive, slowest, and most realistic. Test systems for the following effects will be reviewed: mutagenicity and carcinogenicity, immunotoxicity, pulmonary toxicity, neurotoxicity, and teratogenicity.

Mutagenicity and Carcinogenicity

There are about 80 metals of which at least 20 have compounds with well documented adverse health effects in man. Only a few of these metals have been shown or are suspected to be carcinogenic, e.g., Cr (VI), Ni, and As compounds.

Evidence of carcinogenicity in animal experiments requires, according to the IARC monograph series' definition, "an increased incidence in malignant tumors." The significant aspect of this definition is the biological end result rather than the mode of action of the compound. Carcinogens acting on the genes (genotoxic carcinogens) or through different mechanisms (non-genotoxic or epigenetic carcinogens such as promoters) are covered by this definition. Only genotoxic compounds will be detected in tests based on mutagenic or DNA-damaging properties of chemicals. Most short-term tests for potential carcinogens and mutagens fall into this category. Hollstein et al. (11) have identified over 100 assays that use a variety of cell types in vitro, as well as tests that can be done directly on animals or people. The 23 assays that are presently evaluated in the GENE TOX program (9) fall, broadly speaking, into four categories: a) gene mutation; b) chromosomal aberration; c) DNA synthesis, damage, and repair; and d) cell transformation. We add a fifth category: e) promotion assays.

Gene Mutations

Microbial systems using bacteria, phage, fungi, insects, and mammalian cells in culture all fall into this category.

Microbial systems. Currently the most widely used bacterial assay is the Salmonella/Ames test with or without metabolic activation. Mutations are detected in histidine auxotrophic strains of S. typhimurium as bacterial colonies that grow in the absence of histidine. Over 2600 chemicals have been tested in the assay. Metal carcinogens are usually negative in the system.

The second bacterial mutation system which has been used to screen metal compounds measures reverse mutations in tryptophan auxotrophs of E. coli WP2.

Phage. The induction of lambda prophage might develop into a useful test system for mechanistic studies of metal compounds. DNA-damaging agents often cause the induction of prophage to the active viral form resulting in viral replication and lysis of most cells. Agents interfering with DNA polymerase would not be expected to induce prophage.

Fungi. Three fungi are used extensively: the yeast Saccharomyces cerevisiae (especially strain D3) and the molds Neurospora crassa and Aspergillus nidulans. Tests using these eukaryotic organisms offer the simplest and fastest methods available for examining several kinds of mutagenic effects that are undetectable in prokaryotes. In S. cerevisiae D3 mitotic recombinations induced by chemicals are easily scored: recombinants form red rather than white colonies.

There are many developments in tests using microbial systems. The sensitivity of tester strains has been improved by genetic manipulation and different mammalian metabolic activation systems have been introduced either as an enzyme extract or intact cell or ogans. In host-mediated assays the microbial cells are inoculated into different organs of living mammals and after chemical treatment of the host recovered to be assayed. The genetic testing of body fluids of exposed humans or treated animals is widely being explored.

Mammalian cells in culture. Chinese Hamster lung (V79) and ovary (CHO) cells and mouse lymphoma cells L5178Y are main representatives of this class. Tests in V79 and CHO cells are mostly carried out at genetic loci controlling sensitivity to purine analogs or ouabain resistance. In the former case, mutations are induced in the gene coding for the enzyme hypoxanthine guanine phosphoribosyl transferase (HGPRT). Mutated cells exposed to toxic purine analogs grow unaffected. Normal cells are killed because the toxic analogs are incorporated into cell components. Similarly, in ouabain resistance, mutations in the Na-K ATP-ase gene are induced.

Systems using fibroblast and epithelial cells are still experimental. Developments in the latter category are eagerly awaited: most tissue cultures are based on non-epithelial cell lines, but 90% of human cancers occur in epithelial tissues.

Chromosomal Aberrations

Main representatives of this class are scoring for sister chromatid exchanges (SCEs), the dominant lethal assay, and the micronucleus test.

Sister chromatid exchange. SCE is the exchange of material between the two arms (chromatids) of the chromosomes of an eukaryotic cell. Exchanges are visualized with special staining techniques. The induction of SCE in cultured cells is one of the most sensitive nonbacterial detection systems for genotoxic compounds. SCE is increasingly being used in humans.

Dominant lethal assay. In this assay male rodents are exposed to test substances and mated with untreated virgin females. Dominant lethal mutations in the sperm cells result in the death of the fetus originating from that germ cell.

Micronucleus test. Micronuclei are small extranuclear interphase chromosome fragments that result from chromosome breakage or cellular malfunction during cell division. Young erythrocytes in bone marrow from exposed rodents are mainly used in this test. In principle the test can be used in humans.

DNA Synthesis, Damage, and Repair

Tests in this category are the PolA E.coli and RecA B. subtilis test, the fidelity of DNA synthesis in vitro, and DNA repair in liver hepatocytes.

PolA E.coli and RecA B.subtilis. Bioassays using these organisms measure the capacity of exogenous agents for differential growth inhibition of wild-type and repair-deficient mutants.

Fidelity of in vitro DNA synthesis. Sirover and Loeb (25) developed a cell-free screening system for mutagenic/carcinogenic metals based on the principle that these compounds decrease the fidelity of in vitro DNA synthesis or synthetic polynucleotide templates. Another bioassay in this category measures the fidelity by which natural template (from bacteriophage ΦX 174) is copied in vitro.

Cell transformation

Representatives of cell transformation are assays using hamster or mouse cells and viral infected cells.

Transformation of hamster or mouse cells. Transformation is the chemically-induced conversion of normal cultured mammalian cells into cells with abnormal growth patterns. Whether or not transformed cells are actually malignant can be ascertained by injecting them into whole animals. Assays in this category employ mostly Syrian hamster embryo cells, the hamster cell line BHK-21, and the mouse cell lines C3H 10 $T^{1}/_{2}$ and BALB 3T3.

Viral infected cells. Enhancement of SA 7 virus-induced transformation of hamster embryo cells by potential carcinogens forms the basis of a sensitive assay. Apparently agents capable of causing (in-) direct damage to cellular DNA will increase the frequency of morphological transformations by increasing the integration of viral DNA into cellular DNA at the damage site.

Promotion Assays
In the classical in vivo assay (i.e., mouse skin painting), tumors are elicited by continuous exposure to a non-carcinogenic substance in a system previously exposed to a carcinogen (initiator) at a dose that by itself yields few if any tumors (26). Besides mouse skin, mouse lung and rat liver are also used.

There are few in vitro promotion assays available, and none of them can be considered "established" yet (see below). In the oldest in vitro system, a phorbol ester (potent promoting agent) increased the expression of spontaneously arising transformed cells in a mixed culture of SV40 transformed mouse 3T3 cells or of the mouse cells alone. In newer more realistic systems, mouse C3H 10$T^{1}/_{2}$ or BALB 3T3 cells or hamster embryo cells are initiated with UV light or a chemical initiator followed by exposure to a promoter. Transformed foci show up only in the cultures exposed to both the initiator and the promoter.

Metals and Carcinogenesis / Mutagenesis Assays
Bacterial mutagenesis systems have not been successful in predicting the carcinogenicity of metals (25). The strongest mutagenic effects have been found with dichromate and with square planar platinum compounds. The standard Ames system may not be suitable for metals because of the large amounts of competing salts present in the medium. Several metals are positive in the E.coli WP2 test and the RecA B.subtilis assay. However, the former test is not nearly as "established" as the Ames test, while the latter has a high false positive rate (21). Metal-induced transformation in Syrian hamster cells and enhancement of viral-induced transformation by metal compounds has been observed (25). The former test has high performance characteristics. Transformation comes closer to the production of cancer than any other measure of end point. A problem with the test is the high level of variability in outcome between laboratories (21). The cell-free test systems are promising. They offer the possibility

of systematic mechanistic studies of mutagenesis. Nickel sulfate, cadmium acetate, and potassium chromate have been shown to be promoters in hamster embryo cells initiated with benzo(a)pyrene (22). Arsenite, but not arsenate, strongly enhances UV mutagenicity in E.coli WP 2(23). The latter two examples indicate that one of the reasons that carcinogenic metal compounds behave poorly in mutagenesis tests is that they act as promoters or cocarcinogens rather than as initiators. In vitro test systems offer largely unexplored opportunities for studies of cellular mechanisms, synergistic (18) and antagonistic effects (17), and the role of speciation. Major disadvantages of short-term tests are that the indicator organisms or cells are not comparable to intact mammals in absorption, distribution, and toxicokinetics (10).

The strengths and weaknesses of short-term tests for mutagenicity and carcinogenicity also hold for the effects described below.

Immunotoxicity

The immune system can also act as an important target organ for toxicity. Interactions with xenobiotics may result in three kinds of effects: immunodeficiency, immunosuppression, or immuno-enhancement; alterations in host defense mechanism; allergy or hypersensitivity. Chronic exposure to substances of low toxicity or acute exposure to highly toxic chemicals may alter immunological functions. Immunosuppression may lead to increased susceptibility to infectious diseases, while immuno-enhancement may result in an array of allergic responses and autoimmune diseases.

The complex immune response is, broadly speaking, mediated by three cell types working independently and in concert; thymus-derived lymphocytes (T-cells), bursa (a bird's organ)-derived lymphocytes (B-cells), and macrophages (4). T-cells are involved in cell-mediated immunity (CMI), B-lymphocytes in humoral immunity (B-cells produce antibodies), and macrophages with persistent intracellular organisms.

Standard immune response tests may be carried out in the three basic ways A, C, and D shown in Table 1 (2). Approach B is the most common approach in immunotoxicity testing today. Archer (2) gives examples of new tests for CMI in the categories B and C.

Cell-mediated immunity is routinely assessed by quantitating in vivo delayed hypersensitivity responses (DHR) and in vitro lymphoproliferative responses (LP) following chemical exposure. In humans, DHRs are measured by the intradermal administration of a panel of antigens to which previous exposure is likely. Depressed delayed response at the challenge site after 24 and 48 hours can be interpreted as positive response to one or less antigens and/or reduced skin reactions. Preliminary studies suggest that a depressed DHR corresponds with increased host susceptibility (15). In the LP assay, various mitogens such as metal ions are used to stimulate selectively cell growth measured as tritiated thymidine incorporation into DNA. Depressed LP responses occur in humans with inherited immunodeficiencies as well as in various conditions such as malignancies, aging, nutritional deprivation, and severe stress (15).

Humoral immune responses are measured in many ways such as immunodiffusion, hemagglutination, radioimmunoassay, and enzyme multiplied immunoassay technique. The last method is based on the principle that the activity of an enzyme label may be modulated when antibodies bind to antigens which are in turn covalently bound to the enzyme. Monoclonal antibodies and enzymes with intensely fluorescent products may be used (16). Metals and their compounds responsible for reduced circulating antibody titers are lead, cadmium, methylmercury, arsenic, nickel, chromium, and platinum. Antibody synthesis has been inhibited by lead, cadmium, methylmercury, zinc, arsenic, silica, and magnesium. However, selenium has stimulated both antibody synthesis and circulating antibody titers (12).

Macrophages, especially those from tissue, are widely studied (see below). Phagocytosis of 14 nickel compounds by rat peritoneal macrophages in vitro did not seem to correlate well with sarcoma incidences following administration of the compounds to rats (13).

Pulmonary Toxicity

Pulmonary alveolar macrophages (PAM) have a protecting role against pulmonary disease. PAM cells participate in immunologic response and show several specialized properties, namely, chemotaxis, pinocytosis, phagocytosis, and bacterial-killing mechanisms. They are largely studied in ways A and C given in Table 1. Metal compounds and metal-coated particles such as fly ash have been widely employed in the PAM system (8). Beck et al. (3) have developed a short-term bioassay using hamsters exposed by intratracheal instillation. This route of exposure allows more accurate dosage than inhalation. The lung lavage of exposed animals is characterized using several cellular and enzymic parameters. Changes in the resistance to respiratory infections provide a sensitive system to study health effects of air pollutants. This model indicates the impairment of basic defense mechanisms of the respiratory system by the combined exposure to low concentrations of pollutants (such as airborne metals) and superimposed bacterial infections (7).

Neurotoxicity
The hallmark of exposure to mercury and lead in vivo is neurotoxicity. Neurotoxic effects are increasingly being studied in in vitro systems. (e.g., receptor binding assay). Nerve cells can be grown in tissue culture so that they maintain their normal shape and much of their typical behavior. Electric action potentials can be measured by inserting glass microelectrodes into the cells. Peptides, neurotransmitters, or toxicants can similarly be injected into or near cultured cells to observe their effects. A recent paper on the effects of heavy metals on nerve tissue culture is available (20).

Teratogenicity
Braun et al. (5) have developed an in vitro assay system for the assessment of teratogenic potential of chemicals. The assay identifies teratogens by their ability to inhibit attachment of ascites tumor cells to plastic surfaces coated with concanavalin A. The assay results show a reasonable qualitative agreement with whole animal studies. Doses analyzed ranged over five orders of magnitude. The teratogens zinc chloride and lithium carbonate were inhibitory in the system. Other metals could be tested as well: e.g., methylmercury and mercury and lead salts are teratogens in animal models.

INSTRUMENTAL TECHNIQUES

Recent developments in instrumental techniques such as nuclear magnetic resonance (NMR), positron emission tomography (PET) and photo-acoustic spectroscopy (PAS) offer interesting possibilities for non-invasive studies of the effects of metals. In vivo neutron activation analysis (IVNAA) and X-ray fluorescence do not measure health effects. Yet these techniques provide information on organ and body burden of heavy metals in vivo that can be compared with observed health effects.

NMR

Three developments in NMR are worth mentioning here: in vitro studies of cells, organelles, and perfused organs; in vivo studies using flat surface coils and greatly increased sensitivity through spectrometer design; and the application of special pulse techniques (14).

In Vitro Studies
In vitro studies have concentrated on the nuclei ^{13}C and ^{31}P (14, 24). Metabolism of ^{13}C labeled compounds and phosphate and the measurement of

intracellular pH using the chemical shift of ^{31}P have been carried out in mitochondria, cultured mammalian cells, red blood cells, and perfused heart, muscle, and kidney of rats. Metal nuclei such as ^{59}Co, ^{111}Cd, ^{113}Cd, ^{115}Sn, ^{117}Sn, ^{197}Pt, ^{199}Hg, and ^{207}Pb have been studied in the presence of proteins such as meet-allothioneins and small biologically relevant molecules. Direct observation of these nuclei in in vitro systems is well within reach, although it would still require high doses of metal ion in most cases. Addition of metal ions to in vitro systems employing ^{13}C and ^{31}P as observational nuclei for intracellular processes does not seem to have been explored.

In Vivo Studies

Earlier spectrometers built for chemical studies had cylindrical sample cavities. New spectrometers with flat surface coils allow chemical reactions to be followed in specific tissue regions: the topical NMR. Studies of phosphate metabolism in rats have recently been published (24). Metabolic disorders and xenobiotic effects on metabolic pathways can now be studied non-invasively in intact organisms.

PET

The position of a positron-emitting radionuclide can be determined because the two photons produced upon the annihilation of the positron with an electron are emitted in opposite directions. By positioning detectors around an animal or person, it is possible to determine the line along which a disintegration occurred. The availability of chemicals labeled with ^{11}C, ^{13}N, ^{15}O, and ^{18}F opens up the study of a whole range of physiologically interesting processes. PET has the potential for measurement of sugar, fatty acid, amino acid, and other substrate parameters such as absorption, distribution, and kinetics, as well as receptor concentrations anywhere in the body. Just as in the case of NMR, metabolic effects of xenobiotics can be explored. The two techniques are clearly complementary: PET has high sensitivity but is largely limited to fast kinetics due to the very short half-lives of positron emitting nuclei. NMR has high specificity but its longer observation times probe slower kinetics. In PET, labeled metabolites cannot be distinguished from labeled products; in NMR they can on the basis of chemical shift information (6).

PAS

In PAS a sample cell is irradiated with chopped light. Absorbed light results in heat fluctuations of the gas in the sample cell, which are picked up as vibrations by a sensitive microphone. A PAS spectrum shows adsorbed radiant power as a function of frequency. PAS is essentially a surface technique allowing the use of opaque samples such as cell membranes, skin, and cornea. Absorbing species such as metal ions can be observed directly or indirectly because of their effect on other absorbing species (19).

IVNAA AND X-RAY FLUORESCENCE

Useful information especially on long-term low level exposure to heavy metals can be obtained with these techniques. Cadmium, lead, and mercury levels in such organs as kidney, liver, and bone have been measured. In IVNAA, γ-rays emitted promptly following capture of thermal neutrons are detected. The technique is especially sensitive in the case of cadmium, but detection of mercury is also possible (1, 27). Transportable systems for cadmium detection have been developed (1). In X-ray fluorescence, nuclei absorb γ-rays and X-rays are emitted. Tissue attenuation makes this technique basically a surface technique. Mercury and lead in finger bones and teeth have been measured.

Acknowledgement. Supported in part by the Netherlands Organization for Applied Scientific Research (TNO) and the Interdisciplinary Programs in Health (IPH) at the Harvard School of Public Health through a Cooperative Agreement with the US Environmental Protection Agency, CR 807809.

REFERENCES

(1) Al-Haddad, I.K.; Chettle, D.R.; Fletcher, J.G.; and Fremlin, J.H. 1982. A transportable system for measurement of kidney cadmium in vivo. Int. J. Appl. Rad. Isot. **32**: 109–112.

(2) Archer, D.L. 1982. New approaches to immunotoxicity testing. Env. Health Persp. **43**: 109–113.

(3) Beck, B.D.; Brain, J.D.; and Bohannon, D.E. 1981. The pulmonary toxicity of an ash sample from the Mt. St. Helens volcano (USA). Exp. Lung Res. **2**: 289–301.

(4) Bick, P.H. 1982. Immune system as a target organ for toxicity. Env. Health Persp. **43**: 3–7.

(5) Braun, A.G.; Buckner, C.A.; Emerson, D.J.; and Nichinson, B.D. 1982. Quantitative correspondence between the in vivo and in vitro activity of teratogenic agents. Proc. Natl. Acad. Sci. USA **79**: 2056–2060.

(6) Brownell, G.L.; Budinger, T.F.; Lauterbur, P.C.; and McGeer, P.L. 1982. Positron tomography and nuclear magnetic resonance imaging. Science **215**: 619–626.

(7) Ehrlich, R. 1980. Interaction between environmental pollutants and respiratory infections. Env. Health Persp. **35**: 89–100.

(8) Garrett, N.E.; Campbell, J.A.; Slack, H.F.; Waters, M.D.; and Lewtas, J. 1981. The utilization of the rabbit alveolar macrophage and chinese hamster ovary cell for evaluation of the toxicity of particulate materials. Env. Res. **24**: 345–365.

(9) Green, S.; and Auletta, A. 1980. An evaluation of bioassays in genetic toxicology. Mut. Res. **76**: 165–168.

(10) Hoffman, G.R. 1982. Mutagenicity testing in environmental toxicology. Env. Sci. Technol. **16**: 560A–574A.

(11) Hollstein, M.; McCann, J.; Angelosanto, F.A.; and Nichols, W.W. 1979. Short term tests for carcinogens and mutagens. Mut. Res. **65**: 133–226.

(12) Koller, L.D. 1982. In vitro assessment of humoral immunity following exposure to heavy metals. Env. Health Persp. **43**: 37–39.

(13) Kuehn, K.; Fraser, C.B.; and Sunderman, F.W. 1982. Phagocytosis of particulate nickel compounds by rat periotoneal macrophages in vitro. Carcinogenesis **3**: 321–326.

(14) Levy, G.C., and Craik, D.J. 1981. Recent developments in nuclear magnetic resonance spectroscopy. Science **214**: 291–299.

(15) Luster, M.I.; Dean, J.H.; and Boorman, G.A. 1982. Cell-mediated immunity and its application in toxicology. Env. Health Persp. **43**: 31–36.

(16) Miller, J.N. 1982. Developments in non-isotopic immunoassay. Nature **295**: (5847) xxiii.

(17) Morimoto, K.; Iijima, S.; and Koizumi, A. 1982. Selenite prevents the induction of sister-chromatid exchanges by methyl mercury and mercuric chloride in human whole-blood cultures. Mut. Res. **102**: 183–192.

(18) Önfelt, A., and Jensen, D. 1982. Enhanced mutagenic response of MNU by post-treatment with methylmercury, caffeine or thymidine in V79 Chinese hamster cells. Mut. Res. **106**: 297–303.

(19) Perkampus, H.H. 1982. Photo-Akustik-Spectroskopie an festen Stoffen. Naturwissenschaften **69**: 162–172.

(20) Prasad, K.N. 1982. Tissue culture model to study the mechanism of the effect of heavy metals on nerve tissue. **In** Mechanical Actions of Neurotoxic Substances, eds. K.N. Prasad and A. Vernadakis, pp. 67–94. New York: Raven Press.

(21) Purchase, I.F.H. 1982. An appraisal of predictive tests for carcinogenicity. Mut. Res. **99**: 53–71.

(22) Rivedal, E., and Sanner, T. 1981. Metal salts as promoters of in vitro morphological transformation of hamster embryo cells initiated by benzo(a)pyrene. Cancer Res. **41**: 2950–2953.

(23) Rossman, T.G. 1981. Enhancement of UV-mutagenesis by low concentrations of arsenite in E. coli. Mut. Res. **91**: 207–211.

(24) Shulman, R.C. 1983. NMR spectroscopy of living cells. Sci. Am. **248**: 76–83.

(25) Sirover, M.A. 1981. Effects of metals in in vitro bioassays. Env. Health Persp. **40**: 163–172.

(26) Sivak, A. 1982. An evaluation of assay procedures for detection of tumor promoters. Mut. Res. **98**: 377–387.

(27) Smith, J.R.H.; Athwal, S.S.; Chettle, D.R.; and Scott, M.C. 1982. On the in vivo measurement of mercury using neutron capture and X-ray fluorescence. Int. J. App. Rad. Isot. **33**: 557–561.

Changing Metal Cycles and Human Health, ed. J.O. Nriagu, pp. 237–250.
Dahlem Konferenzen 1984. Berlin, Heidelberg, New York, Tokyo: Springer-Verlag.

BIOCHEMICAL INTERACTIONS OF MERCURY, CADMIUM, AND LEAD

J.H.R. Kägi* and H.-J. Hapke**
*Biochemisches Institut der Universität Zürich
8057 Zürich, Switzerland
**Tierärztliche Hochschule Hannover
3000 Hannover 1, F.R. Germany

Abstract. The chemical basis of the interaction of mercury, cadmium, and lead with cells and tissues is reviewed. Thermodynamic considerations suggest that the preferential targets of these metals in biomolecules are the thiol (mercapto) group and chelating structures containing this ligand. Especially sensitive metabolic targets are enzymes possessing a dithiol configuration at their active site. The preference of the mercapto ligand for polarizable metal ions also plays a crucial role in defense mechanisms that have evolved against these toxic metals. Thus, the ubiquitous tripeptide glutathione effectively protects many intracellular structures by intercepting intruding metal ions through chelation. Similarly, the cysteine-rich protein metallothionein allows for substantial intracellular metal sequestration and is responsible for the development of cellular tolerance towards certain toxic metal ions.

INTRODUCTION

Mercury, cadmium, and lead are poisonous to most forms of life. Both in acute and chronic exposure, these metals interact with many different cellular components, thereby interfering with the normal metabolic functions, causing cellular injuries and, in extreme cases, death of the organism. Thus, toxicological and biochemical research have led to the identification of a large number of potential or real target compounds whose normal activity is inhibited, enhanced, or modified by these agents (33). At the same time, it has also become clear that living organisms have at their disposition a variety of homeostatic mechanisms by which they effectively protect the cellular machinery from damage by the intruding ions. Thus, it is now established that the adverse impact of many metal ions entering a cell is attenuated by their (a) formation of complexes with the ubiquitous intracellular chelating peptide glutathione, (b) sequestration in the cytosolic protein metallothionein, and (c) deposition in water-insoluble intracellular organic and inorganic concretions, in some cases. In addition, there are a number of metabolic and transport mechanisms actively effecting the elimination of metal ions from cells and organisms. Hence, the biological consequences of exposure to toxic elements are to be viewed as a balance of biochemical interactions determined both by chemical variables, such as the nature and the concentrations of the reactants, and by the homeostatic response of the living cell to the metal. In this presentation, we will review some of the chemical and biochemical principles which are pertinent to an understanding of these processes.

SOME CHEMICAL CONSIDERATIONS

An important common chemical feature of the posttransitional elements mercury, cadmium, and lead is the stability of the bivalent state. Thus, unlike iron or copper, these three elements have no strong tendency to participate in redox processes. The reactions of their ions with biological systems are limited essentially to the formation of complexes with nucleophilic ligands of biomolecules. On the "softness" scale of Pearson, Cd(II) and Hg(II) qualify as typically soft acids and Pb(II) as an intermediate acid which have the common tendency of forming strong covalent complexes with soft bases, among them especially with sulfur electron donors. A list of the most prominent monodentate ligands occurring in cells and of the approximate values for the first stability constants of their complexes with Hg(II), Cd(II), and Pb(II) is given in Table 1. For ligands with high pK_a, the protons compete effectively with the metal for binding. It is therefore more meaningful to consider the apparent stability constants calculated for a given pH also included in Table 1 (26). Thus, comparison of the values computed for pH 7 indicates that under these conditions ligation of Cd(II) and Pb(II) to the amino group becomes negligible and that the apparent affinity for the thiolate ligand (RS^-) too is lowered by more than two orders of magnitude. Nonetheless, this strong nucleophile remains by far the most avid monodentate ligand available for binding of the metal ions considered. An additional factor to be taken into account in assessing the effective stability of particular metal complexes under in vivo conditions is the presence or absence of competing ligands. This is of special significance for Hg(II) which forms very strong complexes with the ubiquitously occurring chloride ions (Table 1). Thus, at the concentrations of chloride encountered in animal

Ligand[a]		Approximate stability constant ($\log K_1$)[b]			Apparent approximate stability constant at pH 7 ($\log K_{1,pH\ 7}$)[c]		
		Hg	Cd	Pb	Hg	Cd	Pb
Carboxyl group	(4.6)	5.6[d]	1.8[e]	1.9[e]	5.6	1.8	1.9
Amino group	(9.5)	8.5[f]	2.8[f]	2.0[g]	6.0	0.3	-0.5
Imidazole group	(6.5)	3.8[h]	2.8[i]	<2.3[i]	3.7	2.7	<2.2
Thiol group	(9.4)	12.6[k]	8.0[l]	7.3[m]	10.2	5.6	4.9
Phosphate group	(6.6)	-	2.8[n]	3.2[n]	-	2.7	3.1
Chloride		7.3	2.0	1.6	7.3	2.0	1.6

[a]Figure in parentheses is approximate pK_a value of ligand in proteins.

[b]Unless indicated otherwise, representative values were selected from (27, 28); K_1 is the equilibrium constant (= stability constant) of the reaction $M + L \rightleftarrows ML$, where ML is the 1:1 complex of a metal ion (M) and a ligand (L); $K_1 = \frac{[ML]}{[M][L]}$.

[c]Calculated using the expression $\log K_{1,pH\ 7} = \log K_1 - \log (1 + 10^{-7}/K_a)$, where K_a is the acid dissociation constant of the ligand (26).

[d]$\log K_1$ of acetate.

[e]Measured in gelatin.

[f]$\log K_1$ of ethanolamine.

[g]$\log K_1$ of ammonia quoted in (11).

[h]$\log K_1$ of imidazole.

[i]Measured in bovine serum albumin.

[k]Estimated from the formation of mercaptalbumin, assuming $K_1 = K_2$ (11).

[l]Quoted in (11).

[m]Estimated from cumulative stability constant β_2 of complex with 2-mercaptoethanol assuming $\log K_1 = 0.5 \log \beta_2$.

[n]$0.5 \log K_1$ of diphosphate.

TABLE 1 – Approximate stability constants of 1:1 complexes of mercury, cadmium, and lead with monodentate ligands of biomolecules.

tissues ($>10^{-2}$ M), the apparent stability constants of Hg(II)-complexes with other monodentate ligands are reduced by at least five orders of magnitude, thereby becoming more comparable to those of the corresponding Cd(II)- and Pb(II)-complexes.

Metal complexes of much higher stability are formed, of course, with pluridentate ligands. A number of chelating compounds of biological importance together with the stability constants for their 1:1 complexes with Hg(II), Cd(II), and Pb(II) are listed in Table 2. To allow meaningful comparisons, the apparent stability constants of the chelates computed for pH 7 are also included. Again, the strongest complexes are those with ligands containing a mercapto group, among them the tripeptide glutathione.

Metal complexes of still higher stability are provided by the specific metal-binding sites in metal-chelating proteins. As documented by a number of cristallographically defined systems, such sites are built up of three to six oxygen, nitrogen, or sulfur ligands containing amino acid side chains forming cage-like coordination spheres of defined geometry (18). The magnitude of the

Ligand	Stability constant[a] ($\log K_1$)			Apparent stability constant at pH 7.0[b] ($\log K_{1,pH\ 7}$)		
	Hg	Cd	Pb	Hg	Cd	Pb
Glycine	10.3	4.8	5.5	7.6	2.1	2.8
Histidine	7.9	5.7	6.8	5.4	3.2	4.4
Mercaptoethylamine	-	11.0	11.1	-	6.0	6.2
Cysteine	14.2	-	11.4	9.9	-	7.1
Penicillamine	16.2	10.9	12.4	11.8	6.5	8.0
Glutathione	-	10.5	10.6	-	6.1	6.2
Citrate	10.9	4.2	3.0	10.9	4.2	3.0
Guanosine	-	4.0	3.5	-	1.8	1.3
Diphosphate	-	5.6	6.4	-	3.8	4.6

[a] Representative values for aqueous solutions of low ionic strength selected from (27,28).

[b] Calculated using the expression $\log K_{1,pH\ 7} =$

$$\log K_1 - \log\left(1 + 10^{-7}\left(\frac{1}{K_a^A} + \frac{1}{K_a^B}\right) + 10^{-14}\frac{1}{K_a^A \cdot K_a^B}\right),$$

where K_a^A and K_a^B are the acid dissociation constants of the liganding groups taken from (27,28).

TABLE 2 — Stability constants of 1:1 chelates of mercury, cadmium, and lead with ligands of biochemical importance.

stability constants of these complexes depends on the chemical nature of the metal ion and the ligands and the number of the latter. Thus, for the zinc enzymes carboxypeptidase A and carbonic anhydrase, where the metal is bound to three nonsulfur protein ligands (18), the apparent stability constants for post-transitional metals range at neutral pH in the order of 10^{10} M^{-1} and above (34). For the metal thiolate protein metallothionein in which each metal ion is coordinated to four cysteine residues, the apparent stability constants at pH 7 are as much as 10^{16} M^{-1} and 10^{12} M^{-1} for the Cd(II)- and Zn(II)-complexes, respectively (35). These figures differ by more than 10 and 7 orders of magnitude from the corresponding apparent stability constants of about $4 \cdot 10^5$ M^{-1} and $4 \cdot 10^4$ M^{-1} calculated for the monodentate complexes of mercapto ligands with Cd(II) and Zn(II), respectively (27, 28). Besides the effect on the stability of the complexes, these figures also illustrate the potentiation of metal ion selectivity by chelation. Thus, compared to the only 10-fold difference in the binding constants of 1:1 complexes of monodentate mercaptans with Cd(II) and Zn(II) (vide supra), the metal-binding sites in metallothionein have a 10,000-fold higher affinity for Cd(II) than for Zn(II) (35). This effect is responsible for the preferential accumulation of Cd(II) in metallothionein which is peculiar to this protein (vide infra).

BIOCHEMICAL TARGETS

Because of their tendency to form strong covalent bonds with many biomolecules, Hg(II), Cd(II), and Pb(II) are capable of interacting quite unspecifically with many biochemical structures. Hence, the list of potential targets is large, and there are numerous reports in the biochemical literature on their effectiveness in causing impairment or loss of normal biochemical function. However, in many cases the effects were observed under in vitro conditions using preparations of isolated proteins, nucleic acids, and subcellular particles, and hence, it remains important to assess their in vivo significance, if any (36). The in vivo targets are determined both by the concentrations of the various ligands in the tissues and by the relative magnitude of the stability constants of their complexes with the metal ions under consideration. Thus, in the presence of powerful monodentate ligands such as the mercapto group or of pluridentate ligands, many well documented but less stable complexes of Hg(II), Cd(II), and Pb(II) with amino acids, nucleosides, and nucleic acids or single amino acid side chains other than cysteine in proteins (1) are not formed to any significant degree. Hence, on the basis of the figures given in Table 1, it is obvious that even at the highest concentrations at which toxic metal ions might be expected to occur in cells ($< 10^{-4}$ M), the mercapto group is practically the only monodentate ligand capable of competing successfully for the metals considered.

The universally occurring and most abundant single mercaptan available for complexing Hg(II), Cd(II), or Pb(II) within cells is reduced glutathione. In most tissues the concentration of this peptide ranges between 10^{-3} M and 10^{-2} M (15), sufficient to combine with all posttransitional metal ions likely to enter the cells. Many enzymes containing single -SH groups are also sensitive to these metals but their individual cellular concentrations are much lower than that of glutathione.

More sensitive potential targets for the action of the toxic metals under consideration are enzymes and proteins which contain several mercapto ligands close enough to form chelate structures with the metal. Candidates include the flavin-containing enzymes lipoamide dehydrogenase, glutathione reductase, and thioredoxin reductase which in their fully reduced form contain at their active center vicinal cysteine residues which are known to bind not only Hg(II), Cd(II), and Pb(II) but also arsenite, a known reagent for vicinal thiols (37). It is conceivable, but not proven, that the strong inhibition of respiration in certain cells by Cd(II) and Pb(II) is brought about by their interaction with the vicinal thiol of the lipoamide dehydrogenase component of the pyruvate and α-ketoglutarate dehydrogenase complexes (30, 33). That the same enzyme complexes could be inhibited by binding of the metal to the two juxtaposed mercapto groups of the fully reduced lipoic acid cofactor of the transacetylase and transsuccinylase component of these complexes, respectively, has likewise been suggested (36). Complex formation with vicinal thiols is also postulated to explain the strong inhibition of oxidative phosphorylation observed in isolated mitochondria by Cd(II) and arsenite (14). These effects also manifest themselves in vivo. Thus, acute poisoning of rats with Hg(II), Cd(II), or Pb(II) leads to partial uncoupling of oxidative phosphorylation and decreased respiratory control in the mitochondria isolated from kidney and liver cells of such animals (36).

Other postulated targets for the action of these metals are the multidentate metal-binding sites of metalloenzymes, especially of many zinc enzymes, some of which consist of nitrogen and oxygen ligands. In vitro, the intrinsic Zn(II) of liver alcohol dehydrogenase, carboxypeptidase A and B, alkaline phophatase of E. coli, and carbonic anhydrase (4, 34) can be replaced by a variety of metal ions resulting in altered or abolished enzyme activity. However, because of the strong binding of zinc in these enyzmes, these exchange reactions actually require rather drastic measures such as exposure to low pH or to high concentrations of chelating agents. It remains, therefore, doubtful whether metalloenzymes with strong metal-protein bonds will undergo such an exchange in vivo.

Substitution of metal ions is much more likely in the so-called metal-enzyme complexes where by definition the metal-protein bond is weaker than that in metalloenzymes (34). An important example for an enzyme in which metal poisoning manifests itself by such an effect in vivo is δ-aminolevulinic acid dehydratase, a key enzyme in porphyrin biosynthesis and known to be inhibited quite specifically both in vitro and in vivo by Pb(II). In a recent study on the octameric bovine liver enzyme it was shown that the effect of Pb(II) is brought about by stoichiometric replacement of its intrinsic Zn(II) by Pb(II) (32). The inhibition can be prevented and reversed by Zn(II) both in vitro and in vivo (12). The specific effect of Pb(II) has been attributed tentatively to the abolition of one of the two substrate-binding sites at the active centers of the enzyme (32).

It is important to recognize, however, that not all chemical interactions of the intruding metal ions with cellular components affect the metabolic functions. Indeed, a substantial proportion of the biochemical targets of heavy metals cannot be implicated in any adverse effects but functions as an important part of cellular defense mechanisms. Thus, the very large concentration of reduced glutathione in most living cells (vide supra) serves, among many other important functions, as an efficient buffer for toxic metal ions entering the system. This action explains why many enzymes known to be strongly inhibited in vitro even by traces of Hg(II), Cd(II), or Pb(II), under in vivo conditions often tolerate much larger concentrations of these metals. The most clearly defined and extensively studied cellular defense mechanism against many metal ions is, however, their sequestration by the multidentate polypeptide metallothionein.

METAL SEQUESTRATION BY METALLOTHIONEIN

Considering the high reactivity and toxicity of most heavy metal ions, it is surprising that some of these metals accumulate very substantially in certain tissues without causing adverse effects. Thus, the kidney of healthy human adults is reported to contain median concentrations per kg wet weight on the order of 1 mg mercury (cited in (24)), 1 mg lead (25), and 20 mg cadmium (7). For cadmium, the accumulation is accounted for by the abundance of metallothionein, a ubiquitously occurring cysteine-rich intracellular protein which also binds the essential metals zinc and copper and which was initially discovered as a cadmium-containing protein in equine and human kidney cortex (17, 19). It is the only biological compound thus far known to contain this metal in vivo, and it is responsible for its steady accumulation in the human body throughout the first half of life (7). This low-molecular weight metal- and sulfur-rich protein is now known to occur in various forms and in varying amounts in many different species and tissues throughout the animal kingdom and in some microorganisms (21). The mammalian metallothioneins are characterized by the largest cysteine content (twenty residues per chain weight of 6000) of any known protein and by a highly unique arrangement of these residues in the sequence (Fig. 1). The unusual polypeptide sequence provides for binding of the metal ions in two oligonuclear metal-thiolate clusters containing three and four metal ions, respectively (22). The clusters are thought to be built up of tetrahedral metal-thiolate units forming incomplete adamantane-type ($Me(II)_4S_6$) cage structures (35). They resemble in their steric organization the inorganic zinc blende (ZnS) lattice and are capable of accommodating in vivo not only the group 2B metal ions [Zn(II), Cd(II), and Hg(II)], but also a number of other d^{10} metal ions, among them Cu(I) and, when available, Ag(I), Au(I), Sn(IV), and Bi(III) (21, 31). Pb(II) binds to the apoprotein under in vitro conditions but does not accumulate in vivo in metallothionein.

Of paramount significance to the understanding of the biological role of metallothionein in metal metabolism is the fact that it not only binds metals but that in analogy to the formation of immunoglobulins in response to an antigen its biosynthesis can be induced by many metal ions. Thus, in a variety of organisms and of cell lines, exposure to Cd(II), Zn(II), Cu(II), Hg(II), Ag(I), Au(I),

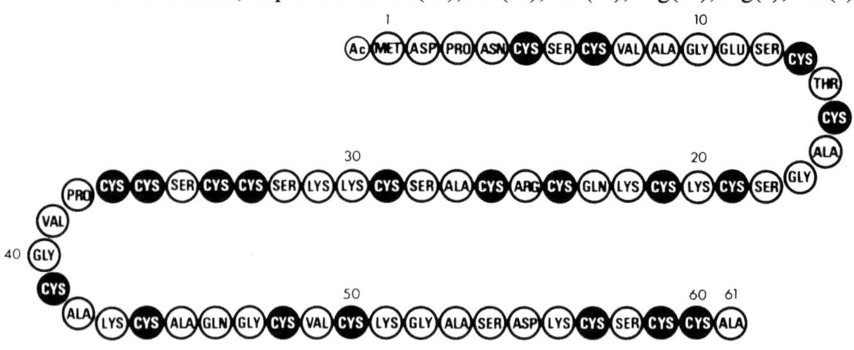

FIG. 1 – Amino acid sequence of human hepatic metallothionein – 2 (adapted from (17)).

Bi(III), Sn(IV), Pb(II), Mn(II), Co(II), Ni(II), Cr(III), and In(III) was found to enhance rapidly and massively the formation of metallothionein (21, 31). The effect involves gene activation with increased synthesis of metallothionein-specific mRNA (6). The induction is most pronounced by Cd(II) and Zn(II), but there are large tissue-dependent differences for the various metals, the biological significance of which remains to be determined. There is also no obligatory link between the ability of a metal ion to induce biosynthesis and its ability to be bound in vivo by the newly formed protein (3, 6). Thus, Mn(II), Cr(III), In(III), and Pb(II), while eliciting increased formation of Zn(II)-metallothionein, were not incorporated into this protein (31).

There can be no doubt that metallothionein plays an important role in short-term metal detoxification and in altering cellular tolerance towards metals (21). Thus, the induction of metallothionein synthesis explains the well-known effect of small amounts of cadmium in preventing, in experimental animals, some of the effects of subsequently given large doses of cadmium or mercury (20). The direct involvement of metallothionein in mitigating the effects of cadmium is also documented by the recent observation that in some cell lines tolerance towards Cd(II) can be acquired by an up to 10-fold amplification of the gene coding for metallothionein (13).

The beneficial role of metallothionein in providing short-term tolerance against certain metal ions by intracellular sequestration is turned into a disadvantage, however, when at high cadmium loads excessively produced metallothionein is released from the cell of its origin into the bloodstream and is taken up by tubular cells in the kidneys where it gives rise to the tubular damage reminiscent of chronic forms of cadmium poisoning in experimental animals and man (21). The basis of the toxicity of Cd-metallothionein in the renal tubule lining cells is not entirely clear. It has been suggested that the protein is itself a nephrotoxin or that it is taken up in the lysosomes and that the metal is released into the cytoplasm of the cells where sequestration by newly induced renal metallothionein may not keep pace with the cytotoxic effects of the released Cd(II) (3). The latter mechanism has been postulated also to account for the nephrotoxic manifestations in copper-poisoned sheep (2).

ROLE OF METALLOTHIONEIN IN INTRACELLULAR METAL DEPOSITS

Although the polypeptide chain of water-soluble cytosolic metallothionein has a short half-life of only a few days, its steady biosynthesis renders this protein a very efficient dynamic trap that provides for a long biological half-life of cadmium (21). In addition, under certain circumstances, metallothionein may also be deposited in intracellular organelles. Thus, in the lysosomes of the liver of human fetuses and newborn calves (21), copper was found to be associated with a particulate high-molecular weight protein now thought to be a polymerized form of metallothionein. Similar deposits were recently shown to occur in the livers of dogs suffering from inherited copper toxicosis, a condition resembling Wilson's disease in man (16). The physiological significance of this deposition, if any, is not clear. However, metallothionein-like proteins have been suggested to play a part in the formation of membrane-bounded metal deposits in mussels (8). These and the inorganic concretions of metals observed in molluscan kidney are likely components of excretory metal pathways for toxic metals in these invertebrates (9). Somewhat reminiscent of these depositions is the formation of nuclear inclusion bodies observed in parenchymal cells of the liver in lead poisoning and in the tubular lining cells of the mammalian kidney both in lead and in bismuth poisoning (10). The inclusion bodies have been reported to contain a low molecular weight acidic protein which has complexed the metal. Since bismuth also induces the biosynthesis of and binds to a metallothionein-like protein (23), its possible involvement in this form of metal deposition deserves further exploration.

Besides the general mechanisms of protection discussed, there are a number of ways by which the cells free themselves of toxic ions. Among the best studied systems are the metal-resistant forms of Staphylococcus aureus which are plasmid-determined and which involve quite specific mechanisms for the various elements (29). The resistance to Hg(II) includes inducible metabolic transformation of inorganic mercury salts and of organomercurials to volatile metallic mercury. The cadmium resistance is brought about by reducing the net uptake of Cd(II) by the resistant cell. The effect is due to a plasmid-encoded Cd(II)-efflux system that actively pumps Cd(II) from the bacterial cell. It is of interest that cadmium-resistance caused by limiting the uptake of Cd(II) from the medium was also recently reported for mutants in a mammalian cell culture (5). The same cell line could also be induced to accelerate metallothionein synthesis, thus attesting to the multiplicity of factors bringing about metal resistance, a prime requirement for the survival of all forms of life.

Acknowledgement. This work was supported by Swiss National Science Foundation Grant 3.207-0.82.

REFERENCES

(1) Blundell, T.L., and Jenkins, J.A. 1977. The binding of heavy metals to proteins. Chem. Soc. Rev. **6**: 139–171.

(2) Bremner, I., and Young, B.W. 1977. Copper thionein in the kidneys of copper-poisoned sheep. Chem. Biol. Interact. **19**: 13–23.

(3) Cherian, M.G., and Goyer, R.A. 1978. Metallothioneins and their role in the metabolism and toxicity of metals. Life Sci. **23**: 1–10.

(4) Coleman, J.E. 1971. Metal ions in enzymatic catalysis. **In** Progress in Bioorganic Chemistry, eds. E.T. Kaiser and F.J. Kézdy, vol. 1, pp. 159–344. New York: Wiley-Interscience.

(5) Corrigan, A.J., and Huang, P.C. 1981. Cellular uptake of cadmium and zinc. Biol. Trace Element Res. **3**: 197–216.

(6) Durnam, D.M., and Palmiter, R.D. 1981. Transcriptional regulation of the mouse metallothionein-I gene by heavy metals. J. Biol. Chem. **256**: 5712–5716.

(7) Friberg, L.; Piscator, M.; Nordberg, G.F.; and Kjellström, T. 1974. Cadmium in the Environment, 2nd ed. Cleveland: CRC Press.

(8) George, S.G.; Carpene, E.; Coombs, T.L.; Overnell, J.; and Youngson, A. 1979. Characterisation of cadmium-binding proteins from mussels, Mytilus edulis (L), exposed to cadmium. Biochim. Biophys. Acta **580**: 225–233.

(9) George, S.G., and Pirie, B.J.S. 1979. The occurrence of cadmium in sub-cellular particles in the kidney of the marine mussel, Mytilus edulis, exposed to cadmium. Biochim. Biophys. Acta **580**: 234–244.

(10) Goyer, R.A., and Cherian, M.G. 1977. Tissue and cellular toxicology of metals. **In** Clinical Chemistry and Chemical Toxicology of Metals, ed. S.S. Brown, pp. 89–103. Amsterdam: Elsevier/North-Holland Biomedical Press.

(11) Gurd, F.R.N., and Wilcox, P.E. 1956. Complex formation between metallic cations and proteins, peptides, and amino acids. Adv. Protein Chem. **11**: 311–427.

(12) Haeger-Aronsen, B.; Schütz, A.; and Abdulla, M. 1976. Antagonistic effect in vivo of zinc on inhibition of δ-aminolevulinic acid dehydratase by lead. Arch. Env. Health **31**: 215–220.

(13) Hildebrand, C.D.; Griffith, J.K.; Tobey, R.A.; Walters, R.A.; and Enger, M.D. 1982. Molecular mechanisms of cadmium detoxification in cadmium-resistant cultured cells: role of metallothionein and other inducible factors. **In** Biological Roles of Metallothionein, ed. E.C. Foulkes, pp. 279–303. Amsterdam: Elsevier North Holland, Inc.

(14) Jacobs, E.E.; Jacob, M.; Sanadi, D.R.; and Bradley, L.B. 1956. Uncoupling of oxidative phosphorylation by cadmium ion. J. Biol. Chem. **223**: 147–156.

(15) Jocelyn, P.C. 1972. Biochemistry of the SH Group. London: Academic Press.

(16) Johnson, G.F.; Morell, A.G.; Stockert, R.J.; and Sternlieb, I. 1981. Hepatic lysosomal copper protein in dogs with an inherited copper toxicosis. Hepatology **1**: 243–248.

(17) Kissling, M.M., and Kägi, J.H.R. 1977. Primary structure of human hepatic metallothionein. FEBS Lett. **82**: 247–250.

(18) Liljas, A., and Rossmann, M.G. 1974. X-ray studies of protein interactions. Ann. Rev. Biochem. **43**: 475–507.

(19) Margoshes, M., and Vallee, B.L. 1957. A cadmium protein from equine kidney cortex. J. Am. Chem. Soc. **79**: 4813–4814.

(20) Nordberg, G.F. 1971. Effects of acute and chronic cadmium exposure on the testicles of mice. With special reference to protective effects of metallothionein. Env. Physiol. (Env. Physiol. Biochem.) **1**: 171–187.

(21) Nordberg, M., and Kojima, Y. 1979. Metallothionein and other low molecular weight metal-binding proteins. **In** Metallothionein, eds. J.H.R. Kägi and M. Nordberg, pp. 41–124. Experientia Suppl. 34. Basel: Birkhäuser Publishing Company.

(22) Otvos, J.D., and Armitage, I.M. 1980. Structure of the metal clusters in rabbit liver metallothionein. Proc. Natl. Acad. Sci. USA **77**: 7094–7098.

(23) Piotrowski, J.K.; Szymanska, J.A.; Mogilnicka, E.M.; and Zelazowski, A.J. 1979. Renal metal binding proteins. **In** Metallothionein, eds. J.H.R. Kägi and M. Nordberg, pp. 363–371. Experientia Suppl. 34. Basel: Birkhäuser Publishing Company.

(24) Schiele, R.; Freitag, E.M.; Schaller, K.H.; Schellmann, B; and Weltle, D. 1981. Untersuchung zur normalen Quecksilberkonzentration menschlicher Organe. Zbl. Bakt. Hyg. I. Abt. Orig. **B 173**: 45–62.

(25) Schroeder, H.E., and Tipton, I.H. 1968. The human body burden of lead. Arch. Env. Health **17**: 965–978.

(26) Sigel, H., and McCormick, D.B. 1970. On the discriminating behavior of metal ions and ligands with regard to their biological significance. Acc. Chem. Res. **3**: 201–208.

(27) Sillén, L.G., and Martell, A.E. 1964. Stability Constants of Metal-Ion Complexes. Special Publication no. 17. London: The Chemical Society.

(28) Sillén, L.G., and Martell, A.E. 1971. Stability Constants of Metal-Ion Complexes. Supplement no. 1. Special Publication no. 25. London: The Chemical Society.

(29) Silver, S. 1981. Mechanisms of plasmid-determined resistance to mercury, cadmium and arsenic in Staphylococcus aureus. **In** Staphylococci and Staphylococcal Infection, ed. J. Jeljaszewics. Zbl. Bakt. Suppl. 10, pp. 1–11. Stuttgart: Gustav Fischer Verlag.

(30) Stein, A.M., and Stein, J.H. 1971. Isolation of the cadmium derivative of lipoyl dehydrogenase. J. Biol. Chem. **246**: 670–676.

(31) Suzuki, Y., and Yoshikawa, H. 1976. Induction of hepatic zinc-binding proteins of rats by various metals. Ind. Health **14**: 25-31.

(32) Tsukamoto, I.; Yoshinaga, T.; and Sano, S. 1980. Zinc and cysteine residues in the active site of bovine liver δ-aminolevulinic acid dehydratase. Int. J. Biochem. **12**: 751–756.

(33) Vallee, B.L., and Ulmer, D.D. 1972. Biochemical effects of mercury, cadmium, and lead. Ann. Rev. Biochem. **41**: 91–128.

(34) Vallee, B.L., and Wacker, W.E.C. 1970. Metalloproteins. **In** The Proteins, ed. H. Neurath, 2nd ed., vol. V. New York: Academic Press.

(35) Vasak, M., and Kägi, J.H.R. 1983. Spectroscopic properties of metallothionein. **In** Metal Ions in Biological Systems, ed. H. Sigel, vol. 15, pp. 213-273. New York: Marcel Dekker, Inc.

(36) Webb, M. 1977. Metabolic targets of metal toxicity. **In** Clinical Chemistry and Chemical Toxicology of Metals, ed. S.S. Brown, pp. 51–64. Amsterdam: Elsevier/North-Holland Biomedical Press.

(37) Williams, C.H. 1976. Flavin-containing dehydrogenases. **In** The Enzymes, ed. P.D. Boyer, vol. 13, pp. 89-173. New York: Academic Press.

Changing Metal Cycles and Human Health, ed. J.O. Nriagu, pp. 251–263.
Dahlem Konferenzen 1984. Berlin, Heidelberg, New York, Tokyo: Springer-Verlag.

STRUCTURAL ASPECTS OF METAL TOXICITY

R.J.P. Williams
Inorganic Chemistry Laboratory, University of Oxford
Oxford OX1 3QR, England

Abstract. The binding and location of metal ions in biological systems is described. The underlying theory indicates that all elements can be toxic. Toxicity is a dose-related concept. The wrong distribution of seemingly harmless elements can be as harmful as that of the obviously poisonous ones.

INTRODUCTION

The changes in cycling of metals in the environment due to man can only be understood on the basis of three pieces of information: (a) an understanding of the cycling and changes in that cycling which occur without man's intervention; (b) the input of metals and their compounds which bind the metals to the environment due to man; (c) the removal of the newly input metals and binding compounds by the environment or man.

NATURAL CYCLING

In itself the cycling of elements within the context of geological and biological systems is not yet understood. The overall processes are irreversible movements of elements from huge stores often in the interior of the earth to the biosphere at the earth's surface. During the course of this movement the elements may go through many chemical forms without the activity of life. The role of life is then both an assistance to and a holding up of the movement of the elements in the overall reaction scheme:

$$\left.\begin{array}{c}\text{Source} \rightarrow \text{Intermediates} \\ \downarrow\uparrow \\ \text{Cycle}\end{array}\right\} \text{Deposit (Sink).} \qquad (1)$$

Different elements have very different progressions through these stages, and it is important first to look at the reaction scheme over geological time and in different regions of the earth. Initially there was no biological component of these movements of elements, but later, living systems became involved. Especially Lovelock (2) has stressed the increasing role of biological systems. In the Gaia Hypothesis (2) it is asserted that the influence of life has been such that the steady state levels of many of the elements in the air, in the sea, in ground water, and it may be true of the top levels of soils, have been fixed by living systems for some long period. This means that in Eq. 1 the steady state levels now observed are largely due to life. Given that life's activity is on this scale, it is obvious that many of the surface deposits (sinks) in Eq. 1 are also the product of biological activity. Such considerations add greatly to the complexity of our problem, for we must know the nature of the complicated biogeochemical steady states, i.e., the states of the intermediates and cycles in Eq. 1, if we are to appreciate the changes man is likely to effect. Biology has energized the distributions which inorganic chemistry alone would have produced. I have no space here to elaborate on Eq. 1, which is based largely on the low-temperature, low-pressure reactions which take place in the biosphere, but major considerations are the effective steady state temperature, redox potential, salt content, and pH. Changes in any of these will cause very considerable influences on the steady state levels of a large number of elements. Examples are apparent in differences between polar and tropical regions, between aerobic and anaerobic life, and between high and low pH zones on the earth. All this living activity stands on top of the equally huge geochemical fluxes due to weathering, volcanoes, and upsurges from the ocean floor. The whole activity is energy driven. It is immediately clear that man throws the whole of this flux into confusion as soon as his energy input into global chemistry is in any way comparable to that of these other activities. This is particularly true since he uses high temperature chemistry. Some element fluxes are more sensitive than others, of course, but all are affected.

As I understand my task at this meeting, it is to consider the influence of man on metal cycles, since we may view other activities as inevitable and occurring at a fixed flux on a short time scale of 10^4 years. Man's influence has arisen within the last two hundred years, since only since then has he changed the nature of the sources in Eq. 1 in a significant way as well as the other terms. This is largely due to the use of a large energy input to sources. The scale of his activities within the context of reaction scheme 1 is such that we must look at each element in turn, describing changes in sources, intermediates, cycling (hold-up), and deposition. My specific task within this frame is to look at the chemical structure of intermediates in Eq. 1 in relation to the toxicity of the elements in biological cycles.

THE BASIS OF TOXICITY

We can state that the toxic action of metals follows *chemical combination* after extraction from known sources. We know that metal ions are Lewis acids. We therefore need to understand the impact of man upon the availability of Lewis acids and their combined centers, Lewis bases, and the related bindings of all such pairs of interactants. Since acid/base reactions are usually in competition, our problem can be pictured first in terms of the multidimensional equilibria

$$M_1L_1 + M_2L_2 \rightleftarrows M_2L_1 + M_1L_2 \qquad (2)$$

for all M and L, and we shall also need to understand the rates of these reactions. Note that some of the reactions are required for life, while some cause only toxic effects, but the balances in Eq. 2 make it possible for all elements to be toxic. Even with the introduction of computers, such an analysis is an impossible task in detail, and we are forced to consider partial model scenarios. Our first concern is an overall description of the biological systems, since they generate the main Lewis bases with which we are concerned.

BIOLOGICAL SYSTEMS AND THEIR STRUCTURES

Elsewhere (5) I have shown that metal ions are partitioned in biology by (a) chemical partners and (b) their kinetics, (c) spatial segregation of M and ML, and (d) timed uptake and release of M and ML (6). (Immediately we see that toxicity will be dependent on chemical, locational, and temporal factors.) Let me define in very gross outline these partitionings.

Chemical Partitioning of Metals: Thermodynamics

In large part, the metal elements are divided through chemical binding by the different biological bases to which they become attached. The division is related to the separation in minerals in the earth and to the old-fashioned Qualitative Analysis Tables for young students, Table 1. The biological bases are usually proteins, and the first class of elements binds best to sulfur rich bases, the second to nitrogen/oxygen bases, the third to oxygen bases alone, and the fourth to water only.

	Sulfur-seeking	Nitrogen-seeking	Oxygen-seeking	Water-seeking
M(I)	H^+, Cu(I)→	←The Proton→	H^+	Na, K, Li
M(II)	Hg, Pb, Cd, Zn		Mg, Ca, Be	
M(III)	Tl, In, Ga		Al	
M(IV)	(Pb)		Sn, Ge, Si	
M(V)	Bi, Sb, As		P	(Oxy-anions of S, N, C, P)
T.M.(II)	Cu to Fe	Cu, Ni, Co, Fe, Mn		
T.M.(III)		← (Co, Fe)	Ln	

The charge on the base and the spin-state and valence of the metal affect many of the divisions of the elements in the center of the table. T.M. = transition metal.

TABLE 1 – Metal partitioning between bases.

The data can be rationalized to a large degree by using the ratio of ionization potential to reach a given ionic state of an element, M^{n+}, relative to the charge/radius ratio. This divides covalent from ionic acids, see Fig. 1. (The work of Williams and Hale (3,7) and Jensen (1) using this ratio has been ignored by biochemists and chemists in favor of more qualitative discussion of hard and soft Lewis acids and bases). In the figure we see that the elements which can compete with one another very effectively are clustered together. In Table 1, competition could be between elements in the same column and is so in chemical equilibria. In biological systems, the elements in columns one and two, except the proton, do not equilibrate, i.e., exchange rates are controlled (see below). The easy exchange of protons leads to important consequences.

The peculiar character of the proton as an acid needs stressing. Much of biological chemistry occurs at neutral pH when the proton is overwhelmingly bound by -O(phenol), N(amine), and -S(thiol) groupings, i.e., all classes in Table 1. Since free H^+ is 10^{-7} molar, these proton bound bases can all become very strong centers for metal-binding, i.e., protons are readily displaced at this pH. In fact, the displacement reaction (Eq. 2) in which we are interested often concerns the proton, and toxicity arises more from proton removal than from metal competition, e.g., Hg(II) reaction with -SH groups rather with Zn(II) centers. As an example, man's interference with pH has led to the release of *aluminum* and poisoning by this metal.

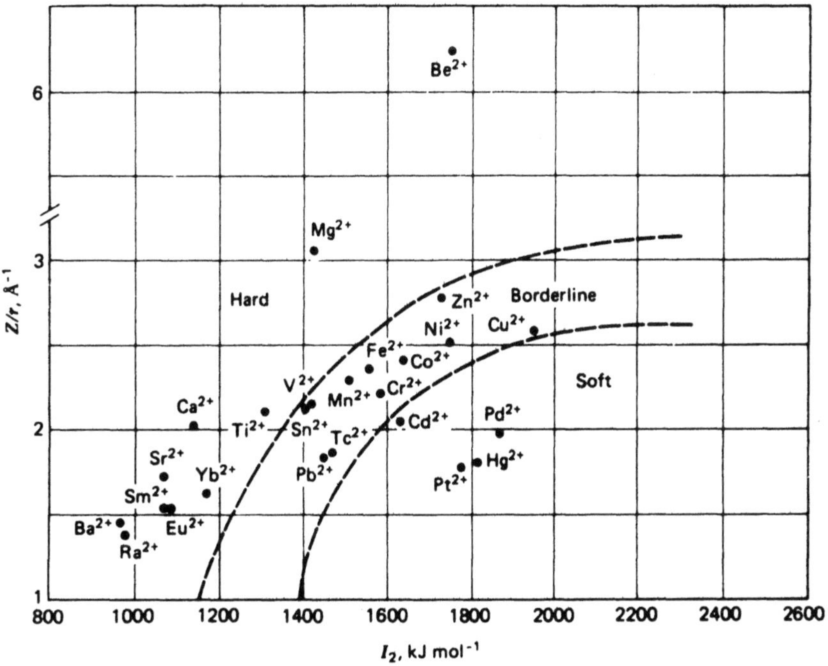

FIG. 1 — A plot of ionization potential, I_2, against charge/radius for divalent ions. Note that the different chemical classes of ions are well-defined (after Jensen (1)). A similar graph for monovalent ions gives Cu, Ag, and Tl as soft while the graph for trivalent ions gives Tl, In, As, Sb, and Bi as soft. The protein is borderline.

Low Coordination (<2)	High Coordination (>3)
H, Ag, Au, Hg, Pt, Pd, and many second and third row transition metal ions (Se, I, Br, P, F)	Ca, (Mg) Zn, Cd, Pb, Sn, Cu, Ni, Co, Fe, Mn, Al, Ga

TABLE 2 — Required coordination numbers for retention.

In order that metals such as those of the first transition series, Mn-Zn, should be strongly (irreversibly) retained by biological macromolecules, three or four bases are required to be deployed in "designed" centers. The proton requires only one base, excluding hydrogen bonding, and such ions as Ag(I), Pt(II),

Au(I), Hg(II), $CH_3Hg(I)$ amongst many other heavy metals, and, unlike the lighter metals, can be retained by just such centers alone. Table 2 makes this point. Chemical competition is not just base-selective but has a selectivity which arises from stereochemical considerations.

Kinetic Chemical Control

Now the divisions in Table 1 are only selective chemical equilibrium effects. It was not possible for biology to obtain the best combination of acids and bases if the rates of the reaction system were not controlled. In biology, the competition is therefore modulated by activation-free energy controls. Elsewhere we have looked at this control and have pointed to three major devices: (a) controlled intake of metals (see spatial control), (b) controlled synthesis of binding proteins, and (c) kinetically controlled transfer of metals and proteins (2). Equation 2 then becomes complicated through the rate of synthesis of proteins, or any other ligand, in a particular reaction vessel (cell compartment) and the rate of transfer of the metal. It is known that synthesis of proteins itself is frequently controlled by the level of metal chelates or metal-proteins, i.e., feedback control. The chemical partitioning is then biased in the sense that selectivity of acid/base partners is enhanced over expectation on simple chemical equilibrium grounds by the application of restrictions on a variety of kinetics. Typically in biology a site "designated" for metal M_1, where M_1 is in the first two columns of Table 1, rarely becomes contaminated with another metal M_2, and the site may be virtually irreversible, specific rather than selective. There are therefore zinc proteins, copper proteins, molybdenum proteins, and iron porphyrin, cobalt corrin, possibly a nickel corrin, and magnesium chlorin complexes, but effectively no cross-reactions. Thus, toxicity amongst these elements rarely arises through competition for a given L, in Eq. 2, between the readily available metals even under conditions of deprivation. (This is not true for metal ions which exchange readily, e.g., calcium.) Excess of a metal

Fast Exchange	Slow Exchange
Na, K	Cu, Zn, Ni
Mg (usually), Ca	Mo, Se
	Most non-metals (covalent)
Cl, F, O (ionic)	H (some)
H (some)	Fe and Mn (some)

N.B. The division is not the same as in inorganic chemistry due to folding of some proteins to make traps. Fe and Mn can be in intermediate exchange as well as slow exchange.

TABLE 3 – Elements in fast and slow exchange.

seemingly leads to binding to proteins which do not normally carry metals, or to the production of stores of metals. A danger is then that these storage devices can be deposited incorrectly. We conclude that deficiency of a required transition metal usually leads to restricted growth but not to the insertion of the wrong metal in its metalloprotein. Excess leads not so much to wrong insertion but to reaction at a new site or storage, generating toxicity. Examples are the deposition of precipitates of iron, and various other metals. Clearly, the situation is very different for the metals which exchange rapidly, e.g., Ca, Mg, Mn, and some Fe (see Table 3 and below).

We now turn to the toxicity factors related to the exchangeable metal ions, Table 3, i.e., those in columns three and four of Table 1. The levels of these elements are high and they do not suffer competition from the trace elements in columns one and two of Table 1. However, the competition between elements such as Be and Mg, or Ca and Al, or Li and Na, could be severe. Understanding of this competition and its effects is particularly difficult since the elements do not have fixed sites, are constantly diffusing from one compartment to another, and the rates of diffusion are time-dependent. All kinds of storage devices are used to assist in the buffering of the concentrations, e.g., blood, bone, and the sarcoplasmic reticulum. Thus, we need to describe compartments before we can handle the new toxicity problems.

The conclusion to this part of my discussion is that since the biological system is fully chemically interactive (Eq. 2), no element is free from the label toxic

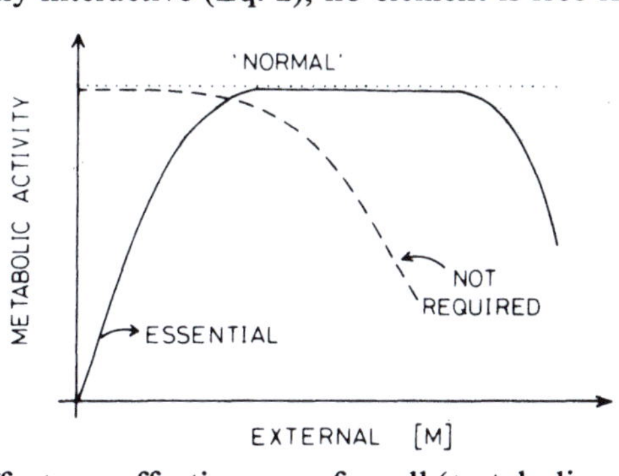

FIG. 2 — The effects on effectiveness of a cell (metabolic activity) of essential and not required elements showing considerable buffering over very different ranges of concentration followed by toxicity. Usual "toxic" elements (not shown) are described as those which produce adverse effects at very low levels of M.

at certain dose levels. However, biological feedback control through synthesis of ligands can look after reasonable variations in concentration of those elements with which it is familiar and which it uses (Fig. 2), although changes in oxygen tension (redox potential), pH, etc., change *all* steady states. (Individual biological systems will be sensitive to these changes in very different ways, and the micro-systems may well be able to evolve and adapt to them readily, but man cannot). The treatment of the equilibria in the sea by Whitfield (4) is a very interesting application of many of these considerations. Apart from control over interactions using chemistry, including synthesis and kinetics, in a single reaction vessel biological systems have two other devices: control through spatial separation using membranes and development in time.

Spatial Partition (Prokaryotes)

Even prokaryotic cells reject some elements as ions, e.g., Na^+ and Ca^{2+} while K^+ and Mg^{2+} are accumulated. The separation is possible due to the properties of membranes (7). Interference with these spatial movements are then forms of toxicity, and the use of Li^+ as a drug may arise in this way. Cs^+ and Tl^+ can mimic K^+ and can affect required movement and steady states in cells.

Other elements are combined naturally and almost totally with proteins and need to be considered in the classes (a) those bound irreversibly, and (b) those bound reversibly (Table 3). In the first group are Zn, Mo, Ni, Co, and much Fe and Mn. In the second group are Mg and Ca, and some Mn and Fe but bound in a different way from those in group (a). Thus some elements are only to be found in irreversibly bound and defined compartments in membranes or fluids, while others undergo intermediate rates of exchange and general redistribution, and yet others undergo rapid exchange everywhere (6). The functions of these three groups of metals have features in common but also show striking differences. All are required for full viability, i.e., they may all be termed essential elements. However, cells require supplies of zinc, molybdenum, and copper for growth, but no external source of them is needed for maintenance of steady states without growth since they are permanently bound. This is not true of elements which suffer losses during cycles due to transfer from compartment to compartment, here in-cell to out-of-cell. Furthermore, cells acquire knowledge of the environment through the rapidly exchangeable ions K^+, Na^+, Mg^{2+}, and Ca^{2+}, which are in almost instantaneous exchange, and these free ions are related to the steady state very quickly. These ions are therefore readily the source of toxicity problems. Finally, cells adapt their internal metabolism to the levels of free iron and manganese slowly, and these free elements are then involved in many steps of intermediate control.

Toxic activity against exchangeable Na, K, Mg, and Ca can be blocking of ion flow or ion binding. As their processing requires energy, the balance is exceedingly delicate. Toxic action against some irreversible iron (and manganese) centers is like that of zinc, but for others it can involve reversible though slow competition and then an effect on flow. Very different toxic effects are that against the production of the proteins required for metal action and that through binding to centers not usually involved with metal chemistry, discussed earlier.

Spatial Partitioning (Eukaryotes)

Turning to eukaryotes, the spatial partitioning is made much more complex in that even the immobilized elements, e.g., zinc or copper, can be put largely into internal organelles and reticula away from the cytoplasm by the movements of metallo-proteins whether or not ions are also moved (6). The complications increased in evolution as cells developed intracellular and cell-cell organization, since cell specialization provided spatially separated functional units which contain very different amounts of different elements. Each

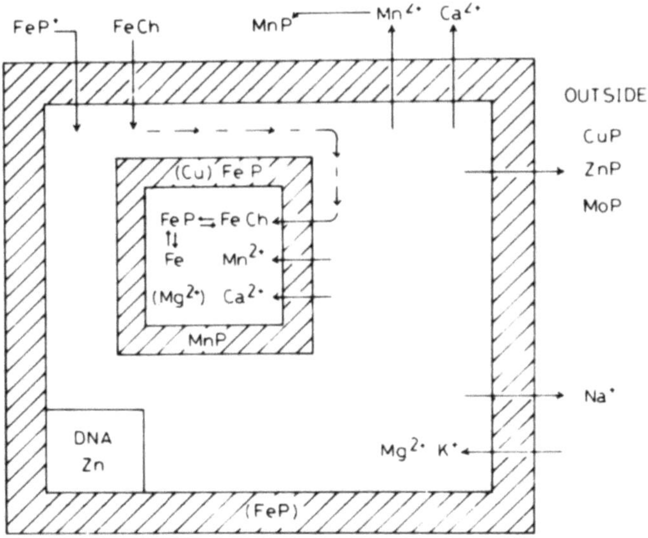

FIG. 3 — A schematic diagram showing element partition into different spatial compartments. Some elements are trapped with proteins, P; some are mobilized by small chelates, Ch; others move as ions. Toxic effects can be through competition for P or Ch, interference with movement, or constraints on synthesis of P or Ch.

organelle or specialized cell has now its own export of ions, and export and import of metallo-proteins (Fig. 3). Even so, the general rule holds that there are three classes of involvement when we consider toxicity (Table 4). However, the level of competition can be very different in different parts of space. For example, the circulating fluids will often carry the highest levels of a foreign or required element. We must then ask about the observed metal distribution by complicated organisms into various compartments. Elsewhere we have shown that the irreversibly bound elements can be distributed into any space through the transfer of their nonexchangeable proteins. Thus copper, molybdenum, and zinc proteins are often used in extracellular fluids, but dissociable iron or manganese proteins are not used in this way. The competition seemingly described by Eq. 2 is therefore quite deceptive, since all the concentration terms can be biased by energized movements of the components

$$\underset{\downarrow}{M_1L_1} + \underset{\downarrow}{M_2L_2} \underset{\text{slow}}{\overset{\text{slow}}{\rightleftarrows}} \underset{\downarrow}{M_2L_1} + \underset{\downarrow}{M_1L_2}$$

pumping of the elements and complexes to compartments.

Toxicity can arise through wrong pumping, i.e., the wrong movement of ML or M. Metals can perhaps use the *carriers* of other metals though they are not eventually incorporated into functional centers.

Toxicity can now arise in special places through the wrong movement of elements. Let us suppose that Pb(II) tends to follow Ca(II). It will then be put into various reticula and organelles but will not affect much the cytoplasm (Table 4). Apart from the fact that it could affect the properties of many storage devices or bind to proteins in the vesicular systems, it could interfere locally with protein or ligand synthesis or element transfer. Toxicity cannot be understood in terms of inorganic exchange equilibria but involves effects on a range of reaction and spatial transfers. The structure of a biological system is not static

	Cytoplasm and Cytoplasmic Membrane	Organelles and Reticula	Body Fluids or Exported
Pro	K, Mg Cu, Zn, Se Fe, Mn, Mo, Co		Na, Ca
Eu	K, Mg Cu, Zn, Se Fe, Mn, Mo, Co	Ca Mn, Fe	Ca, Na Certain Zn, Cu, Mo proteins

TABLE 4 – Cellular division.

but constantly evolving. It is then necessary to see toxicity in relation to the time domain as well as in relation to chemical / spatial equilibrations.

Temporal Partition

If the energy supply to a cell is adjusted at a given time, then immediately the levels of Ca, Mg, H, K, and Na change in different compartments. There is no change in Zn, Cu, Ni, and Mo, although the level of production of the proteins for their internal capture will change very slowly compared with the first process. Such a change can cause toxicity. In the case of Fe and Mn, the situation is more complicated. The production of their protein partners will change very slowly, too, but there will also be quite rapid changes in the free levels of the ions which affect major pathways of metabolism such as oxidation and glycosyl transfer. These two elements have a dual role, and cells have developed intricate mechanisms to recognize and adjust their ionic levels. These mechanisms include complicated reserves (stores), capture devices, and distribution or transport systems. High or low levels of free iron or manganese are toxic. Poisoning of various organelles by one or the other is well documented.

Some of the management schemes for these elements are shown in Eq. 3,

```
                              Feedback
                  (arrow from Use back to Store)
Input        →    Store   ⇄   Distribution   ⇄   Use   →   Waste,
Source
(Ferroamins)      (Ferritin)  (Transferrin
                               Citrate)                        (3)
```

which is another form of Eq. 1. Not only prokaryotes, but eukaryotes use this sophisticated control. It is also used by organelles. Disturbance leads to increased activity, e.g., new production of capture devices. Low levels stop cell division.

THE PERTURBATIONS

The perturbations in Eq. 1 are due to man's large-scale application of energy to new chemistry. In this sense *all* mining and agricultural operations are the source of the risk. Even phosphate and calcium fertilizers have toxic risk. Sewage concentrates impurities. The dangers of nitrogen fertilizers in the form of nitrate in water are obvious. Increase in carbon as CO_2 of the air and change of water content due to deforestation are toxic risks involving carbon, hydrogen, and oxygen cycles. Here we are looking only at metals, but they are greatly affected by these other cycles. Wrong levels of magnesium and iron, two of the most common elements, cause disease in some animals for certain. The problem lies in the fact that relatively delicate balances have been set up following Eq. 1 over very long periods of time, such as 10^6 years, and involving large numbers of elements. Man's activities challenge these balances not just directly, by feeding in extra toxic elements such as lead, cadmium, and mercury, but by altering the fluxes of many elements by changing ambient conditions of the climate, pH, the soil, and so on. If we remember that it is only approximately the top meter of the earth whch is involved in the balance, then man's activity is now very large. We shall have to monitor the movement of many elements carefully for the next hundred years or so in order to take care of toxicities.

SUMMARY

In effect, there have always been two ways in which Eq. 1 mobilized elements: geological and biological. Both use the energy of the inside of the earth or of the sun to cause the element transfer from source to sink. In the process of transfer, intermediates appear which can act as useful or toxic products for life. Oxygen is one such intermediate. Ferric iron in the sea is another. Cadmium is a third. All elements at certain dose levels can be toxic. Man's activity has introduced a third way in which the transfer of elements is brought about, and it must be understood that should he contaminate the earth with a new level of any element, e.g., iron or cadmium, there would be no permanent damage to all forms of life since some organism would come to use it; but from man's point of view it would be a disaster.

REFERENCES

(1) Jensen, W.B. 1980. The Lewis Acid Base Concepts. New York: Wiley-Interscience.

(2) Lovelock, J.E. 1979. Gaia, A New Look at Life on Earth. Oxford: Oxford University Press.

(3) Nieboer, E., and Richardson, D.H.S. 1980. A biologically and chemically significant classification of metal ions. Env. Pollut. **B 1**: 3–26.

(4) Whitfield, M. 1981. The world ocean — mechanism or machination? Interdis. Sci. Rev. **6**: 12–35.

(5) Williams, R.J.P. 1981. Physico-chemical aspects of inorganic element transfer through membranes. Phil. Trans. Roy. Soc. Lond. **B 294**: 57–80.

(6) Williams, R.J.P. 1983. The symbiosis of metal ion and protein chemistry. 22nd International Meeting of Coordination Chemistry, Budapest, Hungary, 1982. Pure Appl. Chem. **55**: 35–46.

(7) Williams, R.J.P., and Hall, J.D. 1966. The classification of acceptors and donors in organic reactions. Struct. Bond. **1**: 249–281.

Changing Metal Cycles and Human Health, ed. J.O. Nriagu, pp. 265–274.
Dahlem Konferenzen 1984. Berlin, Heidelberg, New York, Tokyo: Springer-Verlag.

METABOLISM AND TOXICITY OF METALS

C.-G. Elinder
Dept. of Environmental Hygiene, Karolinska Institute, and
The National Institute of Environmental Medicine
104 01 Stockholm 60, Sweden

Abstract. Uptake, distribution, and excretion of different metals and metal compounds and their toxic effects are highly variable. Most research has focused on a limited number of essential metals (Fe, Zn, Cu, Mn, and Se) and nonessential metals (Pb, Hg, Cd, and As). For these metals metabolism and toxicity in humans and animals are now fairly well established. A number of extensive and comprehensive reviews dealing with these metals are available. Discussion today concentrates not so much on the types of toxic effects that these metals may induce, but rather under which circumstances and at what exposure levels these effects occur. There are, however, several other metals (e.g., Cr, Co, Mo, Ni, Al, Sn, V, and Sb) which may produce severe toxic effects in humans under certain conditions. More research is needed to evaluate the kind of health hazards that may be involved in connection with these less studied metals.

METALS OF INTEREST

There are about 45 different metals utilized in industrial processes to an extent which may lead to their exposure to humans. Workers using, or handling, a specific metal or its compounds are inevitably most prone to experience the toxic effects of such a metal. A certain amount of information on metabolism and toxicity in animals and humans is available for most of these 45 metals.

However, for several of these the information is rather scarce. Figure 1 presents the number of scientific publications dealing with metabolism and toxicity of specific trace metals that have appeared during the last five and a half years (from 1977 to the latter half of 1982). It can be deduced from this figure that about 70% of such publicationss issued in recent years provide information on only three essential metals (Fe, Zn, and Cu) and four nonessential metals (Pb, Hg, Li, and Cd). Apart from the eight metals depicted in Fig. 1, which are considered to be essential to humans, there are four other metals (Na, K, Ca, and Mg) which are essential for all forms of animal and human life. These latter metals are usually included in the term "major elements" (7) and will not be discussed in this paper. Furthermore, Ni is believed to be essential for

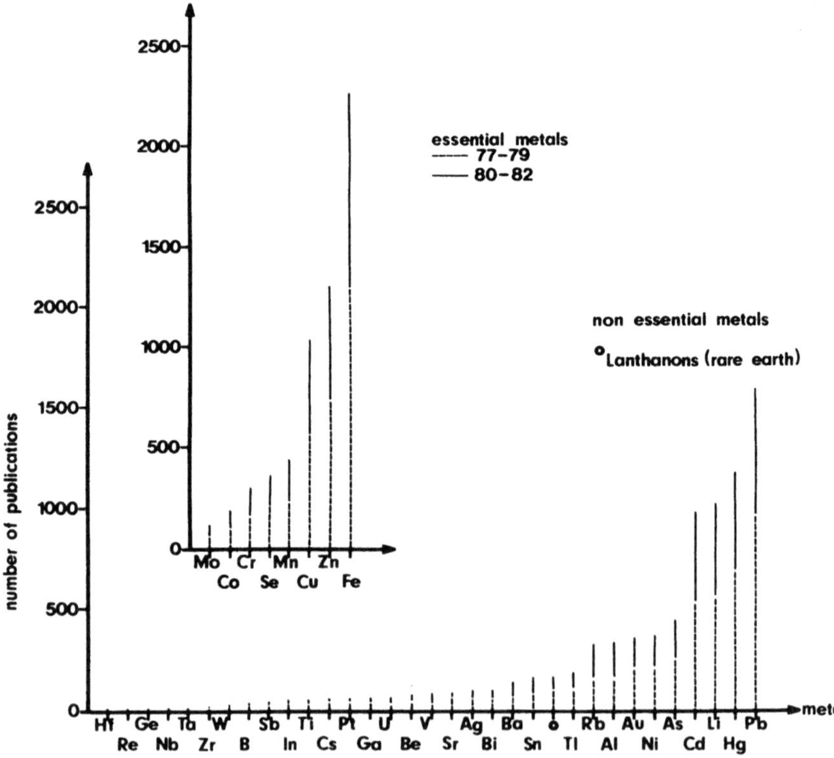

FIG. 1 — Number of publications dealing with metabolism and/or toxicity of metals included in MEDLINE (5) between 1977 and the latter half of 1982.

some animal species but has not yet been established as an essential metal for humans (3, 7). The criteria used to classify a metal as essential or nonessential are described elsewhere (7) and will not be discussed in any detail here.

In the following text, the metabolism and toxicity of essential and some nonessential metals will be reviewed. Of necessity this review is somewhat simplified and superficial, and the reader is therefore referred to recently published comprehensive handbooks which cover most of the metals discussed below (3, 8).

ESSENTIAL METALS

The metabolism of the essential metals is usually adjusted to a fine homeostasis in order to maintain body burdens and concentrations in biological fluids at an optimal level. This can be achieved in various ways: uptake from food may increase, or excretion decrease, when body burdens are low, and vice versa. Iron, zinc, and copper are of considerable clinical interest. Deficiency of iron and zinc in humans is not uncommon. This has occasioned a large amount of research and subsequently published material. For iron and zinc, regulation takes place in the gastrointestinal tract. In the event of iron or zinc deficiency or low body stores, absorption may increase considerably. When the body stores are saturated, or overloaded, most of the zinc and iron in the diet remains unabsorbed and will be eliminated via feces. For copper the uptake from the gastrointestinal tract is always comparably high, and the copper status of the body appears to be regulated by an effective and rapid biliary excretion. Specific transport proteins in plasma exist for iron (transferrin) and copper (ceruloplasmin). Much less is known about the mechanisms of homeostasis for the other essential trace metals. In the case of manganese, both regulation via changes in degree of gastrointestinal absorption and biliary excretion appear to be involved in maintaining optimal levels in the body (11). For selenium, chromium, cobalt, and molybdenum, urinary excretion seems to be the most important route of elimination. In this context it should be mentioned that cobalt should not be genuinely regarded a a true essential metal for humans, since it is essential merely as an integral part of vitamin B_{12}. Due to the homeostasis of essential metals, long-term toxicity from ingestion of these metals has rarely been reported in humans. An exception to this general rule is cobalt, which was the cause of an epidemic outbreak of cardiomyopathy among heavy consumers of beer fortified with cobalt chloride. High concentrations of molybdenum in water and food have been implicated in an increased prevalence of gout.

High peroral doses of essential metals may induce acute intoxication. For example, ingestion of soluble salts of iron, zinc, and copper in doses of several hundred milligrams give rise to nausea, vomiting, abdominal pains and occasionally diarrhea. At very high doses, on the order of several grams, severe systemic effects, as well as fatalities, may occur. The emetic and acute toxic doses for the other essential metals are somewhat lower, i.e., approximately 10 to 100 mg.

Inhalation of relatively high concentrations of copper and zinc oxides and fumes of these metals may take place during, e.g., smelting and welding. Exposure can cause an influenza-like syndrome called metal fume fever shortly after exposure. Long-term effects from inhalation of zinc and copper have not been observed. Inhaled iron oxide particles are retained in the lung for long periods of time and may cause a benign type of pneumoconiosis, termed siderosis of the lung. Chromium is the only essential metal known to be carcinogenic in humans. Inhalation of chromium in its hexavalent form has repeatedly been shown to be associated with cancer of the lung. Probably chromium is essential only in its trivalent form. Manganese is of special interest, since it may cause systemic effects after inhalation. Chronic manganese poisoning due to inhalation of manganese has been reported for several types of industrial processes which involve high exposure to manganese dust. The disorder is characterized by psychiatric and neurologic manifestations closely resembling those found in Parkinson's disease (11). Manganese is also of concern from an environmental health point of view. Pollution of ambient air with manganese has been associated with an increased incidence of respiratory infections in the vicinity of point sources (11).

NONESSENTIAL METALS

"Well Documented Metals"

The nonessential metals that have attracted most attention are lead, mercury, lithium, and cadmium, followed by arsenic (Fig. 1). Most of the reports dealing with lithium focus on the use of lithium salts as an effective drug in the treatment of certain psychiatric disorders, mainly recurrent manodepressive psychosis. Since lithium is readily absorbed, the drug can be given perorally. Chemically, lithium resembles sodium, a fact which allows uniform distribution throughout the body. Elimination takes place mainly via urine. Toxic effects, neuromuscular and central nervous system changes, and signs of renal toxicity are seen as a result of overdosing. Industrial and environmental exposure to lithium has not been reported to be problematical.

Lead, mercury, cadmium, and arsenic are recognized as important industrial hazards and have all caused major environmental health problems as a result of pollution. Comprehensive reviews describing exposure routes, metabolism, and toxic effects in humans are available for lead (6, 10), mercury (9), cadmium (1, 4), and arsenic (12).

Lead. Inorganic lead enters the body mainly by inhalation or ingestion. Exposure to lead via inhalation is caused predominantly by occupational sources. The most important routes of environmental exposure are ingestion of lead-contaminated food, wine, dust, and paint. Elimination of lead is slow and takes place mainly via urine. The dominating toxic effects of inorganic lead are seen as disturbances in the production of heme and in the nervous system. Lead inhibits several enzymes, especially ALA-dehydratase, which are necessary for the normal production of heme. Pronounced and persistent inhibition of these enyzmes will result in anemia. At high exposure levels lead may give rise to dramatic effects on the central and peripheral nervous system, with cramps, unconsciousness, abdominal pains, and polyneuropathies. Comparably high exposure levels may cause impairment in intellectual capacity, especially in small children. It is also possible that relatively low levels of lead have a similar effect. At present there is considerable debate as to the lowest levels of lead in blood which can be associated with the first sign of effects on the central nervous system (6, 10). This problem is discussed in more detail by Needleman elsewhere in this volume.

Mercury. Inorganic mercury vapor is chiefly an industrial problem. It is absorbed effectively via the lung after inhalation, passes through the blood-brain barrier, and subsequently is retained in the brain. Toxic effects of inorganic mercury (mercurism) include gingivitis, tremor, and a mental disorder called erethism. The gastrointestinal absorption of metallic mercury is, however, virtually nil.

When released into the environment, inorganic mercury becomes methylated, forming different types of organomercury compounds, mainly methylmercury. Methylmercury is accumulated in the food chain of aquatic ecosystems and is found at comparably high concentrations in certain larger fish, e.g., pike, shark, and tunafish. Fish consumption constitutes the predominant route for human exposure to organic mercury. One exception is the consumption of seed grain treated with methylmercury. The gastrointestinal absorption of methylmercury is very efficient and the elimination from the body is slow, mainly taking place via the bile. The brain is the critical organ also for organomercury compounds, but the symptoms are somewhat different. The first sign of chronic methylmercury poisoning is parasthesia, notably around the mouth. Later ataxia, dysarthria, and deafness may develop. What is of particular interest when discussing organomercury compounds is their capacity to pass through the placenta into the fetus. Signs of fetal intoxication are severely impaired intellectual capacity and mental retardation. Since fetal intoxication has been recorded in children born to mothers without any sign of mercury poisoning, it is likely that the developing child is more susceptible than the adult (9). Health hazards from exposure to methylmercury is reviewed more thoroughly by Clarkson elsewhere in this volume.

Cadmium. As in the case of lead, the main exposure routes for cadmium are via inhalation and ingestion. Exposure via inhalation occurs primarily under occupational conditions. Smokers are liable to elevated exposure, since cigarettes contain cadmium which is inhaled while smoking. For nonsmokers, various types of food are the major source of exposure, even though gastrointestinal absorption is limited to around 5%. A low molecular and sulfur-rich protein, metallothionein, plays a key role in the metabolism of cadmium (2). This protein firmly binds cadmium ions and is thought to be involved in the pronounced accumulation of cadmium in the liver and kidneys (Kägi and Hapke, this volume). The elimination of cadmium is very slow, and the biological half-time has been estimated to be on the order of decades. The kidney is considered to be the critical organ for long-term exposure to cadmium and is the organ that first reveals signs of toxicity. The first sign of cadmium-induced dysfunction is increased urinary excretion of proteins, especially low molecular proteins, such as β_2-microglobulin and retinol binding protein (see Nogawa, this volume). As a result of high occupational exposure, local effects in the lung in the form of emphysema, have also been seen. Under certain circumstances, long-term cadmium exposure may also give rise to severe bone lesions, or osteomalacia.

One major problem that remains to be solved in connection with cadmium toxicity is the evaluation of the clinical importance of so-called tubular proteinuria. Is this merely a benign sign of tubular dysfunction or does it proceed to severe renal disease? Available data indicate that once tubular proteinuria becomes manifest, it is irreversible in most cases. On the other hand, the disease progresses very slowly. Therefore, severe renal dysfunction can be prevented if exposure ceases. A question under debate is: what is the lowest concentration of cadmium in the kidney that produces the first signs of tubular dysfunction in humans? The best estimate at present is about 200 mg/kg in kidney cortex, suggested by Fribert et al. in 1971 (1, 4).

Arsenic. The metabolism of arsenic is very complex. Many different forms of inorganic and organic arsenic exist, all of which appear to be metabolized somewhat differently. Most soluble arsenic compounds are readily absorbed from the gastrointestinal tract, and most of the arsenic is subsequently rapidly excreted via the urine. To a certain extent, inorganic arsenic becomes methylated in the body before it is excreted. Cases of acute poisoning due to inorganic arsenic are well-known. Long-term peroral exposure to much lower amounts of inorganic arsenic may also cause toxic effects. Consumption of drinking water containing arsenic has resulted in an increased prevalence of vascular disease, polyneuropathies, and cancer of the skin. Exposure to inorganic arsenic via inhalation, predominantly in the form of arsenic trioxide, has repeatedly been shown to increase the risk of lung cancer in smelter workers (12). Organic forms of arsenic are found at comparably high concentrations in certain marine organisms, e.g., shellfish. These organic arsenic compounds are considered to have a very low acute and chronic toxicity.

"Less Well Documented Metals"

Of the remaining metals, some such as nickel, aluminum, beryllium, vanadium, and thallium are known occupational hazards, and others such as gold and antimony have been used therapeutically. For most of the other metals, very little information is available. Only a few of the less studied metals are mentioned below.

Inhalation of insoluble *nickel* compounds, such as nickel hydroxide, has been shown to cause cancer of the lung and upper respiratory tract in occupationally exposed workers, as well as in experimental animals. Nickel absorbed from the lung or from the gastrointestinal tract is excreted rapidly via the urine, and no accumulation takes place in the body. No systemic effects due to nickel exposure have been reported. It should be mentioned that nickel is a common cause of contact dermatitis, but this disease is not usually regarded as a direct toxic effect.

Gold salts given parenterally are used in the medical treatment of rheumatoid arthritis. Toxic reactions, involving skin lesions, liver and kidney damage, and in some cases bone marrow depression, are common as a result of this treatment. Elimination takes place via urine.

Aluminum is an abundant metal which has long been considered virtually nontoxic and nonabsorbable from the gastrointestinal tract. More recent studies have shown that ingested aluminum compounds are to a certain extent absorbed from the gut. The actual absorption rate is unknown. There is a lack of data on metabolism and toxicity of the different forms or compounds of aluminum that might be ingested by humans. Elimination of aluminum from the body takes place mainly via the urine. Persons with kidney dysfunction (uremia patients) may accumulate the metal. Dialysis may be an important contributor to the accumulation of aluminum in the body since water used for dialysis frequently contains significant amounts of aluminum. Severe health effects may develop in the central nervous system as a result of aluminum accumulation in the body, especially in the brain. Typical symptoms include dementia, aphasia, ataxia, convulsions, and characteristic EEG-changes. The disease, which has been named dialysis encephalopathy or dialysis dementia is usually fatal within a couple of years from the date of diagnosis. During occupational exposure to aluminum, effects on the respiratory system are most prevalent. Exposure to dusts during the handling of aluminum salts, fluorides, or oxides has been associated with asthmatic complaints. Aluminum metal is a health risk when it occurs as a fine powder, usually called stamped aluminum powder. When inhaled, this powder may cause a severe lung disease or aluminosis.

Rubidium has several physicochemical similarities to potassium and may to some extent substitute for potassium in the body under experimental potassium deficiency. Very little information on occupational and/or environmental toxicity is available.

Thallium is a highly toxic metal which has caused a number of fatalities. It has been used as an effective rodenticide. It is readily absorbed, giving rise to dramatic and severe toxic symptoms, such as gastroenteritis, and painful polyneuropathies. Following chronic exposure, alopecia (hair loss) is a typical symptom. Environmental contamination with thallium has been observed near certain larger smelters and coal-fired plants.

Ingestion of inorganic **tin** is common since tin is used widely in the canning industry. Gastrointestinal absorption is low. Acute intoxication with symptoms of food poisoning can develop following intake of large doses of tin, on the order of or exceeding 50 mg. Certain man-made organotin compounds, mainly the short chain alkyl compounds, are absorbed efficiently from the gut and can cause severe and fatal intoxications. Inhalation of inorganic tin has caused a benign type of pneumoconiosis, called stanosis.

Vanadium is also poorly absorbed from the gastrointestinal tract. When inhaled, vanadium causes irritation of the respiratory tract, with coughing and bronchitis. A greenish discoloration of the tongue is often mentioned as a characteristic sign of occupational exposure to vanadium, but this may be absent even after prolonged and extensive exposure.

Antimony has several physicochemical similarities to arsenic. It has been used as a drug in the treatment of certain parasitic diseases. The gastrointestinal absorption of antimony is relatively high, and it is eliminated via the urine. Ingestion of large doses via the diet causes food poisoning. Occupational exposure and medical treatment with antimony have been shown to induce adverse effects on the heart with, e.g., ECG changes. When inhaled, antimony also acts as an irritant on the respiratory system.

CONCLUSION

Metals and their compounds display wide variation in terms of their metabolism, distribution and type of effects. So far, most research has concentrated on a limited number of metals. This is natural since the metals studied have produced the most severe environmental and occupational problems or have been of importance from a clinical point of view. There is clearly a need for pertinent information to properly assess the potential health hazards associated with the increasing use and emission of different types of metals and their compounds to the environment.

REFERENCES

(1) CEC. 1978. Criteria (Dose / Effect Relationships) for Cadmium. Report of a Working Group of Experts prepared for the Commission of the European Communities, Directorate-General for Social Affairs, Health and Safety Directorate. London: Pergamon Press.

(2) Foulkes, E.C., ed. 1982. Biological Roles of Metallothionein. Proceedings of a USA-Japan Workshop, Cincinnati, Ohio, March 22–27, 1981. Amsterdam: Elsevier / North Holland.

(3) Friberg, L.; Nordberg, G.F.; and Vouk, V.B., eds. 1979. Handbook on the Toxicology of Metals. Amsterdam: Elsevier / North Holland.

(4) Friberg, L.; Piscator, M.; Nordberg, G.F.; and Kjellström, T. 1974. Cadmium in the Environment, 2nd ed. Cleveland, OH: CRC Press.

(5) MEDLINE. 1981. National Library of Medicine, Bethesda, MD.

(6) Needleman, H.L., ed. 1980. Low Level Lead Exposure: The Clinical Implications of Current Research. New York: Raven Press.

(7) Underwood, E.J. 1977. Trace Elements in Human and Animal Nutrition. New York: Academic Press.

(8) Waldron, H.A., ed. 1980. Metal in the Environment. New York: Academic Press.

(9) WHO. 1976. Environmental Health Criteria 1. Mercury. Geneva: WHO.

(10) WHO. 1977. Environmental Health Criteria 3. Lead. Geneva: WHO.

(11) WHO. 1981. Environmental Health Criteria 17. Manganese. Geneva: WHO.

(12) WHO. 1982. Environmental Health Criteria 18. Arsenic. Geneva: WHO.

Changing Metal Cycles and Human Health, ed. J.O. Nriagu, pp. 275–284.
Dahlem Konferenzen 1984. Berlin, Heidelberg, New York, Tokyo: Springer-Verlag.

CADMIUM

K. Nogawa
Kanazawa Medical University
Uchinada, Ishikawa, Japan

Abstract. Studies on health effects of environmental cadmium pollution have revealed that low molecular weight proteinuria, especially β_2-microglobulinuria, was the earliest sign and itai-itai disease was the most advanced case. Dose-response relationship of environmental cadmium pollution was observed in three different cadmium-polluted areas and this suggested that the maximum allowable intake of cadmium might be very low, compared with nonpolluted levels of cadmium.

INTRODUCTION

Environmental cadmium pollution has been encountered in some countries (3,8,28) and has now become one of the major social and political problems, especially in Japan. Since Japan is the country where health effects caused by environmental cadmium pollution are observed, it is important and useful for evaluating potential threats of cadmium to human health to investigate the present status of environmental cadmium exposure in that country. This paper focuses on the health effects of environmental cadmium pollution, the dose-response relationship, and the maximum allowable intake, which have been derived from human studies mainly performed in Japan.

HEALTH EFFECTS OF ENVIRONMENTAL CADMIUM EXPOSURE

Itai-itai disease is considered to be the most advanced case of chronic cadmium poisoning, although some researchers are strongly opposed to this theory (28).

It is necessary to clarify the clinical picture of itai-itai disease in order to understand the end result of the natural course of chronic cadmium poisoning. There are many reports on the clinical symptoms of the disease. Based on the latest reports (10,26), the present status of itai-itai patients can be summarized as follows: The physical constitution of the patients was small, and deformities of the spine and chest were frequently observed. As for the urinary findings the patients showed increased excretions of total protein, glucose, amino acids, calcium, cadmium, and low molecular weight proteins. The Ca/P ratio in urine was higher in these patients. In regard to the blood findings, patients showed low levels of serum, inorganic phosphorus, potassium, total protein, uric acid, and high levels of chloride, alkaline phosphatase, LDH, BUN, and creatinine. The patients showed normochromic anemia, increased granulocyte counts, and decreased lymphocyte counts. The patients also showed renal tubular dysfunction as well as a moderate disturbance of renal glomerular function. Arterial blood gas analysis revealed low pH, HCO_3, and base excess in the patients. There were no cases of hypertension, and many patients were hypotensive. Most of the patients showed peripheral nerve disturbance. As for the bone changes, looser's zone was found in about one-third of the patients, and the zone was either in the process of healing or had completely healed. Osteoporotic changes were noticed in almost all cases. The main clinical picture of the disease which has been reported to date is in accord with the concept of the Fanconi syndrome. The patients had been treated over a period of a year with vitamin D and other medication. Therefore, there was the suggestion that the side effect of the treatment was one causative factor of the observed syndrome. Since many inhabitants in the area where itai-itai disease is endemic and in the Cd-polluted areas other than the Jinzu River basin showed similar changes in urine, blood, and renal function as those of the itai-itai patient, the side effects of the treatment could be totally ruled out. The inhabitants who were defined as requiring observation showed the same grade of abnormality in urine, blood, and renal function as those of the itai-itai patients (17,18,25). The prevalence of the proteinuria, glucosuria, aminoaciduria, and low molecular weight proteinuria, especially β_2-microglobulinuria, was much higher in the Cd-polluted Jinzu River basin than in nonpolluted areas (7).

Health effects of environmental cadmium exposure were also observed in Cd-polluted districts other than the Jinzu River basin: Kosaka Town in Akita Prefecture (24), Kakehashi River basin in Ishikawa Prefecture (4), Ichi River basin in Hyogo Prefecture (12,14,28), and Tsushima Island in Nagasaki Prefecture (20). Higher prevalences of urinary abnormality such as low molecular weight proteinuria and glucosuria were found in these areas. The inhabitants with abnormal urinary finding showed the same type of renal dysfunction and abnormal blood finding as those of itai-itai patients. In summarizing the available data which were published in Japan, low level exposure to cadmium first induces renal tubular dysfunction indicated by low molecular weight proteinuria, especially β_2-microglobulinuria, followed by abnormal blood findings, and lastly reaching itai-itai disease state. In other words, the first point in the natural course of low level exposure to cadmium is β_2-microglobulinuria and the last point is itai-itai disease. It is not yet clear in what order the symptoms progress and has not yet been demonstrated that environmental cadmium exposure causes hepatic damage, malignant tumor, or hypertension in human beings. Most of the itai-itai patients were hypotensive and did not show abnormal hepatic functions. However, further studies are necessary to determine whether cadmium is related to hypertension in humans, particularly at low levels of exposure.

As for bone effects, there are still arguments as to whether environmental cadmium exposure could cause osteomalacia. Takeuchi and Kajikawa (28) emphasized that itai-itai patients were found only in the Jinzu River basin and no case was found in the cadmium-polluted areas elsewhere. Five cases which were reported by Nogawa et al. (12) in the cadmium-polluted Ichi River basin in Hyogo Prefecture were not found to have osteomalacia by the Research Committee on the Differential Diagnosis organized by the Environment Agency (Chairman B. Takase). The main reason to rule out itai-itai disease was that postmortem examinations did not reveal osteomalacia. It is completely unreasonable to base the diagnosis only upon pathological findings because the patients had shown the same abnormalities in urine, blood, and X-ray of bone as those of itai-itai patients, and the looser's zone was completely healed after treatment with a large dose of vitamin D. One case with osteomalacia which also showed the same urine and blood findings as those of itai-itai patients was found in Cd-polluted Tsushima Island in Nagasaki Prefecture by postmortem examination (27). However, the Histopathological Group of the Cadmium Research Committee organized by the Environment Agency (Chaired by K. Kajikawa) said that itai-itai patients showed pathological anatomical features and clinical data different from those of the patients of Nagasaki Prefecture, and the pathogenesis of the bone lesions in these patients appeared to be different (2). A histological staining method for identifying osteoid tissue developed by Yoshiki was recently used for examining bone lesion of the patients requiring observation in the Jinzu River basin. Mukawa et al. (9) performed a bone biopsy on four patients who required observation. Three of them showed abnormal osteoid tissue increase, and the findings were consistent with osteomalacia. Kitagawa (5) has done nine autopsies of the patients requiring observation and reported that all cases showed abnormal increase of osteoid tissue. These studies indicate that there may be a large number of people who should be treated as actual itai-itai patients in the Jinzu River basin. It should be noted that bone changes caused by environmental cadmium exposure are not as rare as has been believed.

DOSE-RESPONSE RELATIONSHIP OF ENVIRONMENTAL CADMIUM EXPOSURE

Most reviews (19,28) regarding environmental pollution of cadmium in Japan emphasized that the dose-response relationship of environmental cadmium had not yet been demonstrated. Because of the long biological half-life of cadmium in the kidney, it is not easy to clarify this relationship, especially in human subjects. However, there are some reports which have demonstrated its existence in Cd-polluted areas. In the Jinzu River basin where itai-itai disease was first found, Fukushima (1) investigated the relationship between cadmium concentrations in rice and abnormal urinary findings in 1972. Cadmium concentrations in glutinous and nonglutinous village rice were employed as indices of cadmium exposure. As indices of the effects of cadmium on health, the percentages of men and women over 50 years old with proteinuria and glucosuria in each village were used. A close relationship between the cadmium exposure and the health effect was found to exist; the correlation coefficients between them were + .68 for nonglutinous rice and + .82 for glutinous rice. The rice samples analyzed for this study were collected from 77 farm families. The sample size corresponded to one to two rice samples in each village. The number of rice samples may not be enough to represent the levels of cadmium exposure in each village. Nogawa and Ishizaki (11) studied the relationship between the cadmium concentration in rice and abnormal urinary findings in the inhabitants of the Jinzu River basin. Proteinuria and glucosuria were used as indices of the effect of cadmium on health and average rice cadmium concentrations in hamlets as indices of cadmium exposure level. Data on urinary findings were obtained from the epidemiological study in 1967, and the cadmium concentrations in rice were obtained from the survey in 1972. The results for 2741 inhabitants in the cadmium-polluted area, 2150 reference inhabitants, and 583 unpolished nonglutinous rice samples stored in farmhouses were used for this study. There was a close relationship between rice-cadmium concentration and renal effects. That is, the prevalence increased with the increasing rice cadmium concentration. Increased prevalence of proteinuria and proteinuria with glucosuria in the exposed population over 50 years of age was observed at average cadmium concentrations of around 0.30–0.49 μg/g wet weight in unpolished rice, which corresponds to daily intake of 201–290 μg of cadmium.

Nogawa et al. (15) also studied the dose-response relationship for cadmium in itai-itai disease patients and found a close relationship between village average cadmium concentrations and the prevalence of itai-itai disease. Nogawa et al. (16) examined the relationship between urinary cadmium concentration expressed as μg/g creatinine and renal effects of cadmium exposure of 542 inhabitants in the Jinzu River basin. Cadmium concentration in urine was employed as the index of cadmium exposure. Total protein with glucose, β_2-microglobulin, retinol binding protein, and proline served as indices of renal effects. The prevalence of renal effects increased with increasing cadmium concentration in urine and probit linear regression lines could be calculated from the graph. The urinary cadmium concentrations corresponding to 1% prevalence rate of β_2-microglobulinuria was 3.2 and 5.2 μg/g creatinine for men and women, respectively. Kjellström et al. (6) reported that there was a tendency for increased β_2-microglobulin excretion with increased cadmium level in rice and time of residence for inhabitants in the Jinzu River basin. In the Kakehashi River basin in Ishikawa Prefecture, Kawano (4) found a significant correlation between village average cadmium concentration and prevalence of urinary abnormalities such as retinol binding proteinuria and aminoaciduria. He concluded that the so-called threshold limit values in the inhabitants older than 50 years was 0.148–0.246 μg/g in rice. Nogawa et al. (13) performed a more detailed analysis on the same data and found a close dose-response relationship between village average cadmium concentrations in rice or in urine and prevalence of retinol binding proteinuria, and of renal tubular dysfunction when the inhabitants were stratified according to cadmium concentrations in the village average rice or urinary cadmium concentrations. The prevalence of tubular proteinuria increased continuously with nonpolluted levels of cadmium in rice or urine in the inhabitants older than 70 years.

Saito et al. (24) demonstrated that in Kosaka in Akita Prefecture the value of cadmium concentration in unpolished rice multiplied by the duration of residence in years had a close correlation with the concentration of urinary β_2-microglobulin ($r = 0.504$, $P < 0.001$). They noted that farmers habitually eating rice containing 0.66 μg of cadmium generally developed chronic cadmium poisoning in 50 years. The studies in three different cadmium-polluted areas in Japan clearly demonstrate the existence of a dose-response relationship for environmental cadmium exposure. The Research Committee organized by the Environment Agency (21) performed health examinations on the 8858 inhabitants (4658 in Cd-polluted and 4200 in nonpolluted) in seven Prefectures from 1976 to 1978. It was concluded that there existed a relationship between cadmium pollution and glycosuria, low molecular weight proteinuria, and proximal tubular dysfunction in some prefectures only. No relation was observed between urinary excretion of protein and cadmium pollution.

ALLOWABLE INTAKE OF CADMIUM

Based on the epidemiological studies on cadmium-polluted areas in Japan, some researchers have proposed limits for the allowable intake of cadmium or the allowable urinary cadmium concentrations. According to Saito et al. (22,23) consumption of rice with approximately 0.3μg Cd/g, could cause significant urinary excretions of β_2-microglobulin, and daily intakes of 177 μg/day of Cd for 15 years (total body burden 1 g) could result in tubular dysfunction on the basis of the work done in Kosaka in Akita Prefecture. Nowaga et al. (11,16) proposed that the allowable average daily intake of cadmium may be set under 201–287 μg and allowable urinary cadmium concentration under 3.2 μg Cd/g creatinine for men and 5.2 μg Cd/g creatinine for women on the basis of the epidemiological study in the Jinzu River basin in Toyama Prefecture. Based on the epidemiological studies in the Kakehashi River basin, Kawano (4) calculated the maximum allowable cadmium concentration in rice as 0.148–0.246 μg Cd/g. Nogawa et al. (13) reported that health effects increased continuously with increasing average cadmium concentrations in village rice or in urine as a result of normal environmental exposure to cadmium. The last authors doubt whether allowable cadmium concentrations in rice or urine do actually exist. In summarizing the available data in Japan, it should be emphasized that maximum allowable intake of cadmium may be very low. Maximum daily intake may be under 200–290 μg of Cd. Cadmium concentration in rice which would not cause renal tubular damage may be set under 0.30–0.40 μg Cd/g wet weight.

DAILY CADMIUM INTAKE IN NONPOLLUTED AND POLLUTED AREAS

There are some reports which estimate the daily intake of cadmium in unpolluted areas. Three methods were used to make this estimation: the national nutrition census method, the total diet collection method, and the fecal excretion method. The normal daily cadmium intake by Japanese based on the three methods was 16–80 μg and an average of 50 μg/day is acceptable. Recently an extensive study on the intake by, and accumulation of cadmium in, inhabitants in the contaminated districts was performed by the Research Committee organized by the Environment Agency with three prefectural governments (29). Table 1 shows the daily intake of cadmium via the total diet as estimated by two methods in polluted and unpolluted areas. Average daily intake of cadmium ranged from 101–184 μg in polluted areas and 42–69 μg in unpolluted areas by total diet collection method. It should be noted that cultivation of rice was already prohibited in highly polluted areas when the investigation was done. The margin between present intake and the intake that might cause a significant prevalence of renal tubular damage may not be large.

		Total diet collection method		Fecal excretion of cadmium	
District	Designation	No. of samples	Cd intake μg/day	No. of samples	Cd excretion μg/day
Nagasaki Pref.	polluted	10	184 ± 134	30	255 ± 171
	control	10	68.5 ± 37.4	30	42.3 ± 30.9
Ishikawa Pref.	polluted	10	123 ± 95.5	40	149 ± 121
	control	10	41.9 ± 15.6	40	102 ± 83.9
Akita Pref.	polluted	10	101 ± 50.0	37	158 ± 108
	control	10	66.3 ± 34.3	39	60.0 ± 49.8

TABLE 1 – Daily cadmium intake via total diet.

REFERENCES

(1) Fukushima, M.; Ishizaki, A; Sakamoto, M.; and Kobayashi, E. 1973. Cadmium concentration in rice eaten by farmers in the Jinzu River basin. Jpn. J. Hyg. **28:** 406–415 (in Japanese).

(2) Histopathology Group of the Cadmium Research Committee. 1982. A histopathological study on 9 autopsied cases who had lived in the cadmium-polluted areas in Japan — Interim Report. Kankyo Hoken Rep. **48:** 197–203.

(3) Harvey, T.C.; Chettle, D.R.; Fremlin, J.H.; Al Haddad, I.K.; and Downey, S.P.M.J. 1979. Cadmium in Shipham. Lancet **i**: 551.

(4) Kawano, S.; Sakai, Y.; Iida, S.; Nishi, M.; Higashi, S.; Inaki, K.; Matsuda, H.; and Katoh, M. 1976. Results of health effects of cadmium

in inhabitants of the Kakehashi River basin. Kankyo Hoken Rep. **38:** 104–117 (in Japanese).

(5) Kitagawa, M. 1982. Abstract of autopsy report of itai-itai patients in Toyama Prefecture. Kankyo Hoken Rep. **48:** 204–206 (in Japanese).

(6) Kjellström, T.; Shiroishi, K.; and Evrin, P.E. 1977. Urinary β_2-microglobulin excretion among people exposed to cadmium in the general environment. Report No. 1. An epidemiological study in co-operation between Japan and Sweden. Env. Res. **13:** 318–344.

(7) Kobayashi, E. 1982. An epidemiological study on the health effects of environmental cadmium (Part 1). Some urinary findings by sex and age. Jap. J. Pub. Health **29:** 123–133 (in Japanese).

(8) Lauwerys, R.; Roels, H.; Bernard, A; and Buchet, J.P. 1980. Renal response to cadmium in a population living in a nonferrous smelter area in Belgium. Arch. Occup. Env. Health **45:** 271–274.

(9) Mukawa, A.; Nogawa, K.; and Hagino, N. 1980. Bone biopsy performed on women living in the cadmium-polluted Jinzu River basin. Jpn. J. Hyg. **35:** 761–773 (in Japanese).

(10) Nakagawa, S.; Tohda, N.; and Kobayashi, S. 1977. Present status of itai-itai patients — Clinical findings by orthopedic examination. Kankyo Hoken Rep. **41:** 53–60 (in Japanese).

(11) Nogawa, K., and Ishizaki, A. 1979. A comparison between cadmium in rice and renal effects among inhabitants of the Jinzu River basin. Env. Res. **18:** 410–420.

(12) Nogawa, K.; Ishizaki, A.; Fukushima, M.; Shibata, I.; and Hagino, N. 1975. Studies on the women with acquired Fanconi syndrome observed in the Ichi River basin polluted by cadmium. Env. Res. **10:** 280–307.

(13) Nogawa, K.; Ishizaki, A.; and Kawano, S. 1978. Statistical observations of the dose-response relationship of cadmium based on epidemiological studies in the Kakehashi River basin. Env. Res. **15:** 185-198.

(14) Nogawa, K.; Ishizaki, A.; Kobayashi, E.; Inaoka, H.; and Shibata, I. 1975. Studies on renal dysfunction of inhabitants in Cd-polluted district. Jpn. J. Hyg. **30:** 549–555 (in Japanese).

(15) Nogawa, K.; Kawano, S.; and Katoh, T. 1982. A relationship between prevalence of itai-itai disease and village average rice cadmium concentration. Kankyo Hoken Rep. **48:** 149–151 (in Japanese).

(16) Nogawa, K.; Kobayashi, E.; and Honda, R. 1979. A study of the relationship between cadmium concentrations in urine and renal effects of cadmium. Env. Health Perspect. **28:** 161–168.

(17) Nogawa, K.; Kobayashi, E.; Honda, R.; and Ishizaki, A. 1979a. Clinico-chemical studies on chronic cadmium poisoning (Part 1), Results of urinary examinations. Jpn. J. Hyg. **34:** 407–414 (in Japanese).

(18) Nogawa, K.; Kobayashi, E.; Honda, R.; and Ishizaki, A. 1979b. Clinico-chemical studies on chronic cadmium poisoning (Part 2). Results of blood examinations. Jpn. J. Hyg. **34:** 415–419 (in Japanese).

(19) Nomiyama, K. 1980. Recent progress and perspectives in cadmium health effects studies. Sci. Total Env. **14:** 199–232.

(20) Ohtsuka, K.; Yajima, K.; Nishimura, K.; Ogata, H.; Yogo, K.; Higashi, H.; Kuwano, K.; and Kobayashi, S. 1977. Investigation of the health effects of cadmium on the inhabitants in Izuharacho in Tsushima Island Nagasaki Prefecture. Kankyo Hoken Rep. **41:** 139–146 (in Japanese).

(21) Research Committee. 1980. Results of analysis of health examinations on inhabitants of cadmium-polluted areas. Kankyo Hoken Rep. (part 2) **46:** 72–110 (in Japanese).

(22) Saito, H.; Kusakabe, K.; Arikawa, T.; Saito, K.; Furuyama, T.; and Yoshinaga, K. 1978. Health survey of residents in cadmium-polluted area. Kankyo Hoken Rep. **44:** 175–177 (in Japanese).

(23) Saito, H.; Nagai, K.; Shioji, R.; Furukawa, Y.; Arikawa, T.; Saito, T.; and Furuyama, T. 1976. Dose-response relationship between total burden of cadmium and β_2-microglobulin concentrations in urine of the inhabitants in the cadmium-polluted area. Kankyo Hoken Rep. **38:** 78–81 (in Japanese).

(24) Saito, H.; Shioji, R.; Furukawa, Y.; Nagai, T.; Arikawa, T.; Saito, T.; Sasaki, Y.; Furukawa, T.; and Yoshinaga, K. 1977. Cadmium-induced proximal tubular dysfunction in a cadmium-polluted area. Contr. Nephrol. **6:** 1–12.

(25) Shinoda, S., and Yuri, K. 1978. Present status of person requiring observation — Clinical findings by internal examination. Kankyo Hoken Rep. **44:** 119–125 (in Japanese).

(26) Shinoda, S.; Yuri, K.; and Nakagawa, S. 1977. Present status of itai-itai patients — Clinical findings by internal examination. Kankyo Hoken Rep. **41:** 44–52 (in Japanese).

(27) Takebayashi, S. 1979. First autopsy case, suspicious of cadmium intoxication, from the cadmium-polluted area in Tsushima, Nagasaki Prefecture. **In** Cadmium-Induced Osteopathy, eds. I. Shigematsu and K. Nomiyama, pp. 124–138. Tokyo: Japanese Public Health Association.

(28) Tsuchiya, K. 1978. Cadmium studies in Japan. Tokyo: Kodansha LTD and Amsterdam, New York, Oxford: Elsevier / North-Holland Biomedical Press.

(29) Tsuchiya, K., and Iwao, S. 1978. Results of cadmium intake of inhabitants in cadmium polluted areas in Akita, Nagasaki and Ishikawa Prefecture. Kankyo Hoken Rep. **44:** 86–115 (in Japanese).

Changing Metal Cycles and Human Health, ed. J.O. Nriagu, pp. 285–309.
Dahlem Konferenzen 1984. Berlin, Heidelberg, New York, Tokyo: Springer-Verlag.

MERCURY

T.W. Clarkson*, R. Hamada*, and L. Amin-Zaki**
*Division of Toxicology, University of Rochester
School of Medicine, Rochester, NY 14642, USA
**Central Hospital, Abu Dhabi, United Arab Emirates

Abstract. Mercury is highly mobile in the environment. Anthropogenic sources contribute about one-third of mercury emissions to the atmosphere. Once in the atmosphere, long distance transport spreads mercury to distant areas of the globe. Both wet and dry precipitation processes remove mercury from the atmosphere. The oceans and their sediments are believed to be the final reservoir.

Mercury is methylated in sediments and suspended particles in fresh and ocean waters by microorganisms. Methylmercury is avidly accumulated by aquatic organisms and by this means enters the human diet. The acidification of natural bodies of freshwater is statistically associated with elevated concentrations of methylmercury in the edible tissues of predatory fish.

Methylmercury, if ingested in sufficient amounts, will cause damage to the human central nervous system, specially affecting the sensory and coordination functions. Prenatal life is the most sensitive stage of the human life cycle to methylmercury.

The global cycle is affected by direct anthropogenic input to the atmosphere and into bodies of natural water. Increasing evidence now suggests that the acidification of rainwater (acidic disposition) may also affect the global cycle resulting in elevated concentrations of methylmercury in fish. If current trends continue, levels of methylmercury in certain edible species of freshwater fish will become a matter of public health concern.

INTRODUCTION

Mercury exists in a wide variety of physical and chemical forms. It has three oxidation states: Hg° (metallic), Hg_2^{2+} (mercurous), and Hg^{2+} (mercuric mercury). It forms stable organometallic compounds with alkyl groups, e.g., CH_3Hg^+ and CH_3HgCH_3 (monomethyl- and dimethylmercury). This review chapter will include only those chemical physical forms of mercury that are involved in the biogeochemical cycle and the forms to which humans are principally exposed in the environment (methylmercury compounds). Human exposure to the vapor of metallic mercury (mercury vapor) is of public health significance only in occupational settings (46).

MERCURY IN THE ENVIRONMENT

Global Cycles

The global cycles of mercury have recently been reviewed by Nriagu (34) and by the National Academy of Sciences (30). The global cycle of mercury involves degassing of the element from the earth's crust and evaporation from natural bodies of water, atmospheric transport believed to be mainly in the form of mercury vapor and deposition of mercury back onto land and water. Mercury ultimately finds its way to sediments in water, particularly to oceanic sediments where the retention time is very long.

Andren and Nriagu (5) have indicated that the residence time of mercury in the atmosphere may vary from approximately 6 to 90 days. Residence times of mercury in soils is on the order of 1000 years, oceans on the order of 2000 years, and sediments are extremely long — measured in millions of years.

Estimates of the quantities of mercury entering the atmosphere from degassing of the surface of the planet vary widely, but a commonly quoted figure is 30,000 tons per year (Table 1).

In fact, wide estimates have been made of the proportion of the mercury in the atmosphere due to anthropogenic sources; figures from 10% to 80% of atmospheric mercury have been credited to man. The ocean and oceanic sediments are believed to be the ultimate destination of mercury on the global cycle. Estimates of the yearly amounts of mercury finding its way to the ocean indicate that atmospheric deposition accounts for the major amount, approximately 11,000 tons per year with land runoff accounting for about 5,000 tons

Source	Amount Metric tons/year
Natural	
degassing of Earth's crust and agricultural activities	~30,000
Anthropogenic	
worldwide mining	10,000
combustion of coal	3,000
combustion of oil	400-1500
smelting of metal sulfide ores	1,500
steel cement phosphates	500

TABLE 1 — Sources of mercury in the environment 1971 (46).

per year. The role of hydrothermal release of mercury directly into the oceans is not well understood, but in oceans where there is appreciable volcanic activity, for example, in the Mediterranean, hydrothermal release directly into the ocean could be substantial.

The measurement of mercury in extremely low environmental concentrations is frequently close to the limit of detection of many current methods. With this caveat, it would appear that the vastly predominant reservoir for mercury is the ocean water containing on the order of 40 million tons (Table 2). In contrast, the atmosphere and freshwater contain much less. As one might expect, therefore, the impact of man-made release of mercury is much greater on these smaller reservoirs, especially those to which man-made release is direct. Thus the impact on the mercury burdens in the atmosphere in freshwaters is appreciable, whereas it is estimated that oceanic concentrations have not appreciably changed in recent history. For example, it is estimated that the mercury content of lakes and rivers may be increased by a factor of 2 - 4 due to man-made inputs (34).

Reservoir	Mercury Content (metric tons)
Atmosphere	850
Freshwater	2,000
Freshwater biota	400
Ocean water	41,000,000
Oceanic biota	200,000

TABLE 2 — The amount of mercury in some global reservoirs (30).

Chemical and Biochemical Cycles of Mercury

This overall global cycle of mercury is the result of extremely complex physical, chemical, and biochemical processes occurring in the main reservoirs and interfaces between these reservoirs. Most of these processes are poorly understood but, nevertheless, certain very important fundamental discoveries have been made in recent years and these are summarized below. It should be recognized that this is an incomplete picture and that this area is undergoing very active research at the present moment.

The most important single discovery in understanding the chemical and biochemical cycles of mercury in the environment was made by Swedish investigators in the 1960s (for a review see (30, 34)). An intensive investigation into the source of the methylmercury compound in freshwater fish led to the finding that microbial activity in aquatic sediments can result in the methylation of inorganic mercury (18). In the environment, the most probable mechanism involves the nonenzymic methylation of mercuric mercury ions by methyl carboning compound (Vitamin B_{12}) that are produced as a result of bacterial synthesis. However, other pathways both enzymic and nonenzymic may play a role (7).

The methylation of ionic mercury in the environment appears to occur under a variety of conditions; in both aerobic and anaerobic waters, in the presence of various types of microbial populations, both anaerobes and aerobes, and in different types of freshwater bodies such as eutrophic and oligotrophic lakes.

The presence of methylmercury in many species of ocean fish, including the wide ranging species such as tuna and shark, strongly suggests that the methylation process can occur in suspended particles in oceans as well as in freshwater.

The methylation of mercury can result in a formation of either monomethylor dimethylmercury compounds (Fig. 1). The monomethylmercury compound is avidly accumulated by fish and shellfish, whereas the dimethyl compound, having a low solubility and high volatility, tends to vaporize from the water phase to the atmosphere where it may be subjected to photolytic decomposition (Fig. 1).

However, these reactions are not understood in detail, and there does not appear to be general agreement in the literature as to those conditions that favor the formation of monomethyl or the dimethyl form; neither is there complete

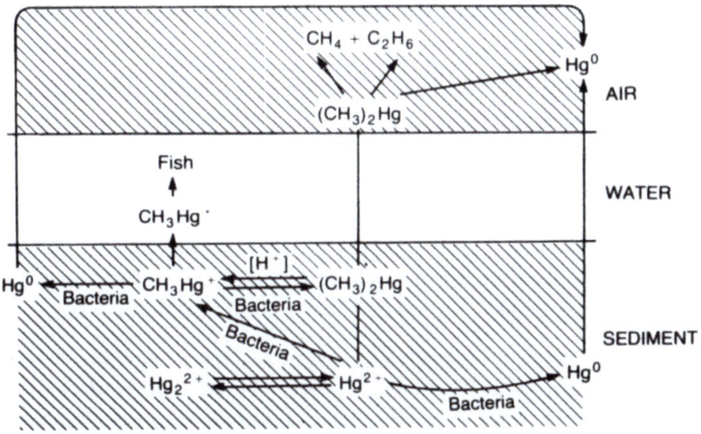

FIG. 1 — A biogeochemical cycle involving bacterial methylation and demethylation of mercury (adapted from Fig. 3.1 of (30)).

agreement as to the extent that the dimethyl species actually vaporizes from the water phase into the atmosphere.

Methylmercury compounds are subject to decomposition in the water phase probably by the action of a variety of microorganisms. The breakdown appears to occur in two enzymically mediated states: first, hydrolase enzymes cleave the carbon mercury bond releasing the methyl group and, subsequently, a reductase enzyme converts the ionic mercury to the metallic elemental form which is then free to diffuse out of the water environment into the vapor phase (Fig. 1). These demethylating microbes appear to be widespread in the environment, occurring in water, sediments and soils and in the gastrointestinal tract of mammals, including humans.

Brosset (21,23) has studied the circulation of mercury in air and natural waters with an emphasis on understanding the mechanisms of the long distance transport and deposition of this element. Mercury is present in the atmosphere, not only as the water insoluble mercury vapor, but also in a water soluble form as well.

Neither the detailed chemical mechanisms nor the kinetics of these processes are understood at this time, for example, the extent to which mercury may be deposited and remitted from water or land surfaces to the atmosphere is still not understood in quantitative terms. Nevertheless, the general picture that emerges is one in which long distance transport of mercury is much more extensive than was originally believed.

Once methylmercury enters the water phase as a soluble compound, it is rapidly accumulated by most aquatic biota and attains highest concentrations in the tissues of such large carnivorous fish as the pike in freshwater and the shark and tuna fish in marine water. Indeed, it is generally believed that the major amount of methylmercury compounds in bodies of water are contained in the biomass of the system. The bioconcentration factors, that is, the ratio of the concentration of methylmercury in fish tissue to concentrations in water can be extremely large, usually on the order of 10,000 to 100,000 (44).

The concentration of methylmercury in fish tissue is of special interest in terms of human exposure to this form of mercury. The aquatic food chain of methylmercury is the main if not the sole source of human exposure barring episodes of accidental discharge or misuse of man-made methylmercury compounds. Thus, factors that affect concentrations of methylmercury in edible fish tissue are of considerable importance in assessing potential human health risks from this form of mercury.

Factors Affecting the Accumulation of Methylmercury in Fish

In general, for any body of water one might expect to see an eventual steady state distribution of methylmercury given the fact that both synthetic and degradation reaction are occurring at the same time. These steady state levels with reference to concentrations in sediment, water, and in the biomass may be expected to be influenced by a wide variety of experimental conditions, perhaps only a few of which have so far been identified. No detailed review will be given in this paper but the reader is referred to other references that give a more specific treatment of this topic (34).

Steady state levels of methylmercury in sediment, water, and biomass are believed to be affected by (a) concentrations of inorganic mercury in sediment and water, (b) oxygen tension, (c) organic matter in water, (d) microbial activity in sediments, and (e) pH of the water. In view of the current interest in the relationship between the use of fossil fuels, particularly coal, and possible acidification of large bodies of freshwater, the influence of aquatic pH on levels of methylmercury in fish will be given special treatment here.

Investigators in Scandinavia were the first to notice the inverse correlation between the concentration of methylmercury in edible tissues of fish and the pH of the water. An example from a recent report (22) is given in Fig. 2.

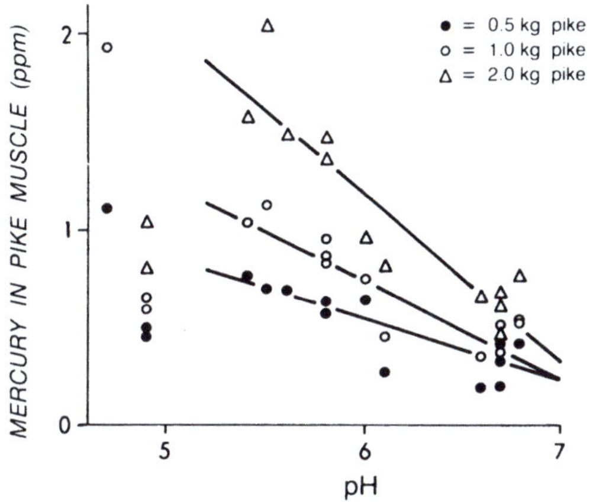

FIG. 2 — The concentrations of total mercury in Pike (muscle tissue) as a function of pH. Regression lines were drawn for three groups of fish distinguished by body weight (0.5, 1.0, and 2.0 kg) over the pH range 5 to 7 (adapted from Fig. 3 of (22)).

The regression lines of mercury in fish muscle versus pH are highly significant for the three sizes of fish that were studied between pH 7 and pH 5. This pH range is significant as it contains the range normally found in most bodies of freshwater. Indeed, elevated mercury levels can now be seen in fish in acidic waters in various parts of the world, including Sweden, Norway, Canada, and certain parts of the United States.

As may be seen in Fig. 2, there is also significant scatter of the data despite the presence of a significant correlation. Clearly a number of factors, in addition to pH, control the accumulation of mercury in fish. In fact, elevated levels of mercury in fish have been correlated with the presence of dissolved organic material, with low water conductivity, with the bioproductivity of the lake, and with the rate of influx of mercury into the aquatic systems. Indeed, a number of covariables may exist in addition to pH and mercury content of fish. For example, the lakes which have a low productivity, i.e., oligotrophic lakes, and low alkalinity tend to be more sensitive to mercury contamination, and mercury accumulation in fish is higher. These conditions also are strongly associated with low pH values so that the effect of pH on mercury concentration in fish might possibly be through an indirect mechanism. For example, correlations between methylmercury in fish based on length versus pH might be confounded by the possibility that fish in acidic lakes grow more slowly. Thus when comparing the same length of fish in an acidic versus a neutral lake, the fish in the acidic lake will be older since it grew more slowly and thus will have a higher mercury content due to age and not due to the lower pH itself.

Therefore, to resolve whether lake pH levels and mercury uptake by the fish actually reflect a cause and effect relationship and indeed whether acidification will actually enhance bioaccumulation of mercury in fish, it is necessary to attempt to understand the effect of acidity on the mercury chemistry, mobilization, and uptake. Field and laboratory studies on mercury cycles have resulted in a number of proposed mechanisms (19). (a) Acidic precipitation may remove mercury from the atmosphere more effectively than neutral pH. (b) The rate of methylation of inorganic mercury by microorganisms is pH-dependent with the maximum rate formation of monomethylmercury occurring at pHs lower than 6. (c) The pH influences the species of methylmercury formed by the microorganisms, a higher pH favoring the formation of the insoluble and volatile dimethylmercury which would not be accumulated readily in fish but rather vaporized into the atmosphere (Fig. 1). (d) A reduced pH should favor the presence of a divalent ionic mercury in water, reducing its presence in complexes and chelates, and therefore making mercury more available for methylation reactions. (e) The total biomass in an acidic lake is usually lower than in a neutral lake. Thus, since most of methylmercury is present in the biomass, one would expect its concentration to be necessarily higher in the biomass (including the fish) in the acidic lake.

Miller and Akagi (29) have found evidence to indicate that an important effect of pH is to mobilize methylmercury absorbed on sediments. Their results, Fig. 3, indicate that, irrespective of the type of sediment, a reduction in water pH shifts methylmercury from the sediment to the water phase. The magnitude of the shift, approximately a factor of 2 for each unit change in pH, would appear to be sufficient to explain the increase in mercury levels in fish in natural waters as a result of pH change, for example, the changes depicted in Fig. 2. Miller and Akagi (29) claim, therefore, that the effect of pH on the equilibrium of methylmercury between water and sediment may be the principal factor responsible for increased mercury levels in fish as a result of a pH change in the aquatic environment. It is, however, an assumption that the results obtained under equilibrium conditions are applicable to natural bodies of water.

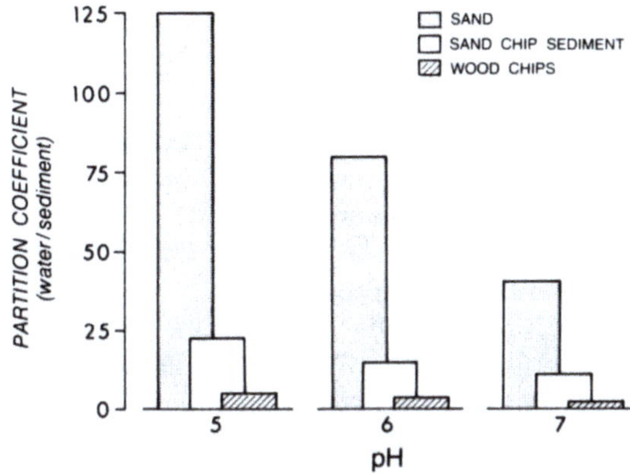

FIG. 3 — The effect of pH on the equilibrium distribution of methylmercury between water and sediment. Mercuric chloride labelled with ^{203}Hg isotope was added to three types of sediment: sand, sand chip sediment, and wood chips. The sediment was equilibrated with samples of natural water from the Ottawa River (Canada) for 21 days. The equilibrium distribution of the naturally-produced methylmercury was measured and is reported above. The values for the water/sediment concentration ratio were multiplied by 10,000 in the vertical scale used in this figure (adapted from Table 1 of (29)).

Mercury Levels in Fish and Anthropogenic Release of Mercury

Mercury levels in freshwater fish may be dramatically affected by the release of mercury into its habitat. The biogeochemical cycles discussed above indicate that inorganic mercury that is released into water systems is subject to methylation and thereby to accumulation by aquatic food chains. The local release of both inorganic and organic mercury compounds to freshwater lakes and streams in a variety of countries have led to extremely high concentrations of mercury, particularly in the large carnivorous fish such as pike. Indeed, the levels in pike have been recorded in excess of 10 ppm in Sweden and in Canada as a result of local contamination (for review, see (46)). In view of the large biological half-times of methylmercury in fish (46), as well as the potentially long residence time of mercury discharged into bodies of water, such local contamination can result in long-term elevation of methylmercury levels in fish.

As indicated previously, the oceans are the major reservoir of mercury, and in terms of total mercury, the oceanic content is so large that anthropogenic emission, even on a century basis, would not likely cause large changes in this reservoir. Indeed, it has been claimed that methylmercury levels in preserved species of oceanic fish, such as swordfish and tuna fish caught early in this century, do not differ substantially from levels of the present time. Nevertheless, it should be recalled that methylmercury will only be a small fraction of the total mercury in oceans, and little is known as yet about the processes that might affect levels of methylmercury in oceanic fish.

Human Exposure to Mercury and Its Compounds

The industrial use of mercury and the release of mercury to the environment by the non-mercury industries (fossil fuel) has resulted in a general increase of mercury in the environment throughout the industrial revolution. This increase is well illustrated in a Swedish study carried out in the late 1960s (for details, see (39)).

Levels of mercury in the general environment and human exposure. The levels of mercury in air, water, and food have recently been summarized and reviewed by the World Health Organization (46) and by the United States Environmental Protection Agency (44). Most of the material presented in this section will be drawn from these two reviews with some additional more recent information from specific papers as deemed appropriate.

Atmospheric levels of mercury and human intake by inhalation. The average concentration of mercury in the ambient atmosphere throughout the world appears to be about 20 ng/m^3. This value, originally reported in Germany in the 1930s, does not appear to have changed substantially. Ranges of concentration in the non-contaminated atmospheres have been reported to go from 0.5 to 50 ng/m^3.

Isolated so-called "hot spots" may have unusually high concentrations of mercury. For example, air levels up to 10,000 ng/m^3 have been found near rice fields where mercury fungicides have been used. Mercury concentrations in the vicinity of mercury mines may be as high as 15,000 ng/m^3.

It is generally accepted that the predominant form of mercury in the atmosphere is mercury vapor present as a monatomic gas. However, as indicated previously, other forms of mercury are also present. Johnson and Bramen (20) in a suburban site in Florida found that approximately 60% of the mercury in the atmosphere was in the form of the vapor, but they also noted that 19% was present as inorganic ionic mercury and approximately 15% as methylmercury compounds. There is general agreement that mercury in the particulate form accounts for only a very small fraction, on the order of 1%, of the total mercury in the atmosphere. Brosset's group, referred to previously, have also found the presence of "soluble" forms of mercury in the atmosphere (21).

The average daily retention of atmospheric mercury in man has been quoted by the USEPA (44) to be approximately 320 ng/70 kg body weight. This is based on the calculation assuming that the average atmospheric concentration was 20 ng/m^3, that the adult male had a ventilation rate of 50m^3, and that 80% of the inhaled rate was retained. Clearly in some urban and industrialized areas, atmospheric mercury concentrations could be higher, but they seem to be unlikely to exceed 50 ng/m^3 which would correspond to a daily retention rate of 800 ng Hg.

Levels of mercury in water and human intake from drinking water. A WHO group (46) concluded that levels of mercury in non-contaminated freshwater were usually less than 200 ng/l. Rainwater appears also to have concentrations generally below 200 ng/l.

The chemical form of mercury in freshwater and drinking water has not yet been well established, mainly because of difficulties of analysis. On theoretical grounds, one would expect mercury to be present mainly as Hg^{++} compounds in well oxygenated water. Generally speaking, methylmercury has not been detected in samples of freshwater. A review by McLean (27) indicated that methylmercury is present at concentrations close to the detection limit of methods. More recently, Kudo et al. (25) have reported that methylmercury compounds account for an average of 30% of the total mercury in samples of natural waters from the Ottawa River in Canada and from river and sewage effluent samples in Japan.

Implications of these findings are not clear with regard to the composition of mercury in drinking water. For example, the effect of water treatment, including chlorination on various forms of mercury, is unknown, but given evidence to the contrary, it will be necessary to assume that an appreciable fraction of mercury in drinking water may be in the form of methylmercury compound.

Levels of mercury in oceanic water are generally well below 300 ng/l. A most recent estimate made by Matsunaga et al. (26) indicates that the reliable value for baseline mercury in nonpolluted waters is approximately 5-6 ng/l. Fitzgerald (13) noted that the median value for oceanic water was approximately 59 ng/l. The form of mercury in oceanic water has not been well established. The high concentration of chloride ion would suggest that at least the inorganic form of mercury would form a complex with chloride.

Human intake from non-contaminated drinking water has been estimated to be generally less than 499 ng Hg/day, assuming two liters of water is consumed by the standard 70 kg adult (44). Even if 30% of the total mercury is present as methylmercury compounds, intake of the latter via drinking water is negligible compared to intake from ingestion of fish and shellfish.

Levels of mercury in flora and fauna and ingestion of mercury by man. The results of a number of extensive surveys (30,38,39,42) indicate that a major distinction should be drawn between fish, particularly the large carnivorous fish, and also shellfish and marine mammals on the one hand and other flora and fauna on the other hand. In general, mercury concentrations in most vegetable and meat samples that have been tested are usually so low as to be near or below the limit of detection of mercury by most analytical methods. Thus figures from surveys carried out by the U.S. Food and Drug Administration indicate that most foodstuffs have total mercury levels of below 20 ng/Hg/g wet weight. Poultry and meat may contain levels of up to 200 ng/g wet weight (30). In fact, in view of the uncertainty of these low numbers, it is impossible to calculate average daily intakes from nonfish food in the United States except that they are low. Indeed, a low intake of mercury from nonfish sources is consistent with the finding that nonfish eaters have the lowest blood concentrations of mercury (39).

Analysis of fish tissue indicates that on the order of 70 to 90% of the total mercury is present as monomethyl mercury compounds (8,39). Levels of methylmercury in the larger predatory oceanic fish such as tuna and shark may reach values up to 5 ppm without having to assume any direct contamination of the fish or the water by human activities (17). For example, over 50% of swordfish tested to date have values in excess of 1 ppm. Observations on canned tuna fish indicate average total mercury concentrations on the order of 0.25 ppm (43,44).

In freshwater, the average concentration in most fish is below 0.2 ppm; however, levels of mercury in pike may frequently exceed 1 ppm, and values as high as 20 ppm have been reported in heavily polluted areas (12). Reference is made to the United Nations Environmental Protection Program (17), to a large survey in the United States on commercial species of fish (43), and to an extensive survey in England (42).

Factors influencing the levels of methylmercury in fish have been discussed briefly in the previous section. It is now well established that two of the most dominant factors controlling levels of methylmercury in edible tissues of fish are the age of the fish and the species. Marine mammals can also accumulate mercury; for example, the livers of seal may obtain very high concentrations of total mercury on the order, in some cases, of 340 ppm, but over 90% of this is in the form of inorganic mercury probably combined in an inert form with selenium (24). Nevertheless, sufficient amounts of methylmercury are found in seal tissue, including liver, so that individuals consuming seal meat, such as the Eskimo, may develop high blood concentrations of methylmercury (14).

The U.S. Department of Commerce (43) conducted a survey of over 25,000 individuals representative of the U.S. population with the intention of determining the intake of methylmercury from fish. On a per weight basis, consumption of tuna far exceeds that of other species of fish in the United States (Table 3). However, since the actual number of users or consumers of tuna fish is also very high, the amount of tuna consumed per user is not dramatically different from the consumption per user of other species of fish. It is of interest to note that the highest consumption per user is of Northern Pike, giving a value of 17.4 g/day, and that other species of freshwater fish, bass and trout, are next highest in per user consumption at about 12 g/day. Tuna falls more or less in the midrange, between 6 and 10 g/day. In fact, the range in per user consumption is quite small being no more than a factor of two, despite the huge range in total consumption of each species. The data from the U.S. Department of Commerce (43) do not allow an accurate calculation of the average daily intake of methylmercury. However, if we assume (a) an FDA figure of 27 g/fish/day as the upper 95% limit of fish intake in the U.S. population, (b) an average value of 220 ng/Hg/g for mercury in tuna, and (c) that 75% of fish consumption is tuna, it follows that 95% of the population consumes less than 4.5 μg/day Hg as methylmercury. Contributions from other fish would be less than 1 μg/day, assuming an average concentration of 0.1 μg/g in fish. Thus it seems likely that 95% of the population will consume less than 5 μg/Hg as methylmercury per day from fish. The average daily fish consumption in the United States is 18.7 g (Table 3), thus the average mercury consumption from fish would be about 3 μg/day for a 70 kg person.

A remarkable feature of Table 3 is the high popularity of tuna in the United States population, indicating that over two-thirds of the population consume tuna fish. The next fish food in order of popularity is shrimp accounting for 21% of the population and flounder 13%. Considerable national differences must exist in patterns of consumption of different species of fish. Individual publications have indicated that certain population groups, such as native peoples in Canada, coastal villagers in Peru, and fishing villages in Samoa and

	Rank	Amount, 10^6 lb/yr	Percent of total by weight	Number of actual users (millions)	Mean amount per user, g/day
Total		2957	100.	197	18.7
Tuna (mainly canned)	1	634	21.4	130	6.1
Unclassified (mainly breaded, including fish sticks)	2	542	18.4	68	10.0
Shrimp	3	301	10.2	45	8.3
Ocean perch[a]	4	149	5.0	19	9.7
Flounder	5	144	4.9	31	8.6
Clams	6	113	3.8	18	7.6
Crabs/lobsters	7	110	3.7	13	10.6
Salmon	8	101	3.4	19	6.7
Oysters/scallops	9	88	3.0	14	7.8
Trout[b]	9	88	3.0	9	12.3
Cod[a]	11	78	2.7	12	8.1
Bass[b]	12	73	2.5	7.6	12.0
Catfish[b]	12	73	2.5	7.5	12.1
Haddock[a]	12	73	2.5	11	8.6
Pollock[a]	15	60	2.0	11	6.8
Herring/smelt					
Sardines	16	54	1.8	10	6.7
Pike[b]	17	35	1.2	2.5	17.4
Halibut[a]	18	32	1.1	5.0	8.0
Snapper	18	32	1.1	4.3	9.3
Whiting	20	25	0.9	3.2	9.7
All other classified		152	5.1		

[a]mainly imports
[b]freshwater

TABLE 3 — Fish and shellfish consumption in the United States (September 1873 — August 1974) (44).

on islands in the Indian Ocean, have very high consumption of fish, as this form of food probably accounts for their total protein intake. The fishing villages in these coastal areas have a pattern of consumption totally dominated, of course, by marine fish and frequently by the higher predatory species such as shark and tuna fish. On the other hand, inland villages close to large freshwater fishing areas, such as in northern Canada, have exclusively an intake of freshwater fish, again fish that tends to be high in the food chain, such as pike, walleye, and bass. Intake of methylmercury in the special population groups having unusually high intake of fish may be 10 to 100 times greater than that in the general population.

UPTAKE, DISTRIBUTION, AND EXCRETION OF METHYLMERCURY IN MAN AND ANIMALS

The pharmacokinetics of methylmercury have recently been reviewed (44,46). Approximately 95% of an oral dose of methylmercury is absorbed whether given as a salt, for example, methylmercury nitrate, or attached to protein, or naturally present in oceanic fish such as tuna fish. Studies on animals also support the conclusion that virtually all of orally ingested mercury is absorbed in the gastrointestinal tract into the bloodstream.

Methylmercury distributes uniformly throughout the body tissues, the distribution phase being complete in approximately 3 days. Differences between tissue concentrations do not normally exceed a factor of about two. The brain contains about 10% of the ingested dose, about 7% is found in the blood compartment. A rough rule of thumb is that in a 70 kg man, approximately 1% of the body burden of methylmercury is found in one liter of blood.

As a consequence of the rapidity with which methylmercury attains steady state distribution throughout the body, one finds that concentration ratios between organs and tissues tend to be constant. Thus the concentration in the red cell is usually ten-fold higher than the simultaneous concentration in plasma and the concentration in the brain in primates, including man, is usually about five times higher than the corresponding whole blood concentration. It is of interest to note that there are very large species differences in brain to blood concentration ratios, for example, in the rat the concentration ratio for brain to blood is 100 to 50 times lower than that seen in primates. The consequence of this is that a given blood concentration in the rat does not have the same toxicologic significance as the same blood concentrations in humans.

Methylmercury readily crosses the placenta and is found in all fetal tissues. At delivery, the concentrations in fetal blood are usually higher in human infants, by about 20%, than the corresponding concentrations in maternal blood.

Methylmercury is accumulated and concentrated in hair at the time the hair is being formed in the hair follicle. The concentration in newly formed hair bears a constant ratio to the simultaneous concentration is blood. This ratio has the value of about 250 but shows some individual variation, usually in the range of 200 to 300. Once incorporated into the hair, the methylmercury concentration remains constant over the remainder of the residence of the hair on the head and even for years after storage of the hair sample. Since hair grows at a rate of approximately 1.1 cm/month, the methylmercury concentrations per centimeter represent the average blood concentration of the month during which that centimeter of hair was being formed. Thus longitudinal analysis of the hair samples centimeter by centimeter offers an effective means of recapitulating previous blood concentrations. This technique has now received wide use and acceptance in studies of populations exposed to methylmercury (46). It is particularly useful for recapitulating blood concentrations during pregnancy (3).

Methylmercury slowly undergoes cleavage of the carbon-mercury bond to release inorganic mercury in the body tissues. This breakdown has been estimated to occur at the rate of about 1% a day on a whole body basis. The ratio of methyl to inorganic mercury in tissues varies from tissue to tissue, as inorganic mercury is very unevenly distributed; it is preferentially accumulated by kidney and the highest fraction is found there in individuals exposed to methylmercury. Brain levels of inorganic mercury are generally low, accounting for about 4% of total mercury in brain and tissue according to animal experiments.

Despite the slow rate of conversion, the formation of inorganic mercury plays a key role in the excretion of mercury from the body in individuals or animals exposed to methylmercury. The fecal pathway accounts for about 80% of total excretion in man, and virtually all mercury in the feces is in the inorganic form. Urine contains about 70% of the total mercury in the inorganic form. Thus, inorganic mercury accounts for virtually all of the mercury excreted from the body in individuals or animals exposed to methylmercury.

The site of breakdown of methylmercury in the body has not yet been identified, but there is convincing evidence that the intestinal flora play an important role (35). Methylmercury is known to be rapidly excreted in bile in experimental animals, and it is believed that part of this is reabsorbed into the blood stream, that part converted by microflora to inorganic mercury which passes into the feces. Evidence of biliary secretion in humans is lacking except that it has been demonstrated that a nonabsorbable thiol resin given by mouth can increase fecal excretion presumably by trapping the methylmercury secreted in bile and preventing its subsequent absorption (11).

Since the distribution of methylmercury within the body is rapid and excretion is slow, the kinetics of elimination from the whole body are first-order; that is to say that they can be described by a single exponential or biological half-time. The biological half-time in man has been measured to be about 70 days. The biological half-time in the blood compartment is somewhat less, but averaging about 50 days and showing large individual differences. The biological half-times in blood have been recorded to be as low as 30 days and as long as 180 days. Some of the blood half-times have been calculated from hair analysis and, in part, this wide range could be due to errors. The reason for this wide range in biological half-times in blood is not understood except that it is known that lactating women have a lower biological half-time than non-lactating women. There is an important research need to understand the factors contributing to the value of the biological half-time in blood and in the whole body. Most observations on animals indicate that the biological half-times in individual tissues do not vary substantially from the biological half-times in blood or in the whole body. There is, however, some controversy with regard to the biological half-time of methylmercury in the brain. Direct evidence in man suggests that the biological half-time in brain is longer than in blood, but not substantially different from whole body values. However, field studies in Japan resulted in the claim that the biological half-time in the brain may be 200 days or more. There is a considerable need to try to resolve these conflicting figures.

As a rough approximation, accumulation and elimination of mercury in man can be represented by a single compartment model in which absorption from food is virtually complete and elimination occurs at the rate of roughly 1% of the body burden per day. This model predicts that the steady state body burden will be attained after a period of constant exposure of approximately one year. At that point, the steady state body burden will be roughly one hundred times the daily intake. No tissue has been found that continues to accumulate mercury throughout the life span of the individual in contrast to accumulation of lead by the skeleton or the accumulation of cadmium in the kidney which seem to take place continuously over most of the individual's lifetime.

EFFECTS AND DOSE-RESPONSE RELATIONSHIPS

Methylmercury Compounds

The human toxicology of methylmercury is well described in the literature due to catastrophic outbreaks in Japan in the 1950s and 1960s and to a large outbreak in Iraq in the winter of 1971–72. Studies on a number of cases of industrial poisoning have also contributed substantially to the understanding of the toxic effects of methylmercury on man (46), primarily, the damaging of the nervous system. The main areas affected are associated with the sensory, visual, and auditory functions and those areas concerned with coordination, especially the cerebellum. The earliest effects are nonspecific symptoms, such as complaints of paresthesia, malaise, and blurred vision. Subsequently, signs appear such as concentric constriction of the visual field, deafness, dysarthria, and ataxia. In the worst cases, the patient may go on to coma and ultimately death. In less severe cases, some degree of recovery occurs in each symptom; this is believed to be a functional recovery that depends on the compensatory function of the central nervous system. The subjective complaint of paresthesia was found to be a permanent symptom with patients exposed in the Japanese outbreak, whereas in the Iraq outbreak paresthesia, in many cases, was transient. The reason for the difference is not known.

Three important characteristics described in the qualitative nature of the effects of methylmercury should be noted. First, the effects are generally irreversible; second, they are selective to the nervous system — usually to specific areas in the central nervous system, and finally, there is a substantial latent period on the order of weeks or months between cessation of exposure and the onset of signs and symptoms.

The irreversible nature of the effects probably results from the fact that methylmercury destroys neurons in the central nervous system which are irreplaceable. The mechanism of the selective effects of the latent period are not too well understood. Syversen, in a series of publications, has claimed that the selective effects refer to the ability of certain cells in the central nervous system to repair the damage initially inflicted by methylmercury (for review see (10)). Thus, those cells capable of repairing injury survive, whereas those cells which lack the facility to repair the damage are the ones that are destroyed. Thus he notes that the small granule cells in the cerebellum lack this repair capacity and are the first cells in that area of the brain to succumb.

The reason for the long latent period is not understood. The mean latent period ranged from 16 to 38 days in Iraq, and in many cases initial symptoms appeared after the cessation of the intake of contaminated bread (3). In the outbreaks in Japan, it is difficult to determine the exact latent period because the starting point of intake is unclear in many cases. In some cases in Japan, however, a very long latent period of up to several years was reported (16,41). Included in these cases were some patients who showed only slight signs and symptoms but later developed the clinical feature of severe poisoning after cessation of eating the polluted fish. In the same period of time the other patients showed more or less improvement or progression (16). Such a long latent period of several years may be partially explained by psychogenic overlay which modifies the symptoms, or subclinical lesions which may be revealed by the aging factor. The long latent period itself, however, cannot be explained by slow accumulation of methylmercury in the brain nor by a slow accumulation of inorganic mercury which has split off from the methylmercury.

Dose-effect and dose-response relationships for methylmercury compounds in man have been presented and evaluated previously (40,46). The most recent analysis of the Iraqi data has been published by Nordberg and Strangert (31–33). This analysis took into account not only the individual range of thresholds to such symptoms as paresthesia but also the interindividual variation in whole body biological half-times. These two distributions were combined to give an overall estimate of the risk of paresthesia for a given steady state daily intake of methylmercury. The results are presented in Fig. 4. Their calculations indicated that an intake of 50 μg/day in an adult person give a risk of about .3% of the symptom of paresthesia, whereas an intake of 200 μg/day would give a risk of about 8% of symptoms of paresthesia. As pointed out by Nordberg and Strangert (32), the background frequency of these nonspecific symptoms such as paresthesia plays a key role in determining the accuracy of the estimates of response of low frequencies. They estimated from the same Iraq data the background frequency of paresthesia of 6.3%. However, there is considerable uncertainty in determining the precise value of the background frequency, and this uncertainty becomes the dominant cause of error at low rates of response.

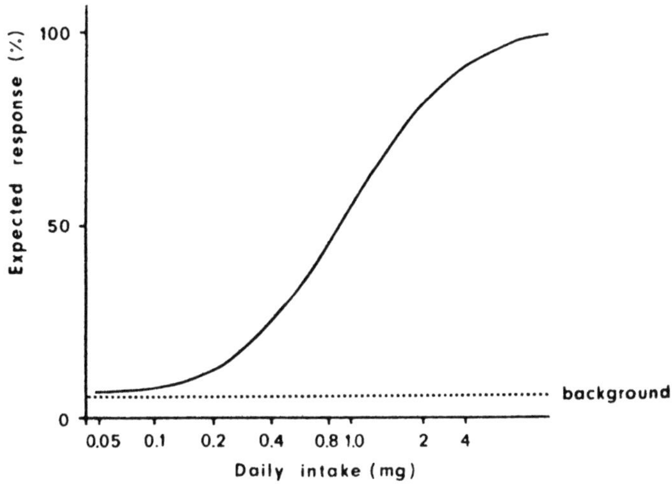

FIG. 4 — The calculated relationship between frequency of paresthesia in adults and long-term average daily intake of methylmercury. The calculations were performed by Nordberg and Strangert (31). The broken line is the estimated background frequency of paresthesia in the population. Data are taken from publications on the Iraqi outbreak of methylmercury poisoning (6,37).

Since the studies on the Iraqi outbreak, a major epidemiological study has been carried out on Cree Indians in Northwestern Quebec exposed to methylmercury in fish (28). The authors claim to find an association in men over age 30 and women over age 40 between a set of neurological abnormalities and the estimated exposure to methylmercury. However, it should be pointed out that this association has been seen by only four of seven observers who reviewed videotaped recordings of the neurological screening tests. The severity of these neurological abnormalities is described as mild or questionable. It was not possible to estimate any threshold body burden or hair levels because this population had been exposed possibly for most of their lives and therefore peak values in previous years were unknown. However, observations on this population over several years indicate that maximum blood concentrations are below 600 ppb and most below 200 ppb (45). A WHO expert group (47), on examining the reports from these studies, raised the possibility that this might be the first example of an endemic disease due to exposure to methylmercury.

Effects on the Developing Central Nervous System

Observations on both human subjects and animals indicate that the developing central nervous system is more sensitive to damage from methylmercury than the adult nervous system. The first indications came from the outbreak of methylmercury poisoning in Minamata, Japan, in the 1950s, when it was found that mothers who were slightly poisoned gave birth to infants with severe cerebral palsy (15). Subsequent studies on animals confirm the increased sensitivity of the fetus. The Iraq outbreak in 1971–72 also reported cases of severe damage to the central nervous system in infants prenatally exposed. More recent follow-up studies in Iraq have indicated a milder syndrome at lower exposure levels. In fact, it has been possible to demonstrate a relationship between the maximum hair level in the mothers during pregnancy and the frequency of abnormalities in their infants.

The clinical picture was dose-dependent. In those who were exposed to high maternal blood levels of methylmercury, the picture was of cerebral palsy indistinguishable from that caused by other factors. Microcephaly, hyperreflexia, and gross motor and mental impairment sometimes associated with blindness or deafness was the main pattern (1). Milder degrees of the infliction were not easy to diagnose during the first few months of life but they became clear with later times. The cases showed mainly psychomotor impairment and persistence of pathological reflexes. The mildest cases had findings quite similar to the findings in the minimal brain damage syndrome (1). The frequency of mental retardation in infants is related to maximum hair levels in the mother. Effects are detectable in prenatally exposed infants at hair levels substantially lower than those associated with the onset of signs in adults.

The McGill Group (28), in their study of Cree Indians exposed to methylmercury in fish, reported an association "...between findings on examination of tone and reflexes in Cree boys and the concentration of methylmercury in the mothers' hair during pregnancy. This association was shown at levels of methylmercury exposure which are very low in relationship to those previously reported to be associated with effects of methylmercury in utero.... These findings were isolated and the variation from normal was mild." The association noted in this study may have been due to chance as their observations on tone and reflexes were part of a number of observations, the rest of which did not correlate with mercury levels.

Methylmercury is excreted in the mother's milk and the concentration is always lower than that in the blood (2). However, when the mother has received substantial exposure to methylmercury, suckling infants may develop blood levels of methylmercury which are in excess of the minimum toxic level for adults (4).

The mechanism of prenatal damage to the central nervous system by methylmercury appears to be entirely different from that seen in adults. Choi et al. (9) reported that methylmercury interferes with the development of the central nervous system in part by interfering with the normal migration of neurons. More recent experimental work by Sager et al. (36) has indicated that one possible molecular target is the protein tubulin. Interaction of methylmercury with tubulin results in a delayed cell division as found in mice undergoing prenatal exposure to methylmercury.

REFERENCES

(1) Amin-Zaki, L.; Elhassani, S.; Majeed, M.A.; Clarkson, T.W.; Doherty, R.A.; and Greenwood, M.R. 1974. Intrauterine methylmercury poisoning in Iraq. Pediatrics **54**: 587–595.

(2) Amin-Zaki, L.; Elhassani, S.; Majeed, M.A.; Clarkson, T.W.; Doherty, R.A.; and Greenwood, M.R. 1974. Studies of infants postnatally exposed to methylmercury. J. Pediatr. **85**: 81–84.

(3) Amin-Zaki, L.; Elhassani, S.; Majeed, M.A.; Clarkson, T.W.; Doherty, R.A.; and Greenwood, M. 1976. Perinatal methylmercury poisoning in Iraq. Am. J. Dis. Child. **130**: 1070–1076.

(4) Amin-Zaki, L.; Majeed, M.A.; Greenwood, M.R.; Elhassani, S.; Clarkson, T.W.; and Doherty, R.A. 1981. Methylmercury poisoning in the Iraq suckling infant: A longitudinal study over five years. J. Appl. Toxicol. **1**: 210–214.

(5) Andren, A.W., and Nriagu, J.O. 1979. The global cycle of mercury. **In** The Biogeochemistry of Mercury in the Environment, ed. J.O. Nriagu, pp. 1–22. Amsterdam: Elsevier/North Holland.

(6) Bakir, F.; Damluji, S.F.; Amin-Zaki, L.; Murtadha, M.; Khalide, A.; Al-Rawi, N.Y.; Tikriti, S.; Dhahir, H.I.; Clarkson, T.W.; Smith, J.C.; and Doherty, R.A. 1973. Methylmercury poisoning in Iraq. Science **181**: 230–241.

(7) Beijer, J., and Jernelöv, A. 1979. Methylation of mercury in aquatic environments. **In** The Biogeochemistry of Mercury in the Environment, ed. J.O. Nriagu, pp. 203–210. Amsterdam: Elsevier/North Holland.

(8) Cappon, C.J., and Smith, J.C. 1981. Mercury and selenium content and chemical form in fish muscle. Arch. Env. Contam. Toxicol. **10**: 305–319.

(9) Choi, B.H.; Lapham, L.W.; Amin-Zaki, L.; and Saleen, T. 1978. Abnormal neuronal migration, damaged cerebral cortical organization and diffuse white matter astrocytes of human fetal brain: A major effect of methyl mercury in utero. J. Neuropath. Exp. Neurol. **37**: 719–733.

(10) Clarkson, T.W. 1983. Methylmercury toxicity to the mature and developing nervous system. **In** Biological Aspects of Metals and Metal-Related Diseases, ed. B. Sarker. New York: Raven Press.

(11) Clarkson, T.W.; Magos, L.; Cox, C.; Greenwood, M.R.; Amin-Zaki, L.; Majeed, M.A.; and Damluji, S.F. 1981. Tests of efficacy of antidotes for removal of methyl mercury in human poisonings during the Iraq outbreak. J. Pharmacol. Exp. Therap. **218**: 714–783.

(12) Femreite, N., and Reynolds, L.M. 1973. Mercury contamination of fish in Northwestern Ontario. J. Wild. Management **37**: 62–68.

(13) Fitzgerald, W.F. 1979. Distribution of mercury in natural waters. **In** The Biogeochemistry of Mercury in the Environment, ed. J.O. Nriagu, pp. 161–174. Amsterdam: Elsevier/North Holland.

(14) Galster, W.A. 1976. Mercury in Alaskan Eskimo mothers and infants. Env. Health Perspect. **15**: 135–140.

(15) Harada, Y. 1966. Study group on Minamata Disease. **In** Minamata Disease, ed. M. Katsuma. Kumamoto: Kumamoto University.

(16) Igata, A.; Niina, K.; Hamada, R.; and Ohkatsu, Y. 1975. The late onset of organic mercury intoxication after exposure. **In** Studies on the Health Effects of Alkylmercury in Japan, pp. 178–179. Japan: Environmental Agency.

(17) IRPTC 1980. Data profile on mercury. International Register of Potentially Toxic Chemicals, United Nations Environment Program, Geneva, Switzerland.

(18) Jensen, S., and Jernelöv, A. 1967. Biosynthesis of methylmercury I. Nordforsk. Biocid. Inf. **10**: 4.

(19) Jernelöv, A. 1980. The effects of acidity on the uptake of mercury in fish. **In** Polluted Rain, eds. T.Y. Toribara, M.W. Miller, and P.E. Morrow, pp. 211–222. New York: Plenum Press.

(20) Johnson, D.C., and Bramen, R.S. 1974. Distribution of atmospheric mercury species near ground. Env. Sci. Technol. **8**: 1003–1009.

(21) KHM 1981. Undersökning av Kvicksilver i Luft och naturliga Vatten. The Swedish Coal-Health-Environment Project. Swedish Power Board, Report No. 5, pp. 1–62, Vallingby, Sweden.

(22) KHM 1981. Kwicksilver Ekosystemet. Swedish Coal-Health-Environment Project, Report No. 192, Swedish State Power Board, pp. 33–51, Vallingby, Sweden.

(23) KHM 1981. Summary Situation Report. The Swedish Coal-Health-Environment Project. Swedish State Power Board, p. 43, Vallingby, Sweden.

(24) Koeman, J.H.; Peeters, W.H.M.; Koudstaal-Hol, C.H.M.; Thoe, P.S.; and DeGoen, J.J.M. 1973. Mercury-selenium correlations in marine mammals. Nature **245**: 385–386.

(25) Kudo, A.; Nagase, H.; and Ose, Y. 1982. Proportion of methylmercury to total amount of mercury in river waters in Canada and Japan. Water Res. **16**: 1011–1015.

(26) Matsunaga, K.; Konishi, S.; and Nishimura, M. 1979. Possible errors caused prior to measurement of mercury in natural waters with special reference to seawater. Env. Sci. Tech. **13**: 63–65.

(27) McLean, R.A. 1980. Mercury transport in the environment — analytical and sampling problems. **In** Polluted Rain, eds. T.Y. Toribara, M.W. Miller, and P.E. Morrow, pp. 151–174. New York: Plenum Press.

(28) Methylmercury Study Group. 1980. Methylmercury study. McGill University, Montreal, Canada.

(29) Miller, D.R., and Akagi, H. 1979. pH affects mercury distribution, not methylation. Ecotoxicol. Env. Safety **3**: 36–38.

(30) National Academy of Science. 1978. An Assessment of Mercury in the Environment. Washington, D.C.: National Research Council.

(31) Nordberg, G.F., and Strangert, P. 1976. Estimations of a dose-response curve for long-term exposure to methyl mercury compounds in human beings taking into account variability of critical organ concentrations and biological half times. A preliminary communication. **In** Effects and Dose-Response Relationships of Toxic Metals, ed. G.F. Nordberg, p. 273. Amsterdam: Elsevier/North Holland.

(32) Nordberg, G.F., and Strangert, P. 1978. Fundamental aspects of dose-response relationships and their extrapolation for noncarcinogenic effects of metals. Env. Health Perspect. **22**: 97–108.

(33) Nordberg, G.F., and Strangert, P. 1982. Risk estimation models derived from metabolic and damage parameter variation in the population. Paper presented at Meeting on Methodology of Evaluation of Chemicals, Rome, Italy, July 1982.

(34) Nriagu, J.O. 1979. The Biogeochemistry of Mercury in the Environment. Amsterdam: Elsevier/North Holland.

(35) Rowland, I.R.; Robinson, R.D.; Doherty, R.A.; and Landry, T.D. 1983. Are developmental changes in methyl mercury metabolism and excretion mediated by intestinal microflora? **In** Reproductive and Developmental Toxicity of Metals, eds. T.W. Clarkson, G.F. Nordberg, and P.R. Sager. New York: Plenum Press, in press.

(36) Sager, P.R.; Doherty, R.A.; and Rodier, P.M. 1982. Effects of methylmercury on developing mouse cerebellar cortex. Exp. Neurol. **77**: 179–183.

(37) Shahristani-al, H., and Shibab, K.M. 1974. Variation of biological half-life of methylmercury in man. Arch. Env. Health **27**: 342.

(38) Simpson, R.E.; Horwitz, W.; and Roy, C.A. 1974. Residues in food and feed. Pest. Monit. J. **7**: 127–138.

(39) Swedish Expert Group. 1971. Methyl mercury in fish, a toxicological-epidemiological evaluation of risks. Nord. Hyg. Tidskr. Suppl. 4.

(40) Task Group on Metal Accumulation. 1973. Accumulation of toxic metals with special reference to their absorption, excretion, and biological half-times. Env. Physiol. Biochem. **3**: 65–107.

(41) Tsubaki, T., and Shirakawa, K. 1977. Neurological aspects of methylmercury poisoning in Niigata. **In** Minamata Disease. Methylmercury Poisoning in Minamata and Niigata, Japan, eds. T. Tsubaki and K. Irukayama, pp. 145–165. Tokyo: Kodansha Ltd.

(42) U.K. Department of the Environment. 1976. Environmental Mercury in Man. A report of an interdepartmental working group on heavy metals. Pollution Paper No. 10. London: HMSO.

(43) U.S. Department of Commerce. 1978. Report on the Chance that U.S. Seafood Consumers Exceeding the Current Acceptable Daily Intake for Mercury and Recommended Regulatory Controls, pp. 1–30. Washington, D.C.: U.S. Department of Commerce, National Marine Fisheries Services.

(44) U.S. EPA 1980. Ambient Water Quality Criteria for Mercury. (EPA 440/5-80-058) Washington, D.C.: EPA, Criteria and Standard Division.

(45) Wheatley, B. 1979. Methyl Mercury in Canada — Exposure of Indian and Inuit Residents to Methyl Mercury in the Canadian Environment, pp. 1–200. Ottawa: Medical Services Branch, Department of National Health and Welfare.

(46) WHO. 1976. Environmental Health Criteria 1. Mercury, pp. 1–131. Geneva: World Health Organization.

(47) WHO. 1980. Consultation to Re-examine the WHO Environmental Health Criteria for Mercury. (EHE/EHC/80.22) Geneva: World Health Organization.

Changing Metal Cycles and Human Health, ed. J.O. Nriagu, pp. 311–322.
Dahlem Konferenzen 1984. Berlin, Heidelberg, New York, Tokyo: Springer-Verlag.

THE HAZARD TO HEALTH OF LEAD EXPOSURE AT LOW DOSE

H.L. Needleman
Depts. of Psychiatry and Pediatrics
University of Pittsburgh School of Medicine
Pittsburgh, PA 15213, USA

Abstract. Knowledge of lead's toxicity has existed since antiquity. Recent studies in children converge on the conclusion that doses of lead below those which produce stark symptoms are associated with neuropsychological deficit. Animal investigations support this assertion.

INTRODUCTION

Man has known and exploited lead's unique properties since antiquity, and for almost as long has been aware of the dangers which accompany its use. This metal is certainly the most thoroughly studied and best understood neurotoxin, yet control of its dispersal into the human environment has been sluggish and marked by protracted controversy and only brief fits of activity.

Current understanding of lead's toxicity is bounded by two extreme positions. At one pole is the assertion that there is no threshold below which lead is innocuous. Opponents hold that lead is an essential trace substance, and that a deficiency of the metal is deleterious to health. In this paper I shall attempt to trace in scant outline the history of scientific knowledge of lead intoxication from ancient times to the present day, and to identify the roadmarks and some of the detours in the progress of our understanding of the significance of low level exposure. I shall also attempt to point out some of the unanswered questions which deserve future study.

Lead's ductility, durability, and its insulating properties caused it to be extensively mined and smelted in ancient Egypt. Nikander, a physician of the 4th centery BC clearly described the symptoms of lead colic, and Dioscorides, in the 2nd century BC, noted that "lead makes the mind give way." In colonial America, Benjamin Franklin noted that the colic of painters and tinkers was due to lead exposure. At the same time, Sir George Baker proved that lead was the cause of the epidemic colic in Devonshire, England, and showed that the source of the metal was the local cider.

At the turn of the century, childhood lead poisoning was described by an Australian ophthalmologist, A.J. Gibson (11). For the next forty years, it was assumed that children who recovered from acute intoxication were free from sequellae. The first investigator to question this conventional wisdom was a Boston neurologist, Randolph Byers. In 1943 Byers (6) followed up a group of children who had ostensibly recovered from frank lead intoxication and found that 19 of 20 were cognitively impaired or behavior disordered. Byers raised for the first time the possibility that some of the learning disorders and behavior problems in American schools were due to undiagnosed lead poisoning.

When screening studies conducted in the 1960s showed that in some areas as many as 10% of "well" children had blood lead levels in the range of 40–50 ug/dl, others began to consider the possibility that a "silent epidemic" of lead intoxication was occurring. It was about this time that a geochemist, Clair Patterson (18), showed that because of unrecognized contamination during the collection and analysis of specimens, earlier measurements of "natural" levels of lead were grossly overestimated. His measurements of pelagic waters, collected with exquisite care, reported much lower concentrations of lead than had been heretofore published. As a result, the relative contamination of waters in contact with man's technological effluents was considerably greater than previously reckoned. Patterson's studies relied on the most meticulous collection and analysis of specimens, under conditions designed to reduce background contaminant levels to a minimum. Few, if any other, laboratories have been capable of approximating the stringency of his methods and the reduction of background contamination that he has been able to achieve.

Because the "natural" levels of lead measured by Patterson and associates are lower by two or three orders-of-magnitude than found by others, the ratio between the "normal" lead burdens found in contemporary urban dwellers and those measured in pretechnological ancients is much larger, in the range of 600:1. It is Patterson's assertion that this increase in body burden is prima facie evidence of intoxication.

Byer's and Patterson's reports, the early screening data, and the burgeoning interest in environmental matters converged, and this shifted attention from the widely acknowledged devasting effects of frank lead poisoning to the possible, unrecognized consequences of exposure at much lower doses. Jane Lin-Fu of DHEW underlined the seriousness of the issue in a widely read review (14), and a large group of epidemiologic studies of children and laboratory investigations of a number of species followed in close succession.

RECENT EPIDEMIOLOGIC STUDIES

The human studies generally took the form of cohort studies in which two groups of children differing in blood lead level were compared on neurologic or psychometric outcomes. Among the earlier reports of merit were three which showed differences between high and low exposure groups. Burde and Choate (8) compared black preschoolers enrolled on the perinatal collaborative disease study of the NIH. Infants who had pica and moderately elevated blood lead levels were compared to non-pica controls. High lead subjects were inferior on psychometric intelligence and behavior. Perino and Ernhart (19) compared a group of black preschoolers with blood lead >40 ug/dl to a group with blood leads <30 ug/dl and found differences on the McCarthy Scales after controlling for socioeconomic status and parental IQ in the multiple regression analysis. Landrigan and colleagues (13) studied children in the vicinity of a lead smelter and found that those with blood lead levels below 40 ug/dl were superior to children with blood levels >40 ug/dl. A number of other studies conducted at about the same time failed to find significant differences between groups (12, 15, 21).

Most of the difficulties in experimental design or execution which affect almost all of the studies to a greater or lesser extent can be subsumed under four categories. First, many studies rely on blood lead as the index of body burden. But blood is a short-term storage system, and lead levels may return to normal once exposure ceases. Subjects may therefore be misclassified if this tissue is the marker. Second, many studies employed group tests or screening tests which do not possess the sensitivity to disclose lesser amounts of deficit. Third, many other factors affect development, and some of these, such as socioeconomic status or health history, may be correlated with lead and thus act as confounders. Finally, bias may be introduced in the self selection of subjects. Mothers who are worried about their childrens' intellectual status, for example, may systematically reject invitation into a study which will examine psychometric function.

My colleagues and I set out to address systematically these difficulties in a retrospective cohort study of first and second grade children in two industrial towns near Boston, Massachusetts (17). We collected shed deciduous teeth, a marker of body burden that persists long after exposure ceases, classified children according to their dentine lead levels, and measured their neuropsychologic function on a large battery of sensitive tests. Confounding was controlled by comparing children on a large number of covariates (N = 39) and

WISC-R	Low Lead $\bar{X}$	High Lead $\bar{X}$	P Value* (2 tail)
Full Scale I.Q.	106.6	102.1	.03
Verbal I.Q.	103.9	99.3	.03
Performance I.Q.	108.7	104.9	.08
Seashore Rhythm Test			
Subtest A	8.2	7.1	.002
Subtest B	7.5	6.8	.03
Subtest C	6.0	5.4	.07
Sum	21.6	19.4	.002
Token Test			
Block 1	2.9	2.8	.37
Block 2	3.7	3.7	.90
Block 3	4.1	4.0	.42
Block 4	14.1	13.1	.05
Sum	24.8	23.6	.09
Sentence Repetition Test	12.6	11.3	.04
Reaction Time Under Varying Intervals of Delay	($\bar{X}$ ± S.D.)	($\bar{X}$ ± S.D.)	
Block 1 (3 sec.)	.35 ± .08 sec.	.37 ± .09 sec.	.32
Block 2 (12 sec.)	.41 ± .09	.47 ± .12	.001
Block 3 (12 sec.)	.41 ± .09	.48 ± .11	.001
Block 4 (3 sec.)	.38 ± .10	.41 ± .12	.01

TABLE 1 – Outcomes which differed between high and low lead groups at $P \leq 0.05$.

controlling in the biostatistical analysis for those five variates which differed between high and low lead groups. When the two groups were compared by analysis of covariance, controlling for socioeconomic status, mother's age at subject's birth, mother's education, family size, and maternal IQ, a number of significant differences were observed. Table 1 summarizes the outcomes which were significantly different between groups. Teachers' ratings of classroom behavior for 2146 students evaluated by teachers blind to the lead levels showed a dose-dependent increase in unfavorable scores with no evidence of a threshold (Fig. 1).

A subsample of these subjects were given a quantitative EEG (3). When power-frequency functions were compared, high lead subjects were found to have less midline alpha content and more midline delta content than their low lead counterparts. Follow-up of these students four years later showed that the teachers' ratings were relatively stable; high lead subjects continued to be graded as less attentive in class. Direct observation of these students by raters blind to their lead levels showed that high lead subjects were off-task a significantly greater amount of time than low lead subjects (2).

Three other investigators have published data since the Boston studies, which also demonstrate effects of low level exposure on children. Winneke and colleagues (27), employing tooth lead as the index of body lead burden, also found decreases in psychometric IQ, attention, and parents' report of behavior. Yule et al. (28) classified London schoolchildren by blood lead levels and found dose-related decreases in IQ, reading scores, and similar responses on the same teachers' rating scale employed by Needleman et al. (17). Ernhart et al. (9) reported no differences in neuropsychologic function as measured by the McCarthy Scales. However, more careful examination of their tables reveals that significant deficits in high lead subjects on the general cognitive, verbal, and motor scales were demonstrated but overlooked.

Taken in sum, these recent studies indicate that exposure to lead, which results in blood lead levels in the range of 13–35 ug/dl, are associated with deficits in psychometric intelligence, attention, auditory and language function, and classroom behavior. Considering the differences in culture of the subjects under investigation, the amount of correspondence between these studies may be considered remarkable. No single study in the group mentioned above is without flaw. Perfect control of exposure or covariates cannot hope to be achieved in epidemiologic investigations. In the real world of children, multivariate space is infinite, and there is always another variable to be suggested and

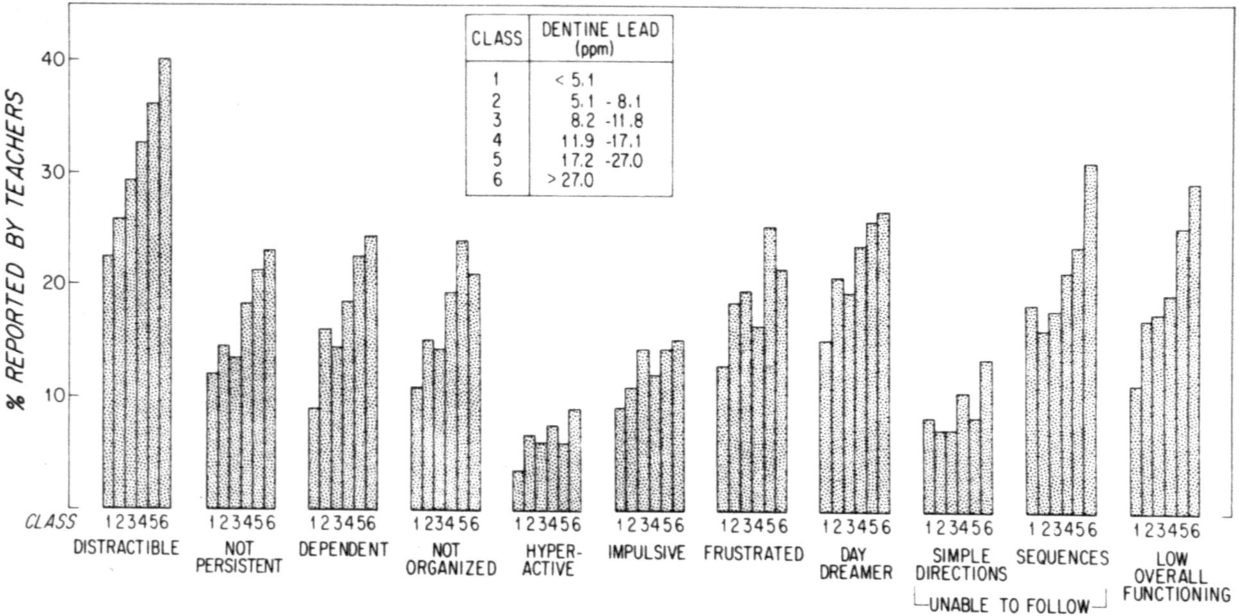

FIG. 1 – Proportion of negative rating on eleven classroom behaviors in relation to dentine lead levels (N = 2146 subjects) (Reprinted from (17)).

controlled. If the studies outlined above are evaluated as a tapestry in relation to each other, each will be seen to supplement the weaknesses of the others.

When the limits to which epidemiologic studies can be driven have been reached, investigators turn to the laboratory where lead can be administered to experimental animals in controlled doses, and experiential, social, genetic, and rearing variables can be brought under relatively systematic control. Impaired learning at low doses of lead has been demonstrated in the rodent (20) and in the primate (5, 23). Delayed appearance of brain cytochromes (4) and delayed synaptogenesis have also been found at low doses (1).

Studies in both the intact animal and in cellular preparations have deepened our understanding of the neurotoxic mechanisms which underlie the altered behaviors observed (26). While considerable work remains to be done before many of the conflicting findings are rationalized, it is clear that lead alters cholinergic and monoaminergic neurotransmission and inhibits brain adenyl cyclase activity. In vivo exposure of rats to lead at extremely low doses alters GABAergic neurotransmission, particularly GABA uptake and release by synaptosomes. Lead appears to be associated with a decrease in GABA transporting sites, probably due to a decrease in GABAergic neurons. It should be noted that diminished GABAergic activity is associated with increased CNS excitability and lowered resistance to seizures, suggesting intriguing connections with the altered behavior reported by teachers and the EEG changes reported earlier.

Another possible route for expression of lead's neurotoxicity has been suggested by Silbergeld (26), who noted that d-ALA may behave as a weak GABA agonist and, by occupying presynaptic GABA receptors, result in diminished GABA release. d-ALA is neuroactive in rodents and is suggested as one of the causes of disordered behavior in patients with acute intermittent porphyria (16). The enzyme d-ALA-D is exquisitely sensitive to lead, and the biochemical profile of lead exposure resembles AIP in many features.

The measured threshold for lead effects in mitochrondrial preparations is 0.1 ug/dl (26). In order to derive a valid relationship between the toxic concentration at the critical tissue site and the external dose or the corresponding blood lead value, considerable work needs to be done on specifying the partition coefficients and the pharmacokinetics of distribution.

THE QUESTION OF LEAD'S ESSENTIALITY

In 1975, Schwarz reported that rats raised on a lead-deficient diet were growth retarded when compared to controls receiving a lead supplement (25). A number of questions have been raised about this study. Lead was administered in two forms, as the subacetate and as nitrate. Differential growth effects were noted, although the two compounds are virtually identical biologically. A more serious reservation grows from the failure to replicate reported by the investigator. This is, in the final analysis, the critical test of any hypothesis.

Reichelmayer-Lais and Kirchgessner (22) have published a series of papers which compare rodents raised on a low lead diet (18 ppb) to controls raised on higher concentrations (1.0 ppm). Deprived animals are reported to show growth deprivation and signs of iron deficiency anemia. These effects appeared remedied when lead was administered. It is important to note that the lead deficient diet was in part produced by treatment with chelating agents (EDTA and APDC). It is unlikely that all of the chelants were removed from the diet, and it is reasonable to suggest that the chelants were responsible for the iron deficiency observed. Furthermore, the real-world significance of any putative trace function for lead is tenuous, given the virtual impossibility of any organism developing a lead deficiency in the 20th century.

UNANSWERED QUESTIONS FOR STUDY

Prenatal Effects of Lead on the Fetus

Lead crosses the placenta and appears in the umbilical cord blood in concentrations closely correlated with the mother's. Umbilical cord blood levels have been shown to decrease over the period of 1978-1981 at a slope of approximately 7% per year (24). The decrease is synchronous with reduced sales of leaded gasoline. Since lead has been shown in in vitro systems to possess mutagenic properties and chromosomal aberrations have been observed in lead workers (7, 10), further study of its mutagenic and teratogenic potential is warranted.

Effects on Aging

The largest pool of lead is in the bone where it is assumed to be metabolically inert. During the later years of life, the bones demineralize. The fate of the lead in this pool, while unknown, raises some intriguing questions for examination. Does some of this lead find its way into the central nervous system? May it be in part responsible for some of the disordered mentation that accompanies, some feel inevitably, the aging process? The systematic study of the mobile lead pool in relation to cognitive processes could prove a rewarding endeavor.

CONCLUSION

This brief began by stating that lead is the best studied neurotoxin. More data from human population studies, from experiments in the intact animal, from cell cultures, and from biochemical systems has been collected and evaluated than for any other CNS toxin. Yet the controversy over whether lesser levels of lead are hazardous continues to smolder. Some, but not all, of the reasons for this protracted debate are scientific: the delivery of an unimpeachable proof from human population studies will likely be a phantom effort. An infinite number of variables affecting brain development exists and no study can control them all. Also, neuropsychologic outcomes do not lend themselves to the same precision in measurement that proteins offer. There is no radioimmunoassay for cognition or attention. And some of the reasons for this debate are political.

Given these limitations, it seems proper for scientists and regulators to devote a portion of their time to the study of what should constitute adequate canons of proof in evaluating this ubiquitous toxin. What criteria should be met to enable science to say, "The danger of not acting is now clear enough, and the evidence of significant toxicity solid enough, to make the next step — removal of lead from the human environment — an inescapable and urgent imperative."

REFERENCES

(1) Averill, D., and Needleman, H.L. 1980. Neonatal lead exposure retards cortical synaptogenesis in the rat. **In** Low Level Lead Exposure: The Clinical Implications of Current Research, ed. H.L. Needleman, pp. 201–210. New York: Raven Press.

(2) Bellinger, D.; Needleman, H.L.; Hargrave, J.; and Nichols, M. 1981. Elevated dentine lead levels and school success. Proceedings of the Biennial Meeting of the Society of Research Child Development, Boston, MA.

(3) Burchfiel, J.; Duffy, F.H.; Bartels, P.; and Needleman, H.L. 1980. The combined discriminating power of quantitative electroencephalography and neuropsychologic measures in evaluating central nervous systems effects of lead at low levels. **In** Low Level Lead Exposure: The Clinical Implications of Current Research, ed. H.L. Needleman, pp. 75–89. New York: Raven Press.

(4) Bull, R.J.; Lutkenhoff, S.D.; McCarty, G.E.; et al. 1979. Delays in the postnatal increase of cerebral cytochrome concentrations in lead-exposed rats. Neuropharmacology **18**: 83–92.

(5) Bushnell, P.J., and Bowman, R.E. 1979. Reversal learning deficits in young monkeys exposed to lead. Pharmacol. Biochem. Behav. **10**: 733–742.

(6) Byers, R., and Lord, E. 1943. Late effects of lead poisoning on mental development. Am. J. Dis. Child **66**: 471–494.

(7) Deknudt, G.; Leonard, A.; and Ivanov, B. 1973. Chromosome aberrations observed in male workers occupationally exposed to lead. Env. Physiol. Biochem. **3**: 132–138.

(8) de la Burde, B., and Choate, M.S. 1975. Early asymptomatic lead exposure and development at school age. J. Ped. **87**: 638–664.

(9) Ernhart, C.B.; Landa, B.; and Schell, N.B. 1981. Subclinical levels of lead and developmental deficit: A multivariate follow-up reassessment. Ped. **67**: 911–919.

(10) Forni, A., and Secchi, G.C. 1972. Chromosome changes in preclinical and clinical lead poisoning and correlations with biochemical findings. Proceedings of the International Symposium of Environmental Health Aspects of Lead (CEC, WHO, EPA), Amsterdam, pp. 473–482.

(11) Gibson, J.L. 1917. The diagnosis, prophylaxis and treatment of plumbic ocular neuritis amongst Queensland children. Med. J. Aust. **2**: 201–204.

(12) Kotok, R., and Herriot, J.T. 1977. Cognitive evaluation of children with elevated blood lead levels. Am. J. Dis. Child **131**: 791–793.

(13) Landrigan, P.J.; Whitworth, R.H.; Baloh, R.W.; Staehling, N.W.; Barthel, W.F.; and Rosenblum, B.F. 1975. Neuropsychological dysfunction in children with chronic low level lead absorption. Lancet **1**: 708–712.

(14) Lin-Fu, J.S. 1972. Under absorption of lead among children — A new look at an old problem. N. Engl. J. Med. **286**: 702–710.

(15) McNeil, J.L.; Ptasnik, J.S.; and Croft, D.B. 1975. Evaluation of long-term effects of elevated blood lead concentrations in asymptomatic children. Arch. Ind. Hyg. Toxicol. (Suppl.) **26**: 97–118.

(16) Moore, M.R.; McGillion, F.B., and Goldberg, A. 1976. Some pharmacological and behavioral effects of D-aminolevulinic acid. **In** Porphyrins in Human Diseases, ed. M. Doss, pp. 148–154. Basel: Karger.

(17) Needleman, H.L.; Gunnoe, C.G.; Leviton, A.; Reed, R.R.; Peresie, H.; Maher, C.; Barrett, P. 1979. Deficits in psychologic and classroom performance of children with elevated dentine lead levels. N. Eng. J. Med. **300**: 584–695.

(18) Patterson, C.C. 1965. Contaminated and natural lead environment of man. Arch. Env. Health **11**: 344–363.

(19) Perino, J., and Ernhart, C.B. 1974. The relation of subclinical lead level to cognitive and sensory impairment in black preschoolers. J. Learn. Disord. **7**: 26–30.

(20) Petit, T.L., and Alfano, D.P. 1979. Differential experience following lead exposure: Effect on brain and behavior. Pharmacol. Biochem. Behav. **11**: 165–171.

(21) Ratcliffe, J.M. 1977. Developmental and behavioural functions in young children with elevated blood lead levels. Br. J. Prev. Soc. Med. **31**: 258–264.

(22) Reichlmayr-Lais, V.A., and Kirchgessner, M. 1981. Zur Essentialität von Blei für das tierische Wachstum. Z. Tierphysiol. Tierernährg. v. Futtermitt. **46**: 1–8.

(23) Rice, D., and Willes, R.J. 1979. Neonatal low-level lead exposure in monkeys (macaccu fasiculoaris): Effect on two choice non-spatial form discrimination. J. Exp. Path. Toxicol. **2**: 1195–1203.

(24) Scanlon, J. 1971. Umbilical cord blood lead concentration. Am. J. Dis. Child. **121**: 325–326.

(25) Schwarz, K. 1975. Potential essentiality of lead. Arch. Rade. Toxikol. (Suppl.) **26**: 13–28.

(26) Silbergeld, E.K., and Howska, R.E. 1980. Neurochemical investigations of low level lead exposure. **In** Low Level Lead Exposure: The Clinical Implications of Current Research, ed. H.L. Needleman, pp. 135–158. New York: Raven Press.

(27) Winneke, G.; Brockhaus, A.; Kramer, U.; Ewen, U.; Kujanek, G.; Lechner, H.; and Janke, W. 1981. Neuropsychological comparison of

children with different tooth lead levels — Preliminary report. **In** Proceedings of the International Conference on Heavy Metals in the Environment, pp. 553–556. WHO Amsterdam, CEP Consultants.

(28) Yule, W.; Lansdown, R.; Millar, I.B.; and Urbanowicz, M. 1981. The relationship between blood lead concentrations, intelligence and attainment in a school population: A pilot study. Dev. Med. Child Neurol. **23**: 567–576.

Changing Metal Cycles and Human Health, ed. J.O. Nriagu, pp. 323–344.
Dahlem Konferenzen 1984. Berlin, Heidelberg, New York, Tokyo: Springer-Verlag.

STANDARD SETTING AND METAL SPECIATION: ARSENIC

R.L. Zielhuis and A.A.E. Wibowo
Coronel Laboratory for Occupational and Environmental Health
University of Amsterdam, 1054 BW Amsterdam, The Netherlands

Abstract. Environmental quality standards are usually based upon concentrations of metals as such, not taking into account essential differences in toxicokinetics and toxicodynamics between different compounds of the same element. There is an urgent need for setting specific standards for different (groups of) metallic compounds; the possibility of total exposure standards is discussed. The present confusion and logical inconsistency between standards is illustrated for arsenic but applies also to other metals.

INTRODUCTION

Different compounds of the same metal may carry widely different health risks. The fractional gastrointestinal absorption of soluble $PbCl_2$ is about four times that of insoluble PbS in fasting humans. Inorganic mercury and short-chain alkylmercury carry essentially different health risks. Differences in water or fat solubility, aggregation state, chemical structure and volatility may determine intake, uptake, distribution, accumulation, biotransformation and excretion, i.e., the toxico*kinetics,* but can also affect the working mechanisms, critical organs and critical effects, i.e., the toxico*dynamics*.

Environmental and occupational standards for metals as regards ambient or workroom air, food, drinking water, or beverages are usually expressed as concentrations of the metal as such, suggesting exposure to the same agent. In that case logical consistency might be expected between quality standards for above mentioned exposure routes. It might also allow the derivation of total exposure standards which are particularly relevant for metals, because man is usually exposed simultaneously to all environmental compartments mentioned. However, no logical consistency appears to exist (46). This discrepancy is due to:

- "*compartmentalization,*" i.e., regulatory agencies which set standards for one compartment, do not cooperate with one another, e.g., the present WHO-drinking water standard for lead is not consistent with the proposed WHO-health-based-occupational exposure limit (44);
- lack of similar *procedures* (43,47), i.e., agencies do not always at first critically evaluate the available toxicity data. A criteria document should allow the establishment of no-adverse-response levels on which recommended health-based quality limits can be based and which only take into account considerations of health protection. This first step should be followed by a second step: the decision making process, in which regulatory agencies together with representatives of industry and of the population at risk (e.g., workers' unions) take into account both the health-based recommended limits and socioeconomic, political, and technological constraints (43,47). Too often the first step is never taken; the present CEC and WHO drinking water standards are not based upon criteria documents (however, WHO has announced such a document for the near future (22));
- lack of *metal speciation data:* the total metal concentrations are analyzed; different metal *compounds* are not measured. Two international conferences of CEC-WHO on "Heavy Metals in the Environment" (11,12) emphasized the urgent need for relevant metal speciation information; various studies were presented on, e.g., methylation of metals in natural waters, sediment, soil, and solid waste. To protect human health, metal speciation is particularly needed when different compounds of the same metal, which differ in metabolism and/or toxicity, are present in food, air, water, etc.

We have examined the logical consistency of various arsenic (As) standards as an example. The objective was to answer the following three questions:

1. Can As be treated as an entity, or is it necessary to distinguish between (groups of) As-compounds?

2. Is it possible to derive health-based total exposure limits for As?
3. Is it possible to derive logically consistent health-based quality limits for As?

The toxicology of As cannot be reviewed in this paper; recent reviews have been published by, e.g., NAS (25) and WHO (45).

PRESENT QUALITY STANDARDS

Table 1 lists present quality standards for As in ambient air, workroom air, drinking water, and the acceptable daily intake (ADI) from food. It should be noted that with the exception of workroom air, all limits refer to total As as such. Official documentation does not exist for the drinking water standards; WHO/FAO (41,42) did not base their tentative acceptable daily intake from food on any validated background information. The US Cancer assessment group (33,35) proposed as drinking water quality standard 0.02 µg As/1, based upon linear extrapolation, accepting a lifetime skin cancer risk of 1.10^{-5}.

The standards for ambient air and drinking water can be compared to the actual As-concentrations as reported inter alia by NAS (25) and WHO (45): in

Country	Medium and As levels	Source
	Ambient air	
USSR	3 µg As/m^3	(26)
	Workroom air	
US	TLV=200 µg As/m^3-twa for As and soluble compounds; no TLV for As in As_2O_3 production	(1)
	10 µg As/m^3 twa-15 min (OSHA)	(27)
USSR	300 µg As/m^3 (As_2O_3 and As_2O_5)	(14)
F.R. GERMANY	200 µg As/m^3 (As_2O_3, As_2O_5, As-acid and -salts)	(14)
THE NETHERLANDS	500 µg As/m^3-twa for As and its compounds 200 µg As/m^3-twa for As_2O_3 production	(14)
	Drinking water	
CEC, WHO, US, USSR	50 µg As/l	(46)
	Food	
WHO/FAO	50 µg/kg b.w./d (ADI)	(41, 42)

TABLE 1 – Present quality standards for arsenic.

ambient air concentrations of up to a few ng As/m^3, except in the vicinity of sources of emission; in drinking water up to 3.4 mg As/1 in areas with high geochemical As-sources (e.g., Taiwan and South America), but in most industrialized countries < 10 µg As/1; the mean intake of total As by food may vary from 10–30 µg As up to about 200 µg As/day; within CEC mineral waters, used as beverage, contain an average of 21 µg As/1 (48).

THREE GROUPS OF ARSENIC COMPOUNDS

Physicochemical properties, toxicokinetics, and toxicodynamics clearly indicate that at least three groups of As compounds have to be distinguished:

1. water soluble inorganic compounds *(As-a)*, inter alia: As fluoride, pentaoxide, trioxide, Mg/Na/K arsenite, Na/K arsenate;
2. non/little water soluble inorganic compounds *(As-b)*, inter alia: As, As-selenide, -sulfide, Cd-/Cr-/Co-/Mn-/Ni-/Ti-/Zn-arsenide, Pb-/Ni-arsenite, Fe-/Pb-/Mg-/Hg-/Ni-/Zn-arsenate;
3. organic As compounds *(As-c)*, in marine food (crustacea, algae) and maybe as organic metabolites in other foods *(As-c1)*; moreover, there exist man-made compounds (drugs, pesticides) *(As-c2)*, which for the sake of simplicity will not be discussed in this paper. Also, potential differences in health risks between inorganic As^{3+} and As^{5+} compounds will not be discussed.

In workroom and ambient air, groups As-a and As-b, in drinking water As-a, and in food As-c1 predominate. The critical organs and effects are for: As-a — skin cancer; As-b — lung cancer; and As-c1 — hardly any health hazard. Health-based recommended exposure limits have to be set separately, taking into account both the As-group and the compartment, if cancer is regarded as critical effect.

TOXICOKINETICS

Fractional Absorption

In man the fractional gastrointestinal absorption (Fo) of As-a and As-c1 is ≥ 0.8; milk is reported to decrease the absorption of As (45) in contrast to Pb and Hg. As^{5+}-a is absorbed somewhat less than As^{3+}-a. As-c1 is almost completely absorbed in humans. Respectively, a fraction of 0.5 (As^{5+}-a), 0.8 (As^{3+}-a), and almost 1.0 (As-c1) was recovered in urine a few days after a single dose (13).

* Abbreviations. **Fo**: gastrointestinal fractional absorption; **Fr**: respiratory fractional absorption; **Io**: oral intake; **Ir**: respiratory intake; **RV**: respiratory volume; **Uo**: gastrointestinal uptake; **Ur**: respiratory uptake.

Human data on pulmonary absorption are hardly available. Intratracheally injected As-a (labelled Na-arsenate) was almost completely absorbed in rats. In lung cancer patients who inhaled a labelled aerosol ($d < 5\ \mu m$, Na-arsenate) the deposited fraction was 0.32 and 0.62 of the dose; clearance was fast. We assume an uptake (Ur) = Ir x 0.3–0.6 (dep.) x 0.8 (Fr) = 0.2–0.5 x Ir for As-a. Bennett (4) assumed 35% deposition and Fr = 0.8; this results in Ur = 0.28 x Ir. Man may inhale As-a and As-b simultaneously, particularly in occupational exposure; the compounds are usually not measured separately. Because As-b will be less well absorbed, we assume total Ur = 0.2 x Ir for As-a + b.

Apparently dermal uptake may also take place for some As-a compounds, as seen in a few case reports of As-poisoning (45).

Distribution and Accumulation

Animal studies indicate considerable species-specificity for As-a: in rats the biological half-life ($T\frac{1}{2}$) in blood after i.v. injection was 60–90 d. In man (i.v. injection) 99% disappeared from blood within 24 h, 0.1% with $T\frac{1}{2}$ = 30 h; a third phase with $T\frac{1}{2}$ = 200 h was observed after one week. As-a is distributed to all organs, particularly liver, kidney, and skin. In thirty-five smelter workers and nine controls the concentrations in kidney and liver (autopsy) were increased 50 and 90%, respectively, in the workers (As-a); in the lung the levels (As-b) increased about six times (6). In liver, kidney, and skin As^{3+}-a may achieve higher levels than As^{5+}-a, in correspondence with increased fractional absorption of As^{3+}-a.

As-cl compounds quickly disappear from the tissues. In rats As-cl in Hizikia (average daily intake in Japan 10 μg As-cl/kg b.w.) was equally well absorbed from the gastrointestinal tract as As-cl from crab meat; urinary excretion occurred more quickly than after intake of As-a (As_2O_3).

Biotransformation

In recent years it has been shown that in man As-a, to a large extent, is metabolized into monomethylarsonic acid (MMAA) and subsequently into dimethylarsinic acid (DMAA) (38,45). Table 2 presents this data (total As in urine (AsU) is 100%). However, As-cl compounds are excreted practically unchanged. In two subjects arsenobetain (from crab meat) was not metabolized (10).

Excretion

As-a and As-cl compounds are predominantly excreted through the urinary tract. In four subjects who ingested 125, 250, 500, or 1000 μg As/day

(Na-arsenite) for five days, a first-order rate of excretion occurred during the first six to eight days; the T½ over nine days after exposure increased with the dose from 39 h (125 μg As/d) to 59 h (1000 μg As/d). A linear relationship was observed between dose and excreted amount of MMAA, but the relative excretion of DMAA decreased from 500 μg As/d onwards, maybe because biotransformation of MMAA to DMAA took place more slowly than that of As-a to MMAA (8). Other authors estimated a T½ for total AsU = 24–96 h, usually about 48 h (18), 10 h for As-a, and 30 h for methylated As (13). Within 85 h after dosage most As has been excreted, 80 % within 61 h. After ingestion of 2 mg As^{3+}-a, 30% was recovered in urine within 24 h; after ingestion of 0.8 mg As^{3+}-a for six days an equilibrium was achieved in which 70% of the daily intake was excreted in urine per 24 h (24). In six subjects ingesting labelled As-a, 58% was excreted in urine within 5 d, 65.9% with T½ = 2.1 d, 30.4% with T½ = 9.4 d, and 3.7% with T½ = 38.4 d (31,36).

Much less As is excreted through other pathways: two subjects lost 2 μg As/h by transpiration in a warm humid environment; skin desquamation is reported to yield 0.1–0.2 μg As/24 h, hair < 0.6% of dose, breast milk 3 μg As/l: in total a few μg As/24 h (45).

Roughly 70–80% As-cl is excreted within 3 d. In eight subjects ingesting 100 g plaice (Pleuronectus platessa) containing about 8 mg As, 65–80% of As was excreted in urine within 24 h, average 75.4% (69.2–85.3%) within five days (23). In fifteen subjects ingesting As-cl from fish (15 mg As), 76.2% was recovered in urine within eight days, 50% within 24 h; only 0.33% was recovered from the feces (37).

Source	No. of subjects	As compounds	period of sampling	% of total AsU As-a	MMAA	DMAA
(5)	4	?	?	As^{3+}-a 8.4	6	66
(13)	1	arsenite in wine	61 h	As^{3+}-a 20*	16*	62*
(36)	6	arsenic acid	5 d	27	21	55
(7)	3	arsenite	14 d	25	25	50

* original data multiplied by $\frac{1}{Fo}$.

TABLE 2 — Fractional distribution of As compounds in urine after oral intake of water soluble inorganic As compounds (As-a).

Biological Monitoring

Measurement of urinary excretion (AsU) is the method of choice for estimating total intake of As-a, if no As-cl has been ingested recently. Usually the concentration does not exceed 50 μg As/l (40 μg As/g creat.), in most cases it is below 20 μg As/l (16 μg As/g creat.) (16,21). In workers a relationship has been established between respirable or total As in air and total (As-a + MMAA + DMAA) urinary excretion of As (34). However, the workers also used respiratory protection, thus the data cannot be regarded as representative. Moreover, oral ingestion of As may have taken place since much As has been found on the hands of workers (32). A relationship between the As content of drinking water and AsU has also been observed.

INTAKE AND UPTAKE

Man is exposed to three groups of As compounds, through two pathways: the respiratory tract and the intestinal tract (primary and secondary ingestion), mainly from three environmental compartments (air, drinking water and beverages, food). In children, intake from contaminated objects, hands, etc., also has to be taken into account. We present an estimated total uptake for an adult standard man, assuming the following characteristics: body weight 70 kg, non-smoking, no recent ingestion of As-cl, no administration of As-containing drugs, RV = 15 m^3/24 h if not working in industry; RV = 20 m^3/24 h for a worker (10 m^3/8 h work, 10 m^3/16 h leisure time and sleep); water consumption 2 l/d; Fo = 0.8 for As-a+cl; Fo = 0 for As-b; Fr = 0.2 for As-a+b+cl; Ir from ambient air is neglected for the general population. Data on oral intake of *total* As from food + drinking water are presented in Table 3; in these studies no speciation of As compounds took place.

Country	Number of subjects	Median intake total As in μg/d	Range	Source
The Netherlands	100	22	11 - 1000	(26)
Belgium	122	11.5	0.1 - 720	(9)
US	?	16	?	(20)
Canada	?	?	25 - 35	(45)
Japan	?	?	70 - 170	(45)
Austria	?	30 (average)	?	quoted by (26)

TABLE 3 — Oral intake of total As in various countries.

When drinking water contains A μg As-a/1, then Uo-As-a = 2 x 0.8 x A = 1.6 A μg/d, i.e., 0.023 A μg/kg/d. If Io from food is B μg As-a+cl/d, then Uo = 0.8 B μg As/d, i.e., 0.011 B μg/kg/d. When the standard man lives in the vicinity of a source of emission, then Ir for As-a+b has to be taken into account: Ur = 0.2 level in air (C_1 μg/m^3) x 15 = 3.0 C_1 μg As/d = 0.043 C_1 μg As/kg/d. When the man is also exposed during work to C_2 μg As/m^3, then Ur = C_1 x 10 x 0.2 + C_2 x 10 x 0.2 = 2 (C_1 + C_2) μg As/d = 0.029 (C_1 + C_2) μg As/kg/d.

In the Netherlands median Io from food + drinking water is 22 μg total As/d, 90% < 44 μg As/d. This corresponds to median Uo = 17 μg As/d, 90% < 35 μg As/d, assuming that As consists of As-a+cl. Upon introducing the CEC/WHO drinking water standard (50 μg As/1), Uo becomes 98 μg As-a/d, 90% < 115 μg As-a/d. Introducing the WHO/FAO tolerable daily intake (50 μg As/kg), then Uo = 2700 μg As/d, probably mainly As-cl. If exposed to both the CEC/WHO drinking water standard and the tolerable daily intake, median man will have total Uo = 2800 μg As/d. When a worker is exposed to the present Dutch MAC (500 μg As/m^3) (14), then during five work days Ur = 5 x 0.2 x 500 x 10 = 5000 μg As-a+b (ambient air neglected), which corresponds to 710 μg As/d, if spread over the whole week.

Bennett (4) developed a similar tentative model, with somewhat different assumptions: Fr = 0.28, RV = 22 m^3/24 h; in food two thirds of As present as As-cl and one-third as As-a; Fo-0.9; mean retention times for inorganic As 8.2 d, and for organic As 2.6 d; urinary excretion in chronic exposure 70%; ambient air is also taken into account. The author estimated as intake from terrestrial food 15 mg/year, from aquatic food 33 mg/year, from drinking water 0.5 mg/year, and negligible amounts from ambient air (6 ng/m^3 rural air, 20 ng/m^3 urban air), based upon current background levels of As. This contributes about 4 μg As-a/kg b.w. and 5 μg As-cl/kg b.w. for the body levels. The author recognized that considerable uncertainties exist in the estimates and assumptions. These estimates of intake and uptake show that total Ir and Uo in combined exposure through various pathways and from various sources can only be estimated for exposure to As-a and As-cl.

ICRP (19) applied their dosimetric models for the respiratory and gastrointestinal systems to arsenic. It assigned *all* "commonly occurring As compounds" to inhalation class W, i.e., with 10 d < $T\frac{1}{2}$ < 100 d, and fi = 0.50 (fraction of the intake reaching the body fluids). However, these assumptions clearly do not agree with the different kinetics of the three As-groups. One aspect of the

ICRP-inhalation model may be relevant. Assuming inhalation of an aerosol with median aerodynamic diameter = 1 μg, 48% of the nonabsorbed particles with $T\frac{1}{2} < 10$ d, 12% of those with 10 d $< T\frac{1}{2} < 100$ d, and 4% of those with $T\frac{1}{2} < 100$ d are transferred to the gastrointestinal tract. In the case of As-a compounds, one may expect a relatively high extra gastrointestinal uptake (Fo = 0.8), whereas secondary ingestion of As-b compounds probably will be less than 10% with Fo almost zero. The assumption by ICRP that of the As leaving the transfer compartment, 35% goes directly to excretion, 7% is translocated to the liver, 1.5% to the kidneys, 0.5% to the spleen, and 56% is uniformly distributed throughout all other organs and tissues, seems not to be warranted, because ICRP does not distinguish between the three groups of As compounds.

ARSENICUM BALANCE MODEL FOR GENERAL POPULATION

Assuming long-term exposure, during which an equilibrium has been achieved between intake and output, we propose a *tentative* model for the arsenic balance. We assume that 70% of daily Ir + Io for As-a + cl will be excreted per day.

In the Netherlands median Io = 22 μg total As/d (assuming mainly As-a + cl); Uo = 17 μg As/d. Neither data on speciation nor on As-U are available; assuming that AsU is similar to total background AsU levels as measured in Belgium, i.e., 12 μg As/d, then the urinary excretion/24 h amounts to about 70% of Ir. As-loss through other routes will be about 17–12 = 5 μg As/d, which reasonably corresponds to the data mentioned before.

Some data from other countries are available. In groups of twenty-two to forty-six subjects, consuming drinking water with < 6.0 to 393 μg As/l a linear relationship was suggested between log AsU (μg/24 h) and log As (μg/l) in drinking water, when the relatively low intake from food was neglected (39). The absolute levels of AsU and As/l drinking water were reasonably similar. Assuming daily consumption of 2 l water, Fo = 0.8, and 70% urinary excretion, then the absolute levels may be expected to be similar (2 x 0.8 x 0.7 = 1.12). In Alaska AsU was measured in 138 subjects; average Io was estimated to be 32.8 (n = 43), 82.5 (n = 19), 126.1 (n = 33), 265.8 (n = 28), and 386.8 (n = 15) μg As/d with, respectively, AsU = < 20, 20–40, 40–80, 80–200, and > 200 μg As/l (17). For the three middle groups AsU/24 h can be estimated to be, respectively, 45, 90, and 210 μg As/24 h (assuming 1.5 l urine/24 h). According to the model AsU was expected to be 82.5, 126.1, and 265.8 μg/24 h, respectively. The authors estimated individual amounts of water ingested; when they had calculated on the basis of 2 l/d the predicted AsU/24 h would have corresponded even better than the estimated AsU/24 h.

In Belgium the median daily intake (food and beverages) of 124 subjects 30–65 years of age was found to be 11.5 μg As/d, 95% < 261 μg As/d (9). According to the authors (Buchet, personal communication), AsU < 10 μg As/g creat. is seldom exceeded in Belgian subjects who do not ingest marine food. Assuming excretion of 1.7 g creat/24 h, then AsU < 17 μg As/24 h would be expected to be excreted in urine. In forty-one non-occupationally exposed workers (in the US) Io was estimated to be 25–40 μg As/d; AsU was 17.5 μg/1, i.e., about 26 μg As/24 h (34). An estimated Io = 30 μg As/d would correspond to an expected excretion of 17 μg/24 h.

Figure 1 presents the data in a diagram. A linear relationship around the 45° line is observed, which reasonably confirms the tentative model of intake/output. No speciation had taken place; it may be assumed that intake of As-a and As-cl existed.

SPECIAL GROUPS AT RISK

Children

Children have a higher intake of food/kg b.w. than adults. Infants may receive up to 500–800 ml water or breast milk per day. Toddlers may have a relatively large intake from contaminated soil, hands, and related objects; this will lead to a higher intake in areas with industrial emission or with increased natural background levels in the soil.

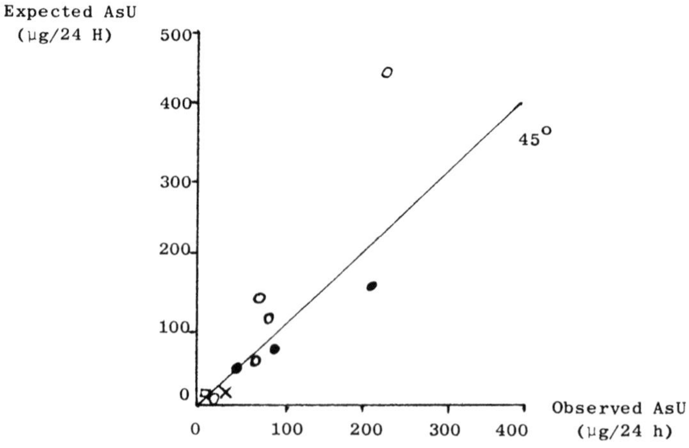

FIG. 1 — Relationship between observed and expected AsU/24 h; references: ○ (39), ● (17), □ (8), and × (34).

In the United States 712 1–5-year-old children living in copper-smelter towns had a geom. average AsU = 18.7 μg As/l (95% confidence interval 17.1–20.5), median 19.8 μg As/l; 20% > 45 μg As/l. In non-exposed children geom. average AsU = 5.8 μg As/l (4.7–17.0), median 7.3 μg As/l, 95-percentile 45 μg As/l (2). Between exposed and non-exposed children the difference in median AsU = 12.5 μg As/l. The median production of urine for 1–5-year-old children is about 30 ml/kg b.w./d; therefore, the extra urinary output of As/d = 0.375 μg As/kg b.w./d. For a child of 15 kg b.w. this corresponds to an extra oral intake of about 8 μg As-a/d.

In Czechoslovakia AsU was measured in children living at various distances from an As-emitting power plant (3): AsU decreased from average 18.9–20.1 μg As/l (maximum 105) at 11 km distance to 8.2 μg As/l (max. 36) at 36 km. The extra urinary As output is 12 μg As/l. Also in this area the extra oral intake of As-a is expected to be about 8 μg As/d. In both studies no distinction has been made between exposure to As-a and As-cl compounds.

Consumers of Mineral Water

Within CEC the average As content of mineral water is 21 μg As/l. The average intake of mineral water was estimated to be 10% of the total intake of water; however, within CEC the intake widely differs between the member states, e.g., the Netherlands 0.5 l/capita/yr compared to France 50 l/capita/yr (48). Regular consumption of mineral water will considerably increase the intake of As-a.

NO-ADVERSE EFFECT/RESPONSE LEVELS

The review presented by WHO (45) allows one to conclude that for induction of *nonmalignant diseases* the no-adverse effect level (NAEL) in long-term exposure will be below 10 μg As-a/kg/d, and certainly below the ADI = 50 μg As/kg/d (mainly As-cl). The following discussion is limited to carcinogenic risks.

The carcinogenic effect in long-term intake of As-a compounds is mainly *skin cancer*. Assuming a daily consumption of 2 l drinking water, a lifespan of 70 years, WHO estimated the lifetime risk from consumption of As in drinking water to be about 25% per mg As-a/l, i.e., 1.10^{-5} per 0.04 μg As-a/l. The EPA Cancer Assessment Group (33, 35) estimated the skin cancer risk to be 1.10^{-5} per 0.02 μg As-a/l. Both estimates, based upon linear extrapolation, agree very well. For our further calculation we therefore use 0.03 μg As-a/l for a risk of 1.10^{-5}. It should be noted that this risk may have been *over*estimated because intake of As-a from food, contaminated hands, objects, etc., in a geochemically polluted environment (such as in Taiwan) was neglected.

As-b compounds are less well absorbed. The carcinogenic effect is mainly *lung cancer*. In one US study on smelter workers (29,30) the extra lung cancer risk was estimated to be +200% (RR=3) in occupational exposure to 50 μg As/m^3 for over 25 years, which corresponds to 1% in exposure to 0.250 μg As-b/m^3 for more than 25 yr. Linear extrapolation to the general population exposed to As-b polluted ambient air suggests an extra risk of +25% in exposure to 1μg as-b/m^3, 24 h/d-lifetime for 70 years. When the background lung cancer risk is assumed to be 3.10^{-2}, then the absolute extra risk will be 0.75% per 1 μg As-b/m^3 (45). For nonsmoking subjects this risk may have been *over*estimated.

The risk of lung cancer as estimated by WHO (45) should be considered as only very *tentative*. Some considerations raise doubt as to its validity:

1. Pershagen (28) observed very large differences in retention of As_2O_3 (0.3 mg As), As_2S_3 (0.5 mg As), and Ca-arsenate (0.5 mg As) in the lungs after intratracheal instillation, four times with an interval of one week, in hamsters. This implies that with decreasing solubility the retention in lung increased and the concentration in the liver decreased and suggests possibly higher local carcinogenic activity of Ca-arsenate and As_2S_3 than of As_2O_3.
2. The same author also observed in copper-smelter workers a kind of multiplicative effect of exposure to As-compounds and smoking of cigarettes. The standard rate ratios for lung cancer death (SRR=1 in non-exposed nonsmokers) was 4.9 in non-exposed smokers, 4.4 in nonsmoking roaster workers, 22.0 in smoking roaster workers; in other As-workers SRR=1.2 in nonsmokers and 14.7 in smokers. However, in the study underlying the WHO estimate, no synergistic effect of smoking was observed. A synergistic effect of smoking might have *over*estimated the risk of exposure to As as such in nonsmokers. Pershagen did not calculate the RR in relation to As-concentrations in air; the speciation of As was not considered. It was known that in the 1940s the average concentration was 0.06 to 2 mg As/m^3 and 15 to 300 mg SO_2/m^3; in the 1960s 0.1 to 0.5 mg As/m^3 and 5 to 10 mg SO_2/m^3.
3. The WHO estimate is only based upon one study; only lung cancer deaths in retired workers were studied. This may have *under*estimated the risk.
4. No distinction was made between workers in different departments, whereas the Swedish study (28) suggested an increased risk particularly in

roaster workers. This again may have *over*estimated the risk; data on As-species were not available.

5. Exposure was estimated from AsU. However, exposure to less well absorbed As-b compounds is difficult to estimate from urinary excretion of As. This may have *under*estimated the risk. Moreover, the coefficient of correlation between levels of AsU and of As in air was 0.53; As-exposure explained less than 30% of AsU.
6. AsU also will be increased by primary hand to mouth transfer of As compounds (32). This may have *over*estimated the risk of respiratory exposure, because AsU also reflects oral exposure to As-a.
7. In previous years, exposure to As may have been higher whereas AsU only measured recent exposure. This may have *over*estimated the risk.

It should be realized that the WHO estimate refers to extra risk, i.e., if the "normal" risk of lung cancer in the general population with a 70 year-exposure is 3%, then one to two extra cases might be expected per 1 μg As-b/m^3 in ambient air per one hundred subjects. This estimate is based upon linear extrapolation.

Recently two new studies were published. The mortality experience of 2802 workers in the same factory on which the WHO estimate was based was reevaluated (15). All the men used in the study worked for one year or more during 1940–1964 at the copper smelter and include both non-retirees and retirees (period of observation 1.1.1941 – 12.31.1976). The airborne As levels (no speciation information) in 1938–1957 were high: up to about 60 mg As/m^3 (small number of samples). The As-U levels dropped in 1973–1975 to about 50% of those in 1948–1952. In the earlier studies (29,30) the AsU levels of 1973 were used; this apparently has *over*estimated the risk. For the new study, estimates of historic exposure were made, still based on AsU levels. Total cancer mortality was significantly elevated (SMR = 129.9, $P < 0.01$); for lung cancer SMR = 198.1 ($P < 0.01$). There was no clear decline in respiratory cancer across successive cohorts. A slight positive relationship was observed between cumulative exposure (AsU x year) and respiratory cancer death. The SMR's were high in the decade immediately following termination of exposure. This suggested a fairly short latency period in some cases and a disappearance of effects after termination of exposure. The difference in mortality rates, when low exposure-intensity jobs were compared with high-intensity jobs, appeared to be small. No effect of simultaneous exposure to SO_2 was observed. The authors concluded that this study strongly suggested that As was acting as a cancer promotor rather than as an initiator.

The second study (40) was carried out in another copper smelter in the United States. The vital status of 1800 men who had worked for at least twelve months before 1.1.1957 was examined. The As operation in the plant ceased after 1964 but the period of observation was 1964–1978. Those in the heavy exposure category had spent at least twenty-four months in the As kitchens, Cottrells, or As roaster. Data on exposure were only available for about half the smelter departments. In departments for which no exposure data were available, exposure was estimated by analogy. The workers were distinguished according to four exposure categories: $\geq$ 5000 μg As/m^3, 500–4999 μg As/m^3, 100–499 μg As/m^3, and < 100 μg As/m^3. Three indices of exposure were used: time-weighted (8 h) estimated over time, ceiling or peak exposure for thirty days, cumulative exposure (μg/m^3 x year). Moreover, three broad SO_2 exposure categories were established, and exposure to asbestos was also qualitatively estimated. An increasing trend from low to high TWA classes was observed: < 100 μg As/m^3 SMR = 138; 100–499 μg As/m^3 SMR = 303 ($P < 0.01$); 500–4999 μg As/m^3 SMR = 375; and $\geq$ 5000 μg As/m^3 SMR = 704. The SMR's for respiratory cancer for men in the low and medium concentrations (< 500 μg As/m^3) were not significantly elevated (129 and 116), in the high concentration (500–4999 μg As/m^3) category SMR = 348 ($P < 0.01$), in the highest ($\geq$ 5000 μg As/m^3) category SMR = 662 ($P < 0.01$). This suggested that a ceiling exposure for thirty days or more was a more important indicator of respiratory cancer risk than TWA exposure. An increasing trend was also observed with increasing cumulative exposure: < 500 μg As/m^3 year no increased risk was observed (SMR = 69); 500–2000 μg As/m^3 year, SMR = 157 (n.s.); 2000–12000 μg As/m^3 year, SMR = 400 ($P < 0.05$); and 12000+ μg As/m^3 year, SMR = 550 ($P < 0.05$). The excess of respiratory cancer could not be ascribed to asbestos exposure but a possible role of SO_2 could not be ruled out. There was little difference in the SMR's for smokers and nonsmokers. The authors concluded that in this plant, As exposure (speciation was not considered) was strongly associated with lung cancer. The data suggested that at exposure < 100 μg As/m^3, maybe even < 500 μg As/m^3, the SMR's were only slightly increased. This study was not based upon indirect assessment of As exposure by means of AsU levels; however, the data on levels in air were limited. Nevertheless, this study again suggests that at low As exposure levels in copper smelters, an increased cancer risk may not exist. However, because of the limited data on exposure, the results must be interpreted with caution.

Both studies suggest that linear extrapolation for assessment of the lung cancer risk may not be the most appropriate method. When As compounds should be confirmed to act as a promoter, then a threshold-extrapolation may be feasible. However, too many uncertainties exist to reject fully the most conservative linear extrapolation method as applied by WHO.

Organic As-c1 compounds are probately hardly toxic and not carcinogenic.

This short review of health risks demonstrates that As compounds should be treated as at least three distinct groups with widely different health risks. Although calculation of total uptake of As-a + c1 compounds, to a certain extent, appears to be possible on the basis of total AsU, the widely different health risks demonstrate that such an approach is not acceptable from the viewpoint of health protection. Moreover, within the groups As-a, As-b, and As-c different compounds may carry different health risks, quantitatively (As-a and As-b) as well as qualitatively (As-c).

We estimated intake and uptake for a standard man. Even if the estimated health risks have been overestimated, the CEC/WHO quality limits for drinking water (50 μg As/l) widely exceed the derived estimates of 0.02–0.04 μg As/l; this will be even more so in the case of consumption of mineral water. The same applies to most present MAC-TLV's. The WHO/FAO acceptable daily intake (50 μg As/kg b.w.), according to present knowledge, may only be acceptable if it refers to As-c1 compounds.

DISCUSSION AND CONCLUSIONS

Limited Objective

This paper has a limited objective which is to demonstrate the necessity for considering groups of As-compounds and not just *total* As in deriving environmental and occupational quality standards. The paper does not fully review As toxicology. Moreover, inter- and intra-individual variability was hardly discussed., relying mostly on "standard" man and child. The intake of arsenic by smokers was also not discussed.

The paper uses arsenic as *an example*. The need for considering speciation also applies to other metals. Recent data demonstrate the necessity to distinguish between different compounds of chromium and nickel because of their difference in toxicokinetics and -dynamics. Some water insoluble Cr and Ni compounds are carcinogenic, and there even exists a difference in carcinogenic potency between some cristalline and non-cristalline Ni compounds with the same chemical structure. The metabolic models and risk assessments of some metals may partly have to be rewritten. In some cases, simple separation of groups of compounds on the basis of their solubility, not requiring elaborate chemical speciation, might already be of much help.

The Validity of Numeric Data

Many numerical data have been presented, either measured as such or estimated. The majority of available data still pertain to *total* As, without any regard to the speciation of As compounds. Usually only mean or median values were used, which did not allow any insight into the distribution. Several assumptions of intake, uptake, and excretion had to be made, whereas the actual data varied within and between subjects and groups. Therefore, it should be stressed that the data are *tentative;* they only provide a "guesstimate." Much study is still needed on levels in various environmental media, rate of intake, fractional absorption, and excretion for specific (groups of) compounds separately. This appears to be the only rational way to derive reliable estimates of health risks, to set valid environmental quality limits, and in this way to protect health.

Environmental Quality Standards

It is necessary to establish separate quality standards for:

1. *Water soluble inorganic As compounds* (As-a) in drinking water, mineral water, food, (ambient air), workroom air, and for total intake;
2. *Water insoluble inorganic As compounds* (As-b) in workroom air, (ambient air), (food);
3. *Organic As compounds* (As-c1) for marine food as total daily intake (and in ambient air).

The environmental compartments indicated between brackets make a negligible contribution, except in specific exposure situations.

When possible attempts should also be made to differentiate between various compounds within groups As-a, As-b, and As-c. Similar measures may have to be taken, e.g., for Cr and Ni compounds.

Risk Assessment

All risks assessed only refer to health-based risks, not taking into account socioeconomic and technological constraints.

Drinking water and food, As-a. Consumption of 0.03 μg As-a/l drinking water has been estimated to carry a risk of 1.10^{-5} in lifetime exposure. This corresponds to a daily uptake of about 0.05 μg As-a. However, the risk may have been overestimated. The estimate is based upon absence of As-a in food, ambient air, and workroom air. Moreover, linear extrapolation as such may also have overestimated the risk.

Occupational exposure, As-a. WHO (45) estimated the prevalence of skin cancer to be increased 5% per 10 g total oral intake of As-a in lifetime exposure.

Assuming Uo = Ur = 0.2 Ir, then total lifetime Ir = 40 g. Extrapolating to a 40 hour workweek, 220 workdays/yr, 25 year exposure, RV = 10 m^3/8 h, this 5.10^{-5} increase in skin cancer risk is expected to occur in occupational exposure to 40 x 1:25 yr x 1:220 d x 1:10 m^3/8 h x 1000 = about 1 μg As-a/m^3, when no As-a is ingested. This risk again may have been *over*estimated, because of the very conservative linear extrapolation, and because of neglect of exposure apart from drinking water. However, in occupational exposure intake may also occur from contaminated hands. This *under*estimates the risk, if based only upon respiratory exposure.

Environmental exposure, As-b. WHO (45) estimated an extra risk of +1% lung cancer death in occupational exposure to 0.25 μg As/m^3 for over 25 years; this probably mainly refers to As-b compounds (smelter workers). This estimate which still is tentative and approximative, certainly needs reassessment as soon as new data are available. It corresponds to a +1% risk for the general population per 1 μg As-b/m^3 in ambient air in lifetime exposure. Recent data (15,40) suggest that linear extrapolation may overestimate the risk, too.

Oral intake of As-c1. The tentative acceptable daily load of 50 μg As-c1/kg (41,42) probably does not carry any health risk.

Total exposure. Today it is feasible to suggest a standard for total oral + respiratory exposure only for water soluble As compounds. For these compounds total uptake should not exceed 0.05 μg As-a/d, if the estimated skin cancer risk is considered socially accepted, and linear extrapolation is accepted as appropriate. However, this method may have overestimated the risk.

CONCLUSIONS

The questions brought forward in the introduction can be answered as follows:

1. Arsenic (As) cannot be treated using only total As concentrations. From the viewpoint of health protection one has at least to distinguish between water soluble (As-a), water insoluble (As-b), and organic As compounds present in marine food (As-c1).
2. It is not possible to derive health-based total exposure limits for arsenic, except maybe for inorganic water soluble As compounds.
3. For As-a and As-b compounds, a health-based recommended limit cannot be established, but only a carcinogenic risk assessment is feasible.
4. The present quality limits/standards for various environmental compartments are not logically consistent.
5. Similar conclusions apply for other metals as well.

REFERENCES

(1) American Conference of Governmental Hygienists (ACGIH). 1982. Threshold Limit Values for 1982. Cincinnati: ACGIH.

(2) Baker, E.L.; Hayes, C.G.; Landrigan, P.J.; Handke, J.L.; Leger, R.T.; Housworth, W.J.; and Harrington, J.M. 1977. A nationwide survey of heavy metal absorption in children living near primary copper, lead and zinc smelters. Am. J. Epidemiol. **106:** 261–273.

(3) Bencko, V., and Symon, K. 1977. Test of environmental exposure to arsenic and hearing changes in exposed children. Env. Hlth. Perspect. **19:** 95–101.

(4) Bennett, B.G. 1981. Exposure commitment assessment of environmental pollutants. MARC Technical Report nr. 23, vol. 1, no. 1. Monitoring and Assessment Research Centre, Chelsea College, University of London, UK.

(5) Braman, R.S., and Foreback, C.C. 1973. Methylated forms of arsenic in the environment. Science **182:** 1247–1249.

(6) Brune, D. 1980. Distribution of 23 elements in the kidney, liver and lungs of workers from a smeltery and refinery in North Sweden exposed to a number of elements and a control group. Sc. Total. Env. **16:** 13–35.

(7) Buchet, J.P.; Lauwerys, R.; and Roels, H. 1981. Comparison of the urinary excretion of arsenic metabolites after a single oral dose of sodium arsenite, monomethylarsonate or dimethylarsinate in man. Int. Arch. Occup. Env. Hlth. **48:** 71–79.

(8) Buchet, J.P.; Lauwerys, R.; and Roels, H. 1981. Urinary excretion of inorganic arsenic and its metabolites after repeated ingestion of sodium metaarsenite by volunteers. Int. Arch. Occup. Env. Hlth. **48:** 111–118.

(9) Buchet, J.P.; Lauwerys, R.; Vandevoorde, A.; and Pycke, J.M. 1981. Evaluation de la quantité moyenne de cadmium, plomb, manganese, cuivre, chrome, mercure, calcium, zinc et arsenic ingéré journellement par la population adulte belge. Arch. Belg. Med. Soc. Hyg. Med. Travail Med. Leg. **39:** 465–490.

(10) Cannon, J.R.; Edmonds, J.S.; Francesconi, K.A.; Raston, C.L.; Saunders, J.B.; Skelton, B.W.; and White, A.H. 1981. Isolation, crystal structure and synthesis of arsenobetaine, a constituent of the Western

Rock lobster, the Dusky shark and some samples of human urine. Aust. J. Chem. **34:** 787–798.

(11) Commission of the European Communities and World Health Organization. 1979. Heavy Metals in the Environment. Edinburgh: CEP Consultants.

(12) Commission of the European Communities and World Health Organization. 1981. Heavy Metals in the Environment. Edinburgh: CEP Consultants.

(13) Crecelius, E.A. 1977. Changes in the chemical speciation of arsenic following ingestion by man. Env. Hlth. Perspect. **19:** 147–150.

(14) Directorate General of Labour (DGA), Inspectorate of Labour. 1982. Advisory Report on MAC for Arsenic. Voorburg: DGA.

(15) Enterline, P.E., and Marsh, G.M. 1982. Cancer among workers exposed to arsenic and other substances in a copper smelter. Am. J. Epidemiol. **116:** 895–911.

(16) Feldman, R.G.; Niles, C.A.; Kelly-Hayes, M.; Sax, D.S.; Dixon, W.J.; Thompson, D.J.; and Landau, E. 1979. Peripheral neuropathy in arsenic smelter workers. Neurology **29:** 939–944.

(17) Harrington, J.M.; Middaugh, J.P.; Morse, D.L.; and Housworth, J. 1978. A survey of population exposed to high concentrations of arsenic in well water in Fairbanks, Alaska. Am. J. Epidemiol. **108:** 377–385.

(18) Hine, C.H.; Pinto, S.S.; and Nelson, K.W. 1977. Medical problems associated with arsenic exposure. J. Occup. Med. **19:** 391–396.

(19) International Commission for Radiological Protection (ICRP). Task Group of Lung Dynamics. 1966. Deposition and excretion models for internal dosimetry of the human respiratory tract. Health Phys. **12:** 173–207.

(20) Jelinek, C.F., and Corneliussen, P.E. 1977. Levels of arsenic in the United States food supply. Env. Hlth. Perspect. **19:** 83–87.

(21) Lauwerys, R. 1980. Principiaux tests biologiques specifiques pour le dépistage d'une exposition excessive à divers toxiques industriels. Cahiers Med. Travail **XVII:** 92–97.

(22) Lewis, W.M., and Smeets, J. 1981. WHO international drinking water standards revised. **In** Water Supply and Health, eds. H. van Lelyveld and B.C.J. Zoeteman, pp. 285–292. Amsterdam: Elsevier Scientific Publishing Company.

(23) Luten, B.J. 1980. Onderzoek naar de uitscheiding van arseen door een aantal vrijwillige proefpersonen na consumptie van schol (Study of the excretion of arsenic by volunteers after consumption of plaice), Report nr. 0-277. Zeist: Central Institute for Nutrition Research-TNO.

(24) Mappes, R. 1977. Versuche zur Ausscheidung von Arsen in Urin. Int. Arch. Occup. Env. Hlth. **40:** 267–272.

(25) National Academy of Sciences/National Research Council (NAS). 1977. Arsenic, Medical and Biologic Effects of Environmental Pollutants. Washington, DC.

(26) National Public Health Laboratory. 1980. Surveillance Programme "Man and Nutrition." Leidschendam, the Netherlands: Ministry Public Health and Environmental Protection; State Supervisory Public Health Service.

(27) Occupational Safety and Health Administration (OSHA). 1980. **In** US Department of Health and Human Services, Summary of NIOSH Recommendations for Occupational Health Standards, nr. GJN16. Cincinnati: NIOSH.

(28) Pershagen, G. 1982. Arsenic and lung cancer with special reference to interacting factors. Stockholm, Sweden: Department of Environmental Hygiene, Karolinska Institute and the National Institute of Environmental Medicine.

(29) Pinto, S.S.; Enterline, P.E.; Henderson, V.; and Varner, M.O. 1977. Mortality experience in relation to a measured arsenic trioxide exposure. Env. Hlth. Perspect. **19:** 127–130.

(30) Pinto, S.S.; Henderson, V.; and Enterline, P.E. 1978. Mortality experience of arsenic exposed workers. Arch. Env. Hlth. **33:** 325–331.

(31) Pomroy, C.; Charbonneau, S.M.; McCullough, R.S.; and Tam, G.K.H. 1980. Human retention studies with ^{74}As. Toxicol. Appl. Pharmacol. **53:** 550–556.

(32) Roels, H.; Buchet, J.P.; Truc, J.; Croquest, F.; and Lauwerys, R. 1982. The possible role of direct ingestion on the overall absorption of

cadmium or arsenic in workers exposed to CdO or As_2O_3 dust. Am. J. Industr. Med. **3:** 53–65.

(33) Sittig, M., ed. 1980. Arsenic. **In** Environmental Health Reviews, pp. 70–79. Park Ridge, NJ: Noyes Data Corp.

(34) Smith, T.J.; Crecelius, E.A.; and Reading, J.C. 1977. Airborne arsenic exposure and excretion of methylated arsenic compounds. Env. Hlth. Perspect. **19:** 89–93.

(35) Stara, J.F.; Kello, D.; and Durkin, P. 1980. Human health hazards associated with chemical contamination of aquatic environment. Env. Hlth. Perspect. **34:** 145–158.

(36) Tam, G.K.H.; Charbonneau, S.M.; Bryce, F.; Pomroy, C.; and Sandi, E. 1979. Metabolism of inorganic arsenic (^{74}As) in humans following oral ingestion. Toxicol. Appl. Pharmacol. **50:** 319–322.

(37) Tam, G.K.H.; Charbonneau, S.M.; Bryce, F.; and Sandi, E. 1982. Excretion of single oral dose of fish-arsenic in man. Bull. Env. Contam. Toxicol. **28:** 669–673.

(38) Vahter, M.; Marafante, E.; Lindgren, A.; and Dencker, L. 1982. Tissue distribution and subcellular binding of arsenic in marmoset monkeys after injection of ^{74}As-arsenite. Arch. Toxicol., in press.

(39) Valentine, J.L.; Kang, H.K.; and Spivey, G. 1979. Arsenic levels in human blood, urine, and hair in response to exposure via drinking water. Env. Res. **20:** 24–32.

(40) Welch, K.; Higgins, I.; Oh, M.; Burchfiel, C. 1982. Arsenic exposure, smoking and respiratory cancer in copper smelter workers. Arch. Env. Hlth. **37:** 325–335.

(41) World Health Organization (WHO). 1967. Arsenic. Technical Report Series no. 373, pp. 14–15. Geneva: WHO.

(42) World Health Organization (WHO). 1973. Arsenic. Technical Report Series no. 532, pp. 49–50. Geneva: WHO.

(43) World Health Organization (WHO). 1977. Methods used in establishing permissible levels in occupational exposure to harmful agents. Technical Report Series no. 601. Geneva: WHO.

(44) World Health Organization (WHO). 1980. Recommended health-based limits in occupational exposure to heavy metals. Technical Report Series no. 647. Geneva: WHO.

(45) World Health Organization (WHO). 1981. Arsenic. Env. Health Crit. **18.** Geneva: WHO.

(46) Zielhuis, R.L. 1982. Standards for chemical quality of drinking water: a critical assessment. Int. Arch. Occup. Env. Hlth. **50:** 113–130.

(47) Zielhuis, R.L., and Notten, W.F. 1979. Permissible levels for occupational exposure. Int. Arch. Occup. Env. Hlth. **42:** 269–281.

(48) Zoeteman, B.C.J., and Brinkman, F.J.J. 1976. Human intake of minerals from drinking water in the European Communities. **In** Hardness of Drinking Water and Public Health, eds. R. Amavis, W.J. Hunter, and J.G.P.M. Smeets, pp. 173–211. Oxford: Pergamon Press.

Changing Metal Cycles and Human Health, ed. J.O. Nriagu, pp. 345–356.
Dahlem Konferenzen 1984. Berlin, Heidelberg, New York, Tokyo: Springer-Verlag.

MODELING EXPOSURE ROUTES OF TRACE METALS FROM SOURCES TO MAN

B.G. Bennett*
Monitoring and Assessment Research Centre
London SW10 0QX, England

Abstract. The exposure of man to trace metals in the environment occurs from transfers along various pathways. The modeling of exposure routes has been developed to bring order and understanding to observations and to allow predictions to be made of the consequences to the receptor from release practices. The general aspects of environmental and metabolic modeling are discussed, and the modeling results for Pb, Cd, Hg, Ni, Sn, As, and Se are presented.

INTRODUCTION

Models are formulated to describe general physical and biological phenomena. They may be highly refined mathematical descriptions or, at the other extreme, they may be little more than rules of thumb, particularly for highly irregular or variable behaviors. In some cases, models may apply to a specific set of conditions, but generally they are intended to reflect broad, representative behavior. Minor differences in conditions and other influencing factors may be overlooked or may be included later when the more generally formulated

* Staff member of the Global Environment Monitoring System (GEMS), United Nations Environment Program.

models are adjusted or adapted to more specific situations. Our interest in models is to describe the transport and behavior of metals in the environment and their metabolism and accumulation in man.

There is a wide range of scope as well as purpose in environmental modeling. Broad models of global geochemical cycling are formulated to understand the major transfer pathways, the sources and sinks, and regional concentrations. Ecological models may continue the multimedia considerations, but on a more regional basis with emphasis on living organisms within the system. More detailed attention may be given to behavior within a single medium, such as atmospheric or aquatic models, and within a single organism, such as metabolic models.

GENERAL ASPECTS OF MODELING

Modeling has a close relationship to monitoring. Monitoring programs provide basic data for input to and validation of modeling efforts. Modeling studies extend the measurement results and also suggest where useful measurements might be made. It has been conventional to separate the main stages of monitoring and modeling. The first stage covers the region from the source to the receptor area. This is source related assessment. The second stage of modeling or monitoring comes closer to the more immediate exposure of the receptor. This may be termed individual related assessment. The movements and habits of the individual are determined as far as these affect the exposure. Modeling of absorption and retention of trace substances may begin with estimates of intake.

Modeling environmental behavior of chemicals begins with the specification of the compartmental arrangement which reflects the system components. For first stage modeling of the regional or global environment, very broad compartment identification may be sufficient: atmosphere, land, freshwaters, ocean. At a later stage, more sensitive areas may be considered, e.g., estuaries or specific ecosystems. Metabolic models also use compartmental arrangements to describe the retention components or the functional components of the organism.

Models can be broadly classed as either time-dependent or time-independent. Time-dependent models give the dynamic behavior of the chemical substance in the environment. First-order kinetics are generally assumed and a system of differential equations describes the changes in amounts of the chemical in the compartment. Solutions may be obtained directly, if possible, or simulation of the transfers occurring may be carried out in small time sequences. This may be referred to as a systems analysis method. It is important that the compartmental arrangement be correctly formulated and the various transfers and interchanges be accurately specified to reflect the proper environmental behavior of the chemical. A large data base is required to evaluate this type of model, but the evaluation is most complete with amounts and transfer rates given for all points and at all times.

A time-independent description of the behavior of a chemical can be made on a more limited data base. In this case, the objectives are to determine the partitioning of the chemical substance along the pathways and to estimate the amounts which ultimately reach the receptor. In the steady states the relationships between compartmental amounts are expressed as ratios, called the concentration factors. The time-independent approach is often referred to as the concentration factor method. A wider application involving cumulative transfer per unit input is called the exposure commitment method (4).

Environmental and metabolic modeling is ultimately concerned with the evaluation of exposures. "Exposure" and the closely related and often interchangeable word "dose" are variously used terms in toxicology and environmental science. Classically, "exposure" may simply refer to the presentation of a substance or physical agent to a receptor and "dose" to the amount administered or taken in via inhalation, ingestion, injection, or absorption through the skin. Both terms have been further used to specify the amount reaching the site of effect. The terms intake and uptake overlap these usages, intake referring to the entry amount and uptake the absorbed amount.

Environmental transfers and biological accumulations depend on the concentrations of the substance and the duration of its presence at the point of interest. The most direct combination of concentration and duration in a definition of exposure is the time integral of the concentration. The concentration may be in an environmental medium, such as air or water, or it may be in the organism or at a specific point in the biological system. This formulation of exposure is particularly useful in describing the movement of chemical substances in the environment. For the general case of transfers described by linear, first-order kinetics, the integrated concentrations and also the integrated fluxes maintain constant relationships.

In terms of effects, this definition of exposure must be used with some caution. The various combinations of concentration and time giving equal exposure, for example, high concentrations of short duration and low concentrations of long duration, may not be associated with equal occurrence of adverse effects.

Other forms of the exposure relationship may be required in the evaluation of specific effects in particular biological systems. Some of these forms may involve exponential or power functions of time (15). It will be necessary for toxicologists to give further consideration to the concentration-time relationship, i.e., the expression of exposure which is most closely associated with the occurrence of effects.

ENVIRONMENTAL MODELING

Environmental models are often used to obtain relationships between the sources of chemical substances and their intake or contact with receptor organisms. A sequence of modeling operations is generally employed; for example, air dispersion or aquatic transport models may be applied to the initial movement of the chemical from the source to the point of interest. Food-chain models may then be used to determine the further transfer of the chemical to the receptors. This general approach has been particularly developed and applied to assess the environmental behavior of radionuclides (8, 19). The methods are applicable to nonradioactive substances as well.

Atmospheric transport models reflect the impact of meteorological factors on the concentrations of trace substances in air. Most commonly, air dispersion models are based on the Gaussian diffusion equation. Dispersion estimates can be made for a wide variety of meteorological conditions, i.e., wind speed and direction and atmospheric stability. A particular result depends on constant conditions and, therefore, there are limits on time and distance validity. Generally, predictions are considered reliable only within a few tens of kilometers from the source. At greater distances, trajectory and grid models may be used. These models can become quite complex, requiring large computer capacity, and must be backed by detailed wind data. However, account can be taken of a wide variety of changing conditions and of complex terrain. A review of air transport models has been published, for example, by Turner (18).

It is necessary to account for various removal processes of chemicals from the air and for transformations which may occur during transport. Deposition is the main depletion process for metals. The deposition velocity, which is the ratio of the aerosol deposition rate per unit area to the average concentration of the aerosol in air, is widely used. Reviews and assessment of deposition measurements for metals have recently been published (10, 11).

Aquatic transport of chemical substances is governed by the general water movements and sedimentation processes. Models are categorized according to whether releases are to rivers, lakes, estuaries, or coastal seas. It may be assumed that relatively rapid mixing occurs in the receiving water body or that equilibrium is achieved. Otherwise, flow characteristics can be modeled if hydrological information is available. Compartmental descriptions of water regions may be employed, for which the turnover time of water in the region is defined and the effective residence time of the substance in the water may be determined. Dynamic models must account for diffusion (mixing) and advection (movement with the water mass) and for sources and sinks including sediment interactions.

Transfers along terrestrial and aquatic food-chains are represented as compartmental sequences with transfer factors or concentration ratios expressing the relationships at steady state. These have been particularly developed and applied to the behavior of the radionuclides in the environment, for example, transfers along the soil-pasture grass-cow-milk-man food-chain. The details of the model depend on the availability of data. The aquatic models often proceed in one step from water to aquatic organism with use of the concentration factor. Linear systems models may also be formulated to describe food-chain transfers. The interlinking transfer rate constants must be evaluated. The solution then gives the concentration variations as a function of time.

METABOLIC MODELING

Metabolic models are used to describe the relationships between intake of chemical substances and their absorption, distribution, retention, transformation, and excretion by the organism. The models may be useful in relating intakes to measured concentrations in body fluids, such as blood and urine. However, their primary use is to calculate the exposures to critical organs, that is, those which accumulate the highest concentrations or are the most liable to damage or dysfunction.

Compartmental models of varying detail and complexity are used to describe metabolic processing of chemicals. The compartments may represent the physical structure of the organism, such as primary organs and tissues, or they may represent the observed components of retention within the body. Retention equations consisting of a series of exponential terms are very useful in reflecting overall metabolism. The parameters are determined from excretion measurements or from analysis of the amounts of the chemical in tissues at various times after an administered amount. The half-times of retention and the effective mean residence times of the chemical substances in the organism can be determined.

Models have been developed which reflect physiological processes following intake. The lung model formulated by the Task Group on Lung Dynamics is widely used (12). The respiratory system is divided into nasopharyngeal, tracheobronchial, and pulmonary compartments. Deposition of inhaled particles in the compartments depends on particle size. Clearance from the compartments is by mucociliary transfer to the gastrointestinal tract or absorption into blood. The model gives the pathways and clearance rates for several classes of particle solubilities.

Multicompartment models have been formulated to represent the metabolism of specific chemical substances, for example, an eight compartment metabolic model for cadmium (13) or the five compartment model for lead (5). Reviews of this and other aspects of metabolic modeling procedures are available (6, 20).

MODELING RESULTS FOR METALS

Application of models to describe the transport and behavior of metals is essentially a process of selection of values for each of the many associated parameters required. There must be a very firm basis underlying this procedure. Background discussion is needed to establish representative values, the uncertainties and variabilities. Each assumption must be fully justified. Clearly, a modeling result cannot stand on its own. However, in attempting to present here a wider survey of results for a number of metals, it will not be possible to do justice to the full procedure. It is anticipated that some of the groundwork is provided in the foregoing papers. Further discussion is available in previously published reports (1–3).

The metals considered in this summary include lead, cadmium, and mercury, which have been widely studied and are toxic at relatively low levels. Nickel and tin are also nonessential elements for man. The metalloids arsenic and selenium are important in terms of their toxicity at higher levels and for the interactions of these elements with metals in the environment and in the organisms.

For most substances, the environmental and metabolic behavior depends on the chemical form. It is thus inadequate to give results for total metals only. Consideration of specific compounds must be given as data become available. Only for mercury and arsenic is a distinction made here between intakes of inorganic (from terrestrial foods) and organic forms (from seafood), which differ in important respects in terms of absorption, retention, and also toxicity.

The important anthropogenic sources of metals include smelting and refining operations, fuel combustion, and waste incineration. Natural sources include windblown dusts, volcanoes, forest fires, and vegetation. On a regional and sometimes global basis, emissions of metals to air from anthropogenic sources may often exceed natural emissions (10, 14).

The greatest source of lead to the atmosphere arises from combustion of leaded gasoline. Fresh exhaust aerosols from automobiles are of submicron size. Away from roadways there is aggregation and attachment to the ambient aerosol with a broad peak in particle size between 0.3 and 1 μm. Increased levels of lead in air near busy roadways are noticeable to distances of about 50m in urban areas and 200m in rural areas. Models to estimate the dependence on distance have been formulated (7).

Metals attached to ambient aerosols are deposited by wet and dry deposition processes. In general, deposition occurs primarily with rainfall at rural sites, but dry deposition is more important at urban sites. For total deposition of the airborne particles, values of the deposition velocity are typically 0.5 cm s^{-1}, though a wider range of values apply to particular conditions and to emissions from specific sources.

Plant uptake of metals depends on the plant species and soil conditions. Generally conditions resulting in less fixation, low organic content, and low pH in soil lead to greater uptake by plants. Uptake of heavy metals is low. In many cases, particularly for lead, there is not a clear distinction between the contributions of root uptake, foliar deposition from air, and direct contact or resuspension of soil particles. It has been demonstrated that over 90% of the lead in ryegrass from a typical background area in Denmark arose from deposition of atmospheric lead on the plant surfaces (16). This contribution is uncertain, but probably somewhat less important for many food crops, and particularly for edible nonleafy parts.

The modeling results presented in Table 1 begin with estimates of intake. In some cases, estimates of dietary intake have been made from food sampling. An effective mean concentration has then been inferred from the estimated total food consumption rate. The intake rates of metals in air or water have been estimated from representative concentrations and average breathing and water consumption rates. The uptake rate to blood is estimated from the fractional absorption, which includes retention in the lungs and absorption to blood following inhalation and absorption through the gastrointestinal tract following ingestion. The concentrations in tissues given in Table 2 are determined by

	Consumption rate (g d^{-1})*	Concentration (μg g^{-1})**	Intake rate (μg d^{-1})	Fractional absorption	Uptake rate (μg d^{-1})
Arsenic					
fish	25	3.6	90	0.9	81
other foods	1200	0.03	40	0.9	36
water	1.2	1.2	1.4	0.9	1.3
air	20	20	0.4	0.3	0.1
				Total	120
Cadmium					
food	1200	0.025	30	0.05	1.5
air	20	30	0.6	0.25	0.2
				Total	1.7
Lead (adult)					
food	1200	0.09	110	0.1	11
food contact	0.06	1000	60	0.1	6
water	1.2	10	12	0.1	1
air	20	2000	40	0.2	8
				Total	26
Lead (child)					
food	750	0.09	70	0.25	18
food contact	0.04	1000	40	0.25	10
dirt on hands	0.02	2400	48	0.25	12
water	1.3	10	13	0.25	3
air	6	1000	6	0.4	2
				Total	45
Mercury					
fish	25	0.4	10	1.0	10
other foods	1200	0.004	5	0.05	0.25
air	20	7	0.14	0.7	0.1
				Total	10.4
Nickel					
food	1200	0.13	160	0.05	8
water	1.2	8	10	0.05	0.5
air	20	20	0.4	0.3	0.1
				Total	8.6
Selenium					
food	1200	0.06	70	0.8	56
air	20	3	0.06	0.2	0.01
				Total	56
Tin					
food	1200	0.17	200	0.02	4
air	20	30	0.6	0.2	0.1
				Total	4.1

* water (ℓ d^{-1}), air (m^3 d^{-1})
** water (μg ℓ$^{-1}$), air (ng m^{-3})

TABLE 1 – Transfer rates of various trace elements to man.

multiplying the uptake rate by the mean retention time, dividing by the organ mass, and accounting for the distribution fraction if the metal is partially retained or distributed to specific organs. In many cases several components of retention have been identified. A weighted effective mean residence time is then used. In some cases the element is assumed uniformly distributed in the body. This is certainly an approximation which should be adjusted when more specific data on organ distribution and retention are available. The estimates are for intake and metabolism of the adult except that, in addition, the two-year-old child is considered for lead. For some metals, such as cadmium, for which there is long-term accumulation, the adult is most at risk. However, absorption and toxicity in the child can often be important, as it is for lead, and efforts must be made to increase the information available on metabolism of metals in children.

Ingestion intake of the metals considered is more significant than inhalation intake. The amounts in food may originate largely from airborne particles, but this must be better quantified. Metals deposited on fruits and vegetables can be removed by washing and peeling, but canning or other processing operations may make additions. Some estimate of this effect is included for lead. An additional contribution to intake of lead by children is from dirt on hands which is transferred to food or mouth. This is obviously quite a variable route of exposure. The lead concentrations in urban areas range from 500 to 1000 $\mu g\, g^{-1}$, with generally somewhat higher levels found indoors and in smaller particles adhering to hands (9).

Lead in blood is partially retained before distribution to other tissues. The concentrations of lead in blood are readily measured. The analysis of lead in shed teeth is also being used in toxicological assessments. In this case, further modeling of the transfers to bone and teeth is required. The estimated concentration (Table 2) of lead in adult blood (± 50%) has typically been reported. The associations between uptake and blood lead levels in the child cannot be completed, as estimates of the residence time of lead in blood of children are unavailable.

Cadmium in the body is retained primarily in the kidneys, which receive an estimated 30% of the amount which reaches the blood. To a first-order approximation, the mean residence time of cadmium in the kidneys is taken as 30 years. The estimated uptake rate of cadmium then results in an eventual concentration in the kidneys of 18 $\mu g\, g^{-1}$. The result would be 50% higher if the cadmium is considered localized in the kidney cortex. Consideration of

	Uptake rate	Distribution	Organ	Mass	Retention time	Concentration
	($\mu g\ d^{-1}$)			(kg)	(d)	($\mu g\ g^{-1}$)
Arsenic	81*	1.0	Body	70	4	0.009
	37**	1.0	Body	70	8	
Cadmium	1.7	0.3	Kidneys	0.31	10950	18
Lead						
(adult)	26	1.0	Blood	5.2	23	0.12
(child)	45	1.0	Blood	2	-	-
Mercury	10*	1.0	Body	70	100	0.015
	0.4**	1.0	Body	70	60	
Nickel	8.6	0.3	Body	70	200	0.007
Selenium	56	0.9	Body	70	140	0.1
Tin	4.1	0.35	Bone	5	350	0.1
		0.15	Body	65	350	0.003

* from intake of organic forms
** from intake of inorganic forms

TABLE 2 – Distribution and retention of absorbed trace elements in man.

age-dependent renal changes can improve modeling results and in particular show declining kidney cadmium concentrations after about the age of 50 years (17).

The estimated concentrations of other elements in the body are in reasonable agreement with measurements, however, in some cases only with the most recent measurements. It is quite noticeable that estimates of intake and body concentrations of metals are declining as analytical techniques are improved. The importance of avoiding sampling contamination and of incorporating quality assurance procedures is now quite generally recognized.

Models and their results obviously must change as understanding of the systems increases. The parameters of models are determined by available knowledge. As new information becomes available, better values of input parameters may be selected or structural changes in the models may be required. There is a danger in holding too firmly to and believing in the models in spite of new information. Measurements are then judged right or wrong based on the preconceived notion. This is certainly not the way. Measurements, if valid and true, are the concrete reality; models are but reflections of current conceptions, never fully fixed and firm.

REFERENCES

(1) Bennett, B.G. 1981. Exposure commitment assessments of environmental pollutants; summary exposure assessments for lead, cadmium, arsenic, vol. 1, No. 1, MARC Report No. 23. London: Monitoring and Assessment Research Centre.

(2) Bennett, B.G. 1981. Exposure commitment assessments of environmental pollutants; summary exposure assessments for mercury, nickel, tin, vol. 1, No. 2, MARC Report No. 25. London: Monitoring and Assessment Research Centre.

(3) Bennett, B.G. 1982. Exposure commitment assessments of environmental pollutants; summary exposure assessments for PCBs, selenium, chromium, vol. 2, MARC Report No. 28. London: Monitoring and Assessment Research Centre.

(4) Bennett, B.G. 1982. The exposure commitment method for pollutant exposure evaluation. Ecotoxicol. Env. Safety **6**: 363–368.

(5) Bernard, S.R. 1977. Dosimetric data and metabolic model for lead. Health Phys. **32**: 44–46.

(6) Camner, P; Clarkson, T.W.; and Nordberg, G.F. 1979. Routes of exposure, dose and metabolism of metals. **In** Handbook on the Toxicology of Metals, eds. L. Friberg, G.F. Nordberg, and V.B. Vouk, pp. 65–97. Amsterdam: Elsevier/North Holland Biomedical Press.

(7) Chamberlain, A.C. 1974. Travel and deposition of lead aerosols. Report AERE-R7676. Harwell: U.K. Atomic Energy Authority.

(8) Commission for the European Communities. 1979. Methodology for evaluating the radiological consequences of radioactive effluents released in normal operations. Joint report by the National Radiological Protection Board and the Commissariat à l'Energie Atomique, Doc. No. V/3865/79-EN, FR.

(9) Duggan, M.J., and Williams, S. 1977. Lead in dust in city streets. Sci. Total Env. **7**: 91–97.

(10) Galloway, J.N.; Thornton, J.D.; Norton, S.A.; Volchok, H.L.; and McLean, R.A.N. 1982. Trace metals in atmospheric deposition: a review and assessment. Atmos. Env. **16**: 1677–1700.

(11) Georgii, H.W., and Pankrath, J., eds. 1982. Deposition of Atmospheric Pollutants. Dordrecht, Holland: D. Reidel Publishing Co.

(12) International Commission on Radiological Protection. 1979. Limits for intakes of radionuclides by workers, ICRP Pub. 30. Oxford: Pergamon Press.

(13) Kjellström, T., and Nordberg, G.F. 1978. A kinetic model of cadmium metabolism in the human being. Env. Res. **16**: 248–269.

(14) Nriagu, J.O. 1979. Global inventory of natural and anthropogenic emissions of trace metals to the atmosphere. Nature **279**: 409–411.

(15) Piotrowski, J.K., and Buchanan, J.M. 1982. Dose-time response functions for toxic chemicals. Env. Monit. Assess. **2**: 139–156.

(16) Tjell, J.C.; Hovmand, M.F.; and Mosbeck, H. 1979. Atmospheric lead pollution of grass grown in a background area in Denmark. Nature **280**: 425–426.

(17) Travis, C.C., and Haddock, A.G. 1980. Interpretation of the observed age-dependency of cadmium body burdens in man. Env. Res. **22**: 46–60.

(18) Turner, D.B. 1979. Atmospheric dispersion modeling, a critical review. APCA J. **29**: 502–519.

(19) United Nations Scientific Committee on the Effects of Atomic Radiation. 1982. Ionizing radiation: sources and biological effects. New York: United Nations.

(20) World Health Organization. 1978. Principles and methods for evaluating the toxicity of chemicals. Part 1. Environmental Health Criteria Nr. 6. Geneva: World Health Organization.

Standing, left to right:
Zdenek Filip, Heinz Tobschall, Arne Jernelöv, Ed Goldberg, Bob Duce, Bill Zoller, Hartmut Heinrichs, Jozef Pacyna, Iain Thornton.

Seated, left to right:
Teruo Asami, Andi Andreae, Kathe Bertine, Patrick Buat-Ménard, Ulrich Förstner.

Changing Metal Cycles and Human Health, ed. J.O. Nriagu, pp. 359–373.
Dahlem Konferenzen 1984. Berlin, Heidelberg, New York, Tokyo: Springer-Verlag.

CHANGING BIOGEOCHEMICAL CYCLES
Group Report

M.O. Andreae, Rapporteur
T. Asami, K.K. Bertine, P.E. Buat-Ménard, R.A. Duce, Z. Filip, U. Förstner, E.D. Goldberg, H. Heinrichs, A.B. Jernelöv, J.M. Pacyna, I. Thornton, H.J. Tobschall, W.H. Zoller

INTRODUCTION

Human society has dramatically altered the face of this planet, a fact which is obvious to anyone who has ever spent some time looking out of the window of an aircraft. Geochemical processes — weathering, erosion, transport — as well as ecological patterns are changing on a global scale. The extraction of mineral and fossil fuels and the processing of these materials represent a selective mode of "weathering" which mobilizes some of the most toxic elements and releases them to the most sensitive compartments of the environment — to the surface waters and the atmosphere.

PERTURBATION OF ELEMENT CYCLES — AN INVENTORY

To assess which elements may be of concern, we have to consider both the change in their biogeochemical cycles by man and their potential threat to human health. Consequently, we proposed four criteria and examined a "laundry list" of elements using these criteria:

1. Has the geochemical cycle of the element been substantially perturbed by man, and on what scales?
2. Is the element mobile in geochemical processes because of either its

volatility or its solubility in natural waters, so that the effect of geochemical perturbations can propagate through the environment?
3. What is the degree of public health concern associated with the element?
4. What are the critical pathways by which the most toxic species of the element can reach the organ in man which is most sensitive to its effect?

In Table 1 we have defined the geographic scale as "global" when the effects of perturbation can be demonstrated at least in large parts of the Northern Hemisphere. Such a global perturbation is most dramatically seen for lead. "Regional" effects refer to a scale of 100–1000 km, while "local" perturbation covers the range of less than 100 km (but larger than the workplace or factory environments which we have not considered here). Significant health effects are usually caused by very strong perturbations at the local level.

Changing land-use patterns and agriculture can mobilize very large amounts of soil materials either as atmospheric dust or as suspended material in rivers. While these processes, which contribute to topsoil loss and desertification, are considered to be of great importance to human welfare, they can be detected only with difficulty in the geochemical record due to the presence of large natural fluxes. Elements which we consider to be mobilized by these processes have been indicated by "c" in Table 1.

As diagnostic environments we have considered those geochemical compartments in which a substantial perturbation for a given element has been documented. Of particular importance here is the atmosphere where, because of the short residence time of most substances and because of the low rates of injection by natural processes for many toxic elements, a perturbation is often most easily detectable. Dated cores of glacial ice and coastal and lake sediments make it possible to use the preindustrial concentration of an element at the same site as an "internal standard" and obtain direct information on historical changes of environmental metal loadings. For many elements, geographical and/or temporal changes in surface and ground waters have been shown to be diagnostic of significant anthropogenic perturbations.

The usually relatively high baseline of metals in soils and the low mobility of metals in this matrix (when compared to air and water) make soils diagnostic for geochemical perturbations on local scales only. Dramatic examples for such perturbations of soils are the presence of 55 ppm Be and 0.5% Cr (VI) in some polluted urban soils in Japan and of 1% lead in agricultural soils in England. Since, in the absence of knowledge on the natural fluxes, a perturbation cannot be rigorously derived from the anthropogenic fluxes alone, we have listed emission studies as evidence only when little or no environmental information was available for an element.

Under "mobility" we have indicated whether an element is volatilized during industrial processes and is therefore likely to enter the atmosphere. We have considered it to be "soluble" if it is likely to be transported over significant distances in the dissolved state in natural waters. The special case of elements which are made soluble under reducing, anoxic conditions has been given a separate symbol (r) in Table 1. In addition to the volatilization under high temperature conditions, several elements may be mobilized by the biological or abiotic formation of organometallic compounds: As, Sn, Ge, Hg, Se, Pb, and possibly Te.

The column on health concerns has been based on information from the Fowler et al. and Kjellström et al. reports (both this volume), that on critical pathways from the Jaworski et al. and Fowler et al. reports (both this volume). Further discussion on these topics can be found in these reports.

Table 1 shows that mobility, especially volatility, is a reasonably good predictor of large-scale perturbation. Significantly, there is also a trend for the most toxic elements to show the most widespread perturbation by man's activities. The table also points to substantial deficiencies in our knowledge of the geochemical cycles of some elements of potential concern, e.g., Pt, Pd, Ti, In, Bi, Be, Ga, and Te.

RESEARCH STRATEGIES AND METHODS

A large concern in understanding geochemical and environmental processes and their changes is the quality of the available data base. Dramatic improvements have been made in this respect in the last two decades through both the development of "clean" techniques (e.g., the procedures used for the study of lead by the Caltech group) and the improvements of analytical detection limits. Much attention still needs to be paid to the sampling process; contamination needs to be avoided, and the sample must be properly taken to be representative. It should be emphasized that the protocols for contamination-free sampling need to be examined for each element individually; a "clean" sample for one element may be "dirty" for another.

The sampling design must be carefully evaluated. What does a sample or group of samples represent? How many samples does it take to characterize a river, a lake, or a rainfall event? How do we preserve the integrity of a sample between collection and analysis?

Element	1 Scale of Perturbation			2 Most Diagnostic Environments	3 Mobility	4 Health Concern	5 Critical Pathway
	global	regional	local				
Pb	+	+	+	A,Sd,I,W,H,So	v,a	+	F,A [8]
V	+	+,c	+	A	9	(+)	A?
As	(+)	+	+	A,Sd,So,W	v,s,a	+	A,W
Sn	(+)	+	+	A,Sd,W	v,a	+ [6]	F
Zn	(-)	+	+	A,Sd,W,So	v,s	E	F
Cd	(-)	+	+	A,Sd,So,W	v,s	+	F
Hg	(-)	+	+	A,Sd,Fish,So	v,a	+ [6]	F,(A)
Sb	(-)	+	+	A,Sd	v,s	(+)	F,W,A?
Cu	(-)	+	+	A,Sd,W,So	v,s	E	F?
Ag	(-)	+	+	A,Sd,W	(v)	(+)	?
Se	(-)	(+)	+	A	v,s,a	E	F
Ge	?	+	+	A,So,W?	v,s,a	(+) [6]	?
Ni	-	+	+	A,Sd	-	E	F,W,A?
Cr	-	+	+	A,Sd,W,Gw	s,v [7]	E	W,F
B	-	(+)	+	A,Sd,Gw	v,s	E	W
K	-	(+)	+	A	s	E	F
Pt	?	?	+	A,Sd	s	(+)	?
Pd	?	?	+	Sd	s	(+)	?
Mo	?	?	+	A,W,So,Sd	s	E	F,W
Tl	?	?	+	Em,So	v,s	(+)	A,F?
In	?	?	+	A,So,Em	v	(+)	?
Bi	?	?	+	A,So,Em	v	(+)	?
Be	?	?	+	A,So,Em	-	(+)	A
Ga	?	?	+	Em	v	(+)	?
Te	?	?	(+)	So	v,a?	(+)	?
Mn	-	c,+	+	A,Sd,W	r	E	A
Fe	-	c	+	A,Sd,W	r	E	F,A,W
Al	-	c	+	A,Sd	-	(+)	W?
Si	-	c	+	A	-	(E)	-
Ti	-	c	+	A,Sd,So	-		-
Co	-	c	(+)	Sd	r	E	F?
Na	-	c,+	+	W,A,So	s	E	F,W
Mg,Ca	-	c	+	A,Sd,Em	(s)	E	
Ba	-	c	+	A,Sd,Em		(+)	F,W
U	-	c	+	A,So,Gw	s	(+)	A,W
Th	-	c	+	A,So,Gw	-	(+)	A
Zr	-	c	+	A	-		-
Y	-	c	(+)	Sd	-		-

TABLE 1 (cont.)

Element	1 Scale of Perturbation			2 Most Diagnostic Environments	3 Mobility	4 Health Concern	5 Critical Pathway
	global	regional	local				
Au	(-)	(-)	+	Sd,A	-	+	-
W	(-)	(-)	+	A	s?		-
Li,Rb, Cs	-	c	c	Sd	s	+	-
Rare Earths	-	c	c	Sd	-		-
Ta,Hf, Sc,Sr	-	c	c	Sd	-		-
Os,Ir, Ru,Rh	?	?	?	-	-		-

1) + = significant perturbation; (+) = possible perturbation; (-) = enriched relative to crustal abundances, but the enrichment may not be anthropogenic; - = no perturbation; ? = not enough information; c = enhanced due to mobilization of crustal materials (soil, dust).

2) A = air; Sd = sediments (coastal, lake); So = soils; I = ice cores; W = surface waters; Gw = groundwaters; H = humans; Em = emission studies (only listed when little geochemical information is available).

3) v = volatile; s = soluble; r = soluble only under reducing conditions; a = mobile as alkylated organometallic species; - = not mobile.

4) + = toxic in excess; (+) = toxic, but little data available; E = essential, but toxic in excess (information gathered from Fowler et al. report, this volume).

5) F = food; W = water; A = air; - = no significant exposure likely (information gathered from Jaworski et al. and Kjellström et al. reports, both this volume).

6) organometallic forms only.

7) hexavalent form volatile and toxic, trivalent form essential.

8) exposure through hand-to-mouth activity is critical for lead in children.

9) enriched relative to crustal abundance from fuel oil combustion (vanadium porphyrins).

TABLE 1 – Perturbations of the geochemical cycles of metals by society. The elements are grouped according to the scale for which such perturbations can be documented.

The analytical procedures must be both accurate and relevant. Standard reference materials are highly desirable for many elements in environmental matrices to assure accuracy. We should use nuclear methods for the determination of the total amount of an element, where possible, to assure that we are not missing important species or forms when using more operationally dependent techniques of sample enrichment and determination. On the other hand, we have to make sure that we determine the environmentally active or biologically available species or forms separately.

The selection of sites and the timing of sampling campaigns is important when we are trying to understand changing biogeochemical cycles. Seasonal changes may be as large or larger than secular ones and could lead to false conclusions if they are not accounted for. The choice of both the "control" site and the "polluted" site is critical. For example, a comparison of the arsenic levels in the upper and lower Rhine indicates the latter to be "polluted," but the lower Rhine has still less arsenic than some "unpolluted" rivers in the Andes mountains. To cope with this natural geochemical variability, it appears most useful to investigate systems with "internal" reference points. Examples for this are ice and sediment cores which make it possible to compare the current state of the system to its preindustrial state. Sediment cores from larger water bodies, e.g., Lake Michigan, the Baltic Sea, or river deltas, may be of special value here, since they integrate over larger areas. On the other hand, water reservoirs provide sensitive and accessible indicators of problems of local concern.

PROCESSES OF MOBILIZATION – EMISSION SOURCES

The production of industrial materials and energy involves the processing of mineral resources at high temperatures, resulting in the release of the volatile elements into the atmosphere. Other emissions result from the leaching of slags, mineral residues, and flue dusts at disposal sites.

Emissions from fossil fuel combustion and metal smelting have been intensively studied in recent years, but large uncertainties persist due to substantial variations in the chemical composition of source materials, in the technology employed, and in the emission control devices utilized. The use of emission factors based on data from the United States to predict emission rates in other regions must be done with great caution.

The progressive industrialization in the developing nations is of concern, since effective emission control is often considered not feasible under the economic circumstances of these countries. Data on the emission characteristics in Eastern Europe, the U.S.S.R., China, and the developing nations need to be gathered and analyzed.

At this time, only limited information is available on the emission of metals from waste incinerators and biomass burning. In the case of oil burning, most of the metals contained in the source material are volatilized, and the emissions can be predicted reasonably well. Waste incinerators receive a highly variable input material and consequently show large variations in their emission characteristics. They appear to be substantial sources for atmospheric tin, antimony, cadmium, and lead, as well as other elements.

In the production of cement, a mixture of shale, limestone, and various additives is exposed to high temperatures. Current practices, including the feedback of precipitator dusts into the input stream, result in spectacular enrichments of some elements in the atmospheric emissions; as much as 2% of thallium have been measured in flue dusts. During failure of the emissions control equipment, large amounts of Tl have been released to the environment, and even under normal operations, as much as sixteen tons/year of Tl are released from cement manufacturing in West Germany, about four times more than is emitted from coal combustion. Emissions of thallium and other elements from cement plants need to be investigated.

The different aspects of biomass burning, including the combustion of fuel wood, forest fires, and agricultural burning, represent one of the major information gaps in the global and regional emission inventories. Fine particulate potassium which is released during biomass combustion can be used as a tracer for these emissions.

In order to understand properly the role of anthropogenic high temperature emissions, we need to further characterize volcanic emissions which represent the major natural source of similar character and which due to their variability in time and space are still inadequately defined. A related issue is the emission of mercury, arsenic, and other elements with volatile species during geothermal power production, which must be expected to become more prevalent as the expansion of this industry continues.

At low temperatures, elements become mobile by dissolution into natural waters and drinking water and by the formation of volatile or finely particulate forms during biological processes. The formation of organometallic species will be discussed in a later section. A largely unresolved question is the role of microorganisms in the release of metals in acid mine drainages from metallurgical wastes and from coal mining wastes. Changes in the physicochemical environment in water-sediment systems can release previously bound elements; examples for this process are the release of cadmium at the estuarine interface and the solubilization of antimony under reducing conditions in sedimentary pore waters and anoxic basins. Mercury is volatilized in cooling towers of power plants, probably because it is reduced to the element or methylated by bacterial activity on the surfaces of the aerators.

Besides characterizing emission sources in a quantitative sense, we must not forget the importance of the qualitative characteristics of the emitted materials. The particle size as well as the solubility and surface free energy of atmospheric emissions determine their transport characteristics and their impact on man and the environment. The materials from high temperature processes are usually quite soluble, and many elements, e.g., tin, which are usually not released into solution from particles of crustal origin can be dissolved quite easily from anthropogenic materials. Consequently, the materials emitted by man's activity often represent forms of elements "new" to biological systems, forms for which the biological defense mechanisms may not be adequate. Examples of such materials are the metal-silicate-oxide spherules which make up most of the fly ash from coal burning, the metallic and metal oxide dusts from smelters, and the high temperature aluminum oxides and silicates (e.g., mullite) produced during combustion processes. When such new forms of elements are emitted, especially if they are in the "respirable" size fraction, even a flux relatively small compared to the natural fluxes of soil elements may be of health concern.

PROCESSES OF MOBILIZATION – INDIRECT EFFECTS

In addition to the direct emissions of metals from man's activities, we are observing indirect effects resulting from processes which change the biogeochemical controls of large-scale natural processes. Agricultural and residential patterns are changing the rates of continental weathering and erosion, and the input of hydrogen ions from acid precipitation changes the chemical characteristics of the weathering process. The acidification of soils enhances the rate of podzolization and consequently the release of many metals which are otherwise either rather immobile or locked into tightly closed loops within the ecosystems. The removal of Co from forest ecosystems may lead to the depletion of this element in forest soils. The mobilization of Al threatens plants, fish, and possibly man. Different ecosystems vary in their sensitivity to acidification; the lowest pH values (about two) and consequently the highest degree of metal leaching are found in spruce forests. Such low pH values must be expected to alter soil microbial ecology, e.g., the prevalence of acidophilic and acid-tolerant microorganisms, which will in turn influence biogeochemical processes in soils.

The composition of groundwaters is altered both by direct injection of metallic pollutants when wastewaters are recharged into the groundwater system, and by changes in soil biogeochemistry resulting from the introduction of excess nutrients from agriculture and organic substances from wastewater injection and from agricultural activities. These processes lead to an increase in the size of the anoxic zone in soils and consequent mobilization of elements soluble under reducing conditions. The opposite effect may be caused by a lowering of the groundwater table because of excess withdrawal, resulting in an enlargement of the oxic zone and release of other metals. Mining activities change the weathering process in large areas and cause the entry of toxic metals, e.g., As and U, into the groundwater system. A similar threat is related to leakage waters from landfills, which are often extremely high in toxic materials and can enter groundwaters.

PRODUCTION AND BEHAVIOR OF ORGANOMETALLIC COMPOUNDS

For a number of elements, organometallic forms, most often methylated species, have been observed in the environment: Hg, As, Pb, Sn, Ge, Se, Sb, and Si (the latter in the form of silicones). While some compounds are clearly anthropogenic, e.g., the butyltins and silicones, the other methylated metal species can be produced both by man and — microbially and abiotically — by nature. Even in view of the considerable research effort which has gone into elucidating the mechanisms by which mercury and other elements are being methylated, much uncertainty persists as to the processes which produce these substances in the environment. What are the methylating agents — biota or abiotic reactions? What are the coenzymes responsible for the biological methyl transfer? What, if any, is the role of methyl iodide in abiotic methyl transfer? What is the source and the concentration of methylmercury in the open ocean? Is there a flux of methylated mercury from the oceans to the atmosphere? Is mercury methylated in soil by humic acids? What elements other than those listed above may become methylated? How does the structure of organometallics relate to their toxicity and persistence? What is the role of methylgroup transfer between different metallic elements (e.g., Sn and Hg)?

These open problems point out many important directions that future research in this field may take. Some suggestions for suitable sites and environments for the identification of "new" methylated elements include marine anoxic sediments, sewage sludges, and the Salton Sea (California) which is an eutrophicated basin with a salinity resembling that of the ocean. Very high concentrations of several methylated elements (As, Ge, S) have been observed here.

With regard to the public health concern associated with these organometallic compounds, each element has to be considered separately. While the organoarsenic compounds in fish and other seafoods are essentially nontoxic, the methylmercury compounds are the only mercury species presenting a significant risk to public health. The absence of a demethylating mechanism for methylmercury in fish (a mechanism which is present in seals, on the other hand) leads to substantial enrichment of this species in large fish. The methylated species of metals low in the periodic table (with a high Z) in which the metal-carbon bond is relatively labile may present a carcinogenic risk resulting from methyl transfer. Silicones are widespread in the environment, but no health risk has been shown for these compounds yet.

CHEMICAL SPECIATION

Since the environmental behavior and toxicity of an element can only be understood in terms of the actual chemical species present, it is essential that we develop and apply techniques which provide us with information on the chemical speciation of metals in the environment. Thermodynamic models may give suggestions as to the possible species to expect, but due to the important role of kinetically controlled processes in biogeochemistry, the actual speciation is often quite different from what can be predicted.

This problem becomes especially complex in heterogenous systems, e.g., soils and aerosol particles. The use of weak extractants to operationally fractionate the forms of metals in soils has not resolved the actual molecular state in which these metals are present. There is an urgent need to develop new approaches.

New techniques for the determination of the chemical state and phase composition in solid materials are now becoming accessible, especially the surface analytical techniques of electron spectroscopy and diffraction. It is to be expected that significant progress will be made when these techniques are applied in atmospheric and soil chemistry.

METAL ASSOCIATIONS AND INTERACTIONS

Most processes release a number of metals at the same time in a more or less characteristic proportion. Examples for this are the release of zinc and cadmium from smelters; of antimony, cadmium, and tin from waste incinerators; and of lead and bromine from automobiles. Such associations are helpful for the fingerprinting of sources of environmental pollutants. On the other hand, they make the epidemiological assignment of a health effect to a given metal extremely difficult. It must be emphasized that these release patterns are not constant but vary with process and usage patterns. Different proportions are seen in different countries, and even in a given country they change as a function of technological and regulatory changes. Before the arrival of stack emissions control, the character of coal combustion particles was alkaline, due to the large amounts of soil element oxides present. Now only acidic gases and particles can escape, leading to a net change in the acidity of the material released.

After the release into the environment, fractionation processes both during transport and biological uptake occur which may separate the metals according to different patterns. In some instances, this may lead to a protective interaction of elements which are associated in their source functions (e.g., the effect of zinc on the cadmium uptake by plant roots), but the complex kinetic control of uptake and biochemical behavior in biota makes generalizations quite difficult here.

TRANSPORT PHENOMENA

The transport of metals through the environment is dependent on their chemical and physical form and on processes which act to change their form during transport. Elements which become very rapidly associated with particles in rivers do not remain accessible to biota for a long time (with the possible exception of filter-feeders) and will not be transported as far as soluble elements. In this context, the addition of complexing agents, such as nitrilotriacetate (NTA), to natural waters through detergents or other emissions is to be viewed with concern. On one hand, the biological effects of some toxic metals will be reduced by their sequestration in complexes, but on the other hand, the increased solubility will lead to more widespread dispersal and to elevated levels in drinking water.

The form of a metal is just as essential in controlling its cycle through the atmosphere. The most abundant form of mercury is gaseous elemental Hg, yet the deposition flux of mercury appears to be controlled by the particulate fraction. Unfortunately, the form of many elements in the atmosphere remains unknown. There appear to be at least two gaseous forms of selenium which are emitted into the atmosphere, one from high temperature processes, which may be Se^0 or SeO_2, and another from the ocean and from plants and soils, maybe a methylated species. The two forms appear to have quite different residence times. It will be essential for an understanding of the selenium cycle to identify these species and to study their chemical behavior in the atmosphere. Uncertainties about the character of atmospheric particulates, e.g., their solubility, hydrophobicity, mineralogy, etc., are in part responsible for the difficulties existing in the parameterization of transport and deposition processes in the atmosphere.

From the "mobile environments" air and water, the metals are scavenged by sediments, soils, and biota. There is a strong influence of the biosphere on this removal process, which may be mechanical, as in the varying efficiency of different tree species for the removal of atmospheric aerosols, or chemical, as in the removal of dissolved metals by biological uptake. When the biological processes are altered by human activities, the removal of pollutants are modified as well. Changes in the abundance of forest cover will change the rate of removal of air pollutants. Eutrophication of lakes and coastal waters can enhance the rate of removal of metals from the surface waters. The degree of burial of this material in the sediments is controlled by the availability of oxygen in the deeper waters, which in turn depends on the flux of organic matter from above. These phenomena point out the intricate balance of biogeochemical processes which control trace metal levels in natural waters. Interesting areas at which to study these phenomena are lakes and coastal environments which receive substantial amounts of metals as well as high nutrient inputs, e.g., the waters around Hongkong and Lake Constance.

In most instances, the metals will be remobilized after having been scavenged. We cannot consider soils and sediments (with the exception of those in the deep sea) as permanent sinks, but rather as intermediate storage sites, or "Zwischenlager," to use a word from nuclear technology. These remobilization processes, which are also subject to biogeochemical control, can make elements available again which were thought to have been safely disposed in the Zwischenlager, e.g., the radioisotopes discharged to the sea from the British and French nuclear establishments, which are being returned to land by marine aerosols. Ignoring such remobilization fluxes may also introduce large errors into the calculation of net fluxes to the sink compartment. When such recycling processes are interrupted by anthropogenic activities, new net fluxes may result; manganese is cycled through the soil-tree-leaf-throughfall-soil cycle in forests without much loss, until a drop in pH by acid precipitation leads to a substantial net removal of Mn from the forest ecosystem.

RELATING BIOGEOCHEMISTRY TO HUMAN HEALTH

Geochemical maps have been prepared for a number of media, e.g., soils and stream sediments, for several geographic areas, among them England, Germany, and California. These can serve as a resource to identify geochemical hot spots in which one can then look for disease information.

Epidemiological maps are also available for some areas, and the relationships between these types of maps need to be further explored. Such relationships are, of course, only empirical but may eventually lead to the discovery of causal relationships.

CONCLUSIONS AND RECOMMENDATIONS

Techniques for the identification of chemical species and physical form of metals in air, water, and soils need to be developed and applied to environmental chemistry. If we are to increase substantially our understanding of the behavior of metals in the environment, we have to be able to identify the molecular species. We can then discuss cycles in terms of species transformations and residence times and of reactions between chemical species. Without this information we are restricted to operational descriptions.

The production and fate of organometallic species in the environment need to be studied, in particular for the elements mercury, tin, and selenium. Again, this investigation must be conducted on the basis of molecular species rather than on operational terms.

Emission inventories should be assembled on the basis of existing data bases, and poorly characterized emissions should be studied. There exists a large body of information with local and national authorities which is not readily accessible to the scientific community. A suitable organization (MARC, OECD, WHO, ECE) should be assigned to collate and publish these data. New data are needed from poorly characterized sources (cement manufacturing, biomass burning, volcanoes) and from the developing countries where large future emissions must be expected.

The classic pollutant elements Pb, Cd, Hg, and As must continue to be studied due to the serious health concern associated with them. At the same time, potential "new" pollutant elements must be evaluated, in particular Tl, In, Bi, Sb, Sn, and Cr. These new elements have been selected on the basis of geochemical as well as public health considerations.

There is a need to improve our understanding of the role of indirect changes of metal cycles resulting from changing biogeochemical processes. Focal points in this research should be the influence of acid precipitation and of the eutrophication of natural waters.

A continued effort is needed to determine the natural baseline of metal concentrations and fluxes in the environment. Without this information, changes by the influence of society cannot be unequivocally demonstrated.

SUGGESTED READING

(1) Andreae, M.O. 1980. Arsenic in rain and the atmospheric mass balance of arsenic. J. Geophys. Res. **85**: 4512–4518.

(2) Buat-Ménard, P., and Chesselet, R. 1979. Variable influence of the atmospheric flux on the trace metal chemistry of oceanic suspended matter. Earth Planet. Sci. Lett. **42**: 399–411.

(3) Byrd, J.T., and Andreae, M.O. 1982. Tin and methyltin species in seawater: concentrations and fluxes. Science **218**: 565–569.

(4) Florence, T.M., and Batley, G.E. 1980. Chemical speciation in natural waters. CRC Critical Reviews in Analytical Chemistry **9**: 219–296.

(5) Förstner, U., and Wittmann, G. 1981. Metal Pollution in the Aquatic Environment, 2nd ed. Berlin: Springer.

(6) Heinrichs, H. 1982. Trace element discharge from a brown coal fired power plant. Env. Technol. Lett. **3**: 127–136.

(7) Heinrichs, H., and Mayer, R. 1980. The role of forest vegetation in the biogeochemical cycle of heavy metals. J. Env. Qual. **9**: 111–118.

(8) Kazantis, G. 1979. Thallium. **In** Handbook on the Toxicology of Metals, ed. L. Friberg. Amsterdam: Elsevier.

(9) Kitagishi, K., and Yamane, I., eds. 1981. Heavy Metal Pollution in Soils of Japan. Tokyo: Japan Scientific Societies Press.

(10) Li, Y.H. 1981. Geochemical cycles of elements and human perturbation. Geochim. Cosmochim. Acta **45**: 2073–2085.

(11) National Academy of Sciences. 1978. The Tropospheric Transport of Pollutants and Other Substances to the Oceans. Washington, D.C.: Printing and Publishing Office, National Academy of Sciences.

(12) Ng, A., and Patterson, C.C. 1981. Natural concentrations of lead in ancient Arctic and Antarctic ice. Geochim. Cosmochim. Acta **45**: 2109–2121.

(13) Nriagu, J.O.; Kemp, A.L.W.; Wong, H.K.T.; and Harper, N. 1979. Sedimentary record of heavy metal pollution in Lake Erie. Geochim. Cosmochim. Acta **43**: 247–258.

(14) Organization for Economic Co-operation and Development. 1977. The OECD Programme in Long-range Transport of Air Pollutants. Measurements and Findings. Paris: OECD.

(15) Thornton, I., and Plant, J. 1980. Regional geochemical mapping and health in the United Kingdom. J. Geol. Soc. Lond. **137**: 575–586.

(16) Wong, C.S.; Boyle, E.; Bruland, K.W.; Burton, J.D.; and Goldberg, E.D., eds, 1983. Trace Metals in Seawater. New York: Plenum.

Standing, left to right:
John Jaworski, Franco Baldi, Dieter Sauerbeck, Hermann Hecht, Fred Brinckmann, Tarzan Legović, Adolf Kloke.

Seated, left to right:
Jerome Nriagu, Al Page, Joan McKenzie, Michael Bernhard, Kurt Wasserman.

Changing Metal Cycles and Human Health, ed. J.O. Nriagu, pp. 375–388.
Dahlem Konferenzen 1984. Berlin, Heidelberg, New York, Tokyo: Springer-Verlag.

ROUTES OF EXPOSURE TO HUMANS AND BIOAVAILABILITY

Group Report

J.F. Jaworski, Rapporteur
F. Baldi, M. Bernhard, F.E. Brinckman, H.P. Hecht, A. Kloke,
T. Legović, J.M. McKenzie, J.O. Nriagu, A.L. Page, D.R. Sauerbeck,
K.J. Wasserman

INTRODUCTION

There are many pathways or routes by which humans are exposed to metallic compounds, and these are changing as our society uses more or less of the metal or changes the chemical form of the metal in the environment (8, 11).

The following discussion considers what pathways are, which are important, which are subject to the influence of natural barriers on metal transfer, and which of these barriers can be circumvented by man's activities. The quality of available data for predicting the relative importance of pathways is assessed as is the usefulness of models in gathering and using the data. Finally, these considerations are used to identify pathways and elements of concern to human health, and recommendations are made for further research to fill in gaps in knowledge.

WHAT ARE PATHWAYS?

An environmental pathway is the route that a substance follows between a source of emission and either a geological depot or a biological receptor (see Fig. 1).

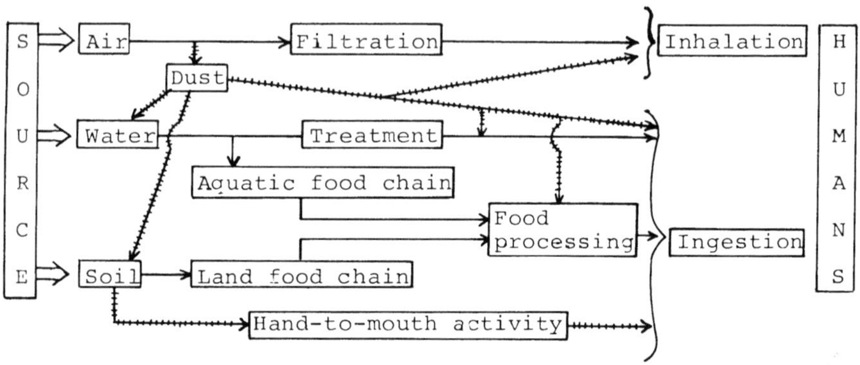

FIG. 1 – Pathways of metals to humans (cross-hatching denotes path of fine airborne fallout).

A given metal may exist in several chemical forms (species), and the pathways described in Fig. 1 can apply to each of these species.

Simple pathways involve only one environmental medium, e.g., a metal emitted into air is transported to the receptor where it is inhaled and subsequently absorbed. More complex pathways involve two or more environmental media and transfer of the metal among the different compartments of the media, e.g., a metal emitted into water may enter the soil via irrigation and from there the terrestrial food chain, or via spray-irrigation may deposit onto crops which are then eaten.

IDENTIFYING IMPORTANT PATHWAYS

For a given metal, the importance of a specific pathway is determined by the flux (the amount per unit time of a substance which is delivered to a site of absorption or deposition), the rate of absorption, and the destination in the body. To identify the critical pathway for a metal species, one must have the following information:

1. The critical effect of the metal species and its location in the body (13);
2. The relative absorption rates of the metal species from the lungs, gut, and skin, as well as the rate of translocation in the body from site of absorption to site of action;
3. The relative fluxes of the metal species via different pathways to the sites of absorption.

Consideration of these factors can become very complex because man's activities not only change the relative fluxes of metals along their pathways but also alter the chemical and physical form of the metals so that their rates of absorption differ from those found in nature (4).

The factors above are comprehended in the term "bioavailability," which is a quantitative parameter expressing the combined effect of environmental mobility, absorption/excretion, transport within the body, and binding to the target site (or site of action). Thus there are two components to bioavailability: (a) outside the organism, factors which increase or decrease exposure and absorption; and (b) inside the organism, factors which increase or decrease distribution and binding to the site of action.

The following characteristics of a substance are considered to augment its bioavailability (Table 1):

1. Small particle size — facilitates long-range air transport and deposition in the respiratory tract as well as deposition onto food;
2. Volatility — vapors are subject to long-range transport and have high rates of respiratory absorption;
3. Aqueous/lipid solubility — depending on the nature of the site of action, bioavailability may be enhanced by one or the other property, e.g., methylmercury and neural tissue; a lipophile and a lipid-rich target.
4. Complexation — complexes having certain molecular sizes or net charges can diffuse (passive transport) across biological membranes more easily than others; in other cases, however, complexation can strongly reduce transfer across biological membranes;
5. Elements or complexes of elements which mimic essential nutrients are more likely to be handled by specific active transport processes and thus have enhanced absorption rates and mobility within the body, perhaps by being bound to particular carrier molecules.

FACTORS AFFECTING PATHWAY FLUX

The flux of a metal species along a pathway is dependent on the factors which diminish or augment the transfer of the metal from one environmental compartment to another. Table 2 lists a number of physical, chemical, and biological processes which can act as partial or almost complete barriers to the movement of metals along their pathway to man. The physical processes involve adsorption, sedimentation, and filtration; the chemical processes involve complexation, precipitation, and redox reactions; the biological processes involve rejection at membranes, restricted distribution among tissues of the organism, biotransformation, and a death or growth inhibition of organisms used as food sources occurring before toxic levels occur in the edible tissues. The

Element Grouping	Major Pathways to Man	- Critical Factors Affecting Bioavailability -		
		Metabolic	Dietary	Receptor
As, Se	air, food, water	biotransformation, replace essential P, S	speciation, interactions	physiological status
Cd, Zn	food	biotransformation, accumulation, processing	"	"
Hg	food	"	speciation, stage of maturity	individual variation, pregnancy
Pb, Sn	air, food, water	" and abiotic methylation	speciation, processing	stage of maturity, race, pregnancy
Fe, Co, Ni, Mn	food	absorption, form metallo-enzymes	speciation	pregnancy, stage of maturity, diseases

Notes on speciation: a) Methylated metals tend to be more soluble than their inorganic analogs. They tend to be more lipophilic which enhances their neurotoxicity. Methylated species sometimes undergo reductive elimination of the methyl group with an anion, resulting in the precipitation of the metal in a reduced oxidation state; the physiological significance of this reaction is unknown. b) Ingested arsenic(III) can be recovered in urine within 12 hours as methylarsonic(V) acid, i.e., oxidative methylation has occurred. c) Methyltins are found in human urine, but this could be from ingested methyltin, not methylation. d) Cd and Zn readily form metallo-organic bonds; Pb, in addition, also forms metal-carbon bonds.

TABLE 1 – Some operational considerations for bioavailability to humans.

Environmental medium	Transfer affected	Type of effect[a]
air	long-range transport	P: settling of larger particles, filtration of air
water	sediment/water	P: adsorption, filtration C: precipitation
	water/biota	C: formation of poorly absorbed complexes, interelement interaction B: rejection at gill membrane
	biota/man	C: as above P: avoiding ingestion of contaminated organs or tissue (selective food processing)
soil	soil/soil solution	P: adsorption onto clay C: binding by humic substances, formation of precipitates
	soil solution/root	C: intermetallic interactions B: root membrane rejection
	plant/man or plant/animal	C: formation of precipitate or adsorbate in roots (no translocation) B: death or lack of productivity before toxic level reached in edible tissue P: food processing (rejecting contaminated tissues)
	animal/man	B: as above P: as above

[a] P = physical; C = chemical; B = biological.

TABLE 2 – Barriers to increased flux of metals along environmental pathways.

physical and chemical barriers are passive, but the biological barriers are mostly active (i.e., metabolic energy is required). In some cases a biological process such as biotransformation may act as a barrier to accumulation of a toxic substance by one organism, but as a result may circumvent barrier mechanisms in other organisms (e.g., biomethylation of Hg is a detoxification mechanism in microorganisms but results in increased uptake of the Hg by the aquatic food chain). Therefore, barrier mechanisms should be considered with reference to a particular target organism.

Terrestrial Food Chains

Most trace metals when added to soil do not enter the soil solution but become fixed, adsorbed, precipitated, and/or chelated depending on the soil type and the specific element. The chemical reactions involved are poorly understood; but, it is known that the immobilization is due to the presence of hydrous oxides, clay, and humus in soil as well as the influence of pH and redox potential. Cationic metals are more strongly fixed in neutral soils whereas anionic metals are less mobile in the acid range. Metals differ in degree of fixation. In consequence, the metals Pb and Cr (also Hg, As, Be, and Sn) are so strongly fixed in near-neutral arable soils that they are hardly absorbed by plants. Transfer coefficients in the range of 0.01–0.1 (plant/soil) illustrate this behavior (see Kloke et al., Table 6, this volume). Soils may therefore accumulate considerable quantities of these metals without this being reflected in increased exposure to humans. However, this does not imply that soils may be loaded with metals indefinitely without harmful effect to the quantity and quality of the food supply (see below).

In contrast, light and/or strongly acidic soils exhibit increased cation availability so that some of the Al, Pb, Cr, and other elements can be absorbed by plant roots. However, another effective barrier may then come into play because often very little of these elements is translocated to the above-ground (edible) parts. The same effect also occurs for transfer to storage roots, tubers, and bulbs. Hence for soil-derived Al, Pb, Cr, and Hg there are two effective barriers which strongly reduce the passage of these elements along the food chain to humans.

Another group of elements (Cd, Zn, Cu, Ni, and Tl) remain somewhat more soluble in soils and thus more available to plant roots. Although translocation from root to shoot is somewhat limited, this barrier is not sufficient to prevent all food and pasture plants from absorbing excessive levels of at least Cd and Tl from contaminated soils. The corresponding soil-to-plant transfer coefficients for these two elements are in the range of 1–10 (Kloke et al., this volume). The particular problem for Cd and Tl (and in some cases also for Cu) is that concentrations which are below the toxic threshold for the plants themselves may already be toxic to the organisms which eat them. In contrast, Zn, Ni, and

Cu are generally more toxic to the plants than to the animals eating them. For these elements crop yields tend to drop well before their metal contents become hazardous to the organisms consuming the plants. Although this represents an effective barrier mechanism, it must be remembered that it occurs at the expense of the quality or quantity of the food supply and thus is an undesirable situation.

Little is known about the dissolution and fractional uptake of metals from ingested plant material. A number of plant constituents such as pectate or phytate are expected to reduce strongly the bioavailability of metals to monogastric animals, whereas in the rumen of cattle insoluble sulfides may form leading to decreased intestinal absorption. Nevertheless, experiments in which soluble metal salts are fed show that after absorption most trace metals accumulate preferentially in a few organs or tissues leaving the other tissues relatively uncontaminated. For example, Cd concentrations in animal tissues decrease respectively from kidney, liver, bone, brain, and muscle while the decreasing order for Pb is bone, kidney, liver, lung, spleen, and muscle. Thus, the selective consumption of muscle rather than organ meat can be construed as another barrier for metal transfer to humans because muscle remains relatively uncontaminated even when pasture soils contain excessive amounts of metals.

In summary, the soil-plant-man pathway is a much more critical route for metals than is the soil-plant-animal-man route; but, fortunately, plants discriminate against most soil-derived toxic metals except Cd and Tl.

Aquatic Food Chains

The chemistry of metals added to natural waters is in many respects identical to that encountered in considering the soil/soil solution system, only, in this case, the aqueous phase is predominant. The same metals are associated predominantly with the solid phase because of adsorption or precipitation processes (sediment/water barrier). Those metals that are soluble (Ni, Cu, Zn, Tl, Cd) are generally more toxic to fish than to humans, i.e., fish will die or cease reproducing before tissue levels become high enough to be toxic to humans consuming the fish. Cd in molluscs and crustaceans and Hg in molluscs and finfish are exceptions. Cd may accumulate to high levels in the gastrointestinal (but not muscle) tissues of molluscs and crustaceans, and, unless these tissues are excluded from the diet, a significant increase in human exposure to Cd may occur. Hg when methylated is so effectively absorbed from water and deposited in muscle tissue by fish that bioaccumulation occurs in spite of the extremely low levels found in natural waters. Prolonged consumption of aquatic organisms contaminated by Hg or Cd can lead to undesirable accumulations or even poisoning. Methylation of As and Se also circumvents the sediment/water/biota barriers; but, even if high levels of these elements are found in fish they are usually present in a chemical form that is rapidly excreted by humans so that the risk of accumulating harmful levels of these substances is low.

FACTORS OVERCOMING PATHWAY BARRIERS

The natural barriers against excessive human exposure to metals can be circumvented or short-circuited as a consequence of human activity.

Fine particles produced by high temperature sources (see Andreae et al., this volume) and emitted as dust into the air can contaminate the outside and inside (home) environment as well as food and drink prior to, during, and after processing. The metals associated with these particles are generally concentrated at the surface in relatively soluble form. Children can ingest and absorb excessive amounts of metals, especially Pb, from household and playground dust during normal hand-to-mouth activity. Thus terrestrial food chain barriers and air filters which do not trap fine particles are bypassed by these fine particles (see crosshatched pathways in Fig. 1; see also Landrigan and Bridbord, this volume).

Smoking is a high temperature source which introduces fine particulate directly into the lungs (where absorption is high) of the smoker and persons nearby as well as contributing to the general fine particle level of the indoor environment. Smokers have significantly higher body levels of at least As, Cd, Ni, and Pb (little information is available for other elements).

Pesticides, fertilizers, and mineral supplements all represent substances that are used either to stimulate growth and food production or to eliminate organisms which would reduce the quality or efficiency of food production. Pesticides containing Cu, Zn, Mn, As, or Pb may bypass the soil/plant barrier by direct deposition onto food plants. Metal contaminants in fertilizers may increase the concentrations of undesirable metals which are available from the soil (e.g., Cd in phosphates; Zn, Cu, Ni, and Cd in sewage sludge). Similarly, metal contaminants in mineral supplements for animals bypass both soil/plant and root/shoot barriers, while those in mineral supplements for humans will bypass the plant/animal/man barriers as well.

Contaminated consumer products may contribute to metal levels in household dust (e.g., Pb in paint; Pb, Bi, and Cd in solder fume) or contaminate food and drink by direct contact (e.g., Pb in soldered cans and ceramic glazes;

organotins in polyvinyl chloride). The terrestrial and aquatic food chain barriers do not operate at this level.

Man not only introduces metals into the environment in such form or quantity that natural barriers are overcome, but also many of the consequences of man's activities change the effectiveness of the barrier processes themselves. Acid deposition from fossil fuel combustion decreases soil pH and shifts the metal equilibrium away from the adsorbed or precipitated phase to the dissolved phase (i.e., in the soil solution). In acid soils (pH < 4.7) Al becomes toxic to a number of plants and its concentration increases markedly in acid ground waters.

The large-scale use of salt to clear ice from roadways has significantly increased NaCl concentrations of surface waters in parts of the United States and Canada. Continued increases of Na^+ may be expected to begin adversely affecting the blood pressures of those drinking the water, while increased Cl^- would be expected to increase the mobilization of Hg, Pb, Cd, and other elements from sediments by the formation of complex anions.

DATA QUALITY

Having discussed pathways and bioavailability, it is necessary to discuss how these concepts may be put together in a quantitative manner to describe the effects of metals on human health.

Modelling

The most important step in assuring the quality of environmental data is to determine what information one really needs to know about the system so that this may be measured most directly and efficiently. Models of systems are goal-oriented approximations of reality and as such help one to identify which initial parameters are measured as well as how these measurements help answer the original questions about the system (3, 5, 6, 10, 12).

Some of the common questions addressed by environmental models are: (a) What is the critical pathway of the substance? (b) If specific changes occur in the emission patterns of a substance, what will be the trend and rate of change of the critical pathway and total exposure? (c) At what points in the pathway can regulation or remedial action decrease the risk of adverse health effects?

Rather than give a textbook treatment of modelling, the following remarks are presented:

1. The type of model chosen is a function of what questions are asked about the system.

2. The type of model chosen will determine what raw data is required. Furthermore, the sensitivity of the mathematical equation describing the model to changes in the respective parameters will determine the precision and accuracy needed for those parameters, i.e., for some, great precision and accuracy will be required, while for others, order-of-magnitude results will suffice.
3. The model should be verified by empirical measurement and should not be extrapolated beyond the range of validity of the input parameters.
4. Laboratory experiments should be used to extend the model beyond present environmental levels and fluxes so as to permit extrapolation and trend analysis.

Data Gathering

After the model has been constructed, considerations of sampling, sample handling, analysis, and data handling come into play. The publication "OECD Principles of Good Laboratory Practice" provides a good general presentation of factors to be considered in gathering data. A few points are emphasized below.

Sampling

1. Sampling must be performed so as to consider critical or vulnerable subpopulations.
2. Sufficient data must be supplied on sampling methods to permit an assessment of whether or not the sample is representative.
3. Integrated sampling methods must be used in order to permit the study of transfer among compartments, e.g., many so-called data banks do not permit the correlation of metal levels in sediment, water, and biota simply because samples were not taken in the same place at the same time, this being because the samples were taken without any conceptual model in mind.

Sample Handling

Wherever possible, sampling and sample handling should be done so as to preserve the chemical species present at the time of sampling.

Analysis

1. Few laboratories have proper clean-room facilities to perform low-level background analyses; more centers should be set up with this capability.
2. The stable isotope tracer technique is a very powerful method for accurately and precisely determining contributions of the different pathways to total intake as well as for determining metabolic parameters for the metals

once they have been absorbed into the body. More of these studies should be undertaken.

3. The certification of methods and laboratories using well-characterized standards and referee laboratories is especially important for low-level analyses, e.g., Pb in blood and Hg in water, and because this has not been done many incorrect data have entered the published literature. This points out an important responsibility of journal editors which often has not been fulfilled.

Data Handling

Data should be made available in a format permitting trend analysis (time, geographic, etc.) as well as other statistical analyses.

Reference Materials

Materials certified to contain certain levels of stipulated metals in homogeneous natural or synthetic matrices are issued by a number of expert organizations. The International Organization for Standardization (ISO) Committee on reference materials (9) has issued a Directory of Certified Reference Materials listing over 300 worldwide sources of supply and suggested usage categories, e.g., environmental, botanical, clinical chemicals, biological, foodstuffs, pharmaceuticals, nuclear/radioisotopes. A very useful compilation of metal concentration data has been published for biological and environmental standard reference materials produced by the U.S. National Bureau of Standards (7). Similar comprehensive inventories for standards produced by other organizations are lacking; however, some reviews are available (1, 2). It should be noted that only *total* metal concentrations are provided in available certified standards. Data on the speciation of metals in such reference materials are not yet available.

ELEMENTS OF CONCERN FOR HUMAN HEALTH

A number of characteristics contribute to the labelling of an element as being a potential or present environmental problem (examples are given):

1. The element has already produced adverse effects in humans or in animals (Pb, Cd, As, Hg, Tl, Be, Cr, Cu, F, Mn, Mo, Ni, Sb) — animals are often used as indicator organisms.
2. The element produces adverse effects (metabolic interference, mutation, cancer) at very low levels (methyl-Hg, Ni, Cr, As).
3. The environmental concentration of the element is expected to increase with the use of existing technology (Al, Cd, Ni, Cr).

4. New or developing technologies are expected to increase emissions, cause emissions in new geographic locations, or emit new species of the element (In, Be, Mn, Ga, As).
5. The rate of exposure or the rate of accumulation is very high (Cd, methyl-Hg, Pb, Na).
6. The dietary habits or geographic location, etc., of a certain subpopulation may make them more likely to have a higher rate of exposure or of absorption (As, Pb, Cd, Hg, Na, Ni, Tl).
7. In certain regions the environmental levels of the element result in body levels near the critical level (Pb, Cd, As, Hg, Na).
8. Certain elements may affect the quality and quantity of the food supply available to humans resulting in food shortages (Cu, Ni, Cr, Zn, Mo, Na).

FUTURE CONSIDERATIONS FOR RESEARCH

A number of trends will have to be evaluated for their impact on future pathways and bioavailability of metals to man, otherwise we shall not be able to anticipate and avoid environmental problems:

1. Mineral deposits have only finite lifetimes and the exhausting of one deposit will mean that either another lower grade deposit will be used (increasing total emissions per kilogram of product unless more stringent controls are instituted) or another metal of perhaps greater or lesser toxicity may be substituted. Similarly, recycling technology has its own set of environmental problems.
2. The mix of fossil fuels used for heating, producing electricity, and in transportation will change with varying cost and availability, resulting in changes of metal emissions (and acid precipitation which may increase the solubility of the metals) on a global as well as regional basis (since combustion produces fine particles).
3. Changing feedstocks for the production of chemicals will also affect the associated waste metal emissions.
4. Deforestation on a large scale and draining of marshes will reduce the number of sinks available for absorbing metal emissions and possibly will mobilize metals already deposited.
5. Increased levels of microorganisms in waterways resulting from improper waste management may increase the potential for biomethylation.
6. Increased urbanization will not only result in large changes of geographic location of peoples, but also will put more at risk from increased exposure to urban pollutants.

7. Salting of waterways and agricultural soils because of poor agricultural practices may markedly affect the mobility and availability of metals.
8. Changing dietary habits (increased vegetarianism, increased consumption of processed foods, etc.) may greatly affect the intake of metals by humans, e.g., mechanically deboned meat contains significant levels of minute bone particles which may greatly increase the levels of Pb and F in that food.
10. Regulatory decisions which result in the substitution of one product or technology for another may result in emissions of a more toxic or less controllable nature, i.e., technology replacement options should be considered in addition to pollutant control options.
11. Factors which affect dietary exposure to metals (age, pregnancy, vegetarianism, diabetes, etc.) and factors which affect absorption from the gut (age, pregnancy, nutritional stress, inborn or infectious disease, and individual variation) should be investigated so that vulnerable subgroups of the population may be identified and protected from excessive exposure to metal compounds.

REFERENCES

(1) Alvarez, R. 1983. Report on reference materials and standard solutions. J.Assoc. Off. Anal. Chem. **66**: 335–337.

(2) Alvarez, R.; Rasberry, S.D.; and Uriano, G.A. 1982. NBS standard reference materials: update 1982. Anal. Chem. **54**: 1226A–1244A.

(3) Bernhard, M.; Buffoni, G.; and Renzoni, A. 1983. Mercury in the Mediterranean Tuna. Why is their level higher than in Atlantic Tuna? A model. Thalassia Jugoslavica, in press.

(4) Bowen, H.J.M. 1979. Environmental chemistry of the elements. New York: Academic Press.

(5) Curtis, E.H.; Beauchamp, J.J.; and Blaycock, B.G. 1977. Application of various mathematical models to data from the uptake of methyl mercury in bluegill sunfish. Ecol. Model. **3**: 127–284.

(6) Fagestrom, T., and Asell, B. 1973. Methyl mercury accumulation in an aquatic food chain. Ambio **2**: 164–171.

(7) Gladney, E.S. 1980. Compilation of elemental concentration data for NBS biological and environmental standard reference materials. Report LA-8438-MS, Los Alamos National Laboratory, Los Alamos, NM.

(8) Hemphill, D.D., ed. Trace Substances in Environmental Health (Series). Proceedings of annual conferences held at the University of Missouri-Columbia. Columbia, MO: University of Missouri Press.

(9) ISO Central Secretariat. 1980. REMCO Directory. Geneva, Switzerland: ISO.

(10) Jørgensen, S.E. 1979. Modelling the distribution and effect of heavy metals in an aquatic ecosystem. Ecol. Model. **6**: 199–222.

(11) Management and Control of Heavy Metals in the Environment (Series). Proceedings of Biennial International Conferences. Edinburgh: CEP Consultants Ltd.

(12) Seip, K.L. 1979. A mathematical model for the uptake of heavy metals in benthic algae. Ecol. Model. **6**: 183–197.

(13) Venugopal, B., and Luckey, T.D. 1978. Metal Toxicity in Mammals, vols. 1 and 2. New York: Plenum Press.

Standing, left to right:
Karl Summer, Simon Silver, Jürgen Kleiminger, Yutaka Kojima, Hans-Jürgen Hapke, Josef Abel, Bob Williams.

Seated, left to right:
Ton Schoot Uiterkamp, Jeremias Kägi, Bruce Fowler, Ellen Silbergeld, Carl Elinder.

Changing Metal Cycles and Human Health, ed. J.O. Nriagu, pp. 391–404.
Dahlem Konferenzen 1984. Berlin, Heidelberg, New York, Tokyo: Springer-Verlag.

STRUCTURE, MECHANISM, AND TOXICITY

Group Report

B.A. Fowler, Rapporteur
J. Abel, C.-G. Elinder, H.-J. Hapke, J.H.R. Kägi, J. Kleiminger, Y. Kojima, A.J.M. Schoot Uiterkamp, E.K. Silbergeld, S. Silver, K.H. Summer, R.J.P. Williams

INTRODUCTION

In order to facilitate assessment of relationships between the chemistry of specific metals and in vivo toxicity, our group decided to utilize an approach beginning with an evaluation of general principles governing simple metal interactions with biological ligands and then applying this information to the evaluation and prediction of metal reactions at higher levels of biological organization. This chemical approach specifically involved first defining expected metal reactivities with known reactive groups of biological molecules and then evaluating effects at the macromolecular, organelle, and intact cell levels. Consideration was then given to the possible roles of these metal-biological interactions in explaining cell and organ toxicities as well as the possible use of this knowledge in explaining observed human health effects and developing new biological indicators of toxicity. It was hoped that this overall approach would provide a logical and cohesive transition between evaluations by Andreae et al. who were concerned with biogeochemical cycling of metals, by Jaworski et al. with routes of human exposure, and by Kjellström et al. with human health effects.

1. THE FUNDAMENTAL CHEMISTRY OF METAL INTERACTIONS WITH REACTIVE GROUPS OF BIOLOGICAL SYSTEMS – A BASIS FOR PREDICTION

On a simple thermodynamic basis, oxygen, nitrogen, and sulfur were considered to be the three primary reactive ligands for metals in biological systems. With the exception of Al, Be, Ge, and Sn, which bind to oxygen, and Mn, which also binds to nitrogen, the vast majority of the metals considered toxic to man are soft Lewis acids that readily form covalent bonds with sulfur. Lewis acids are centers which can act as acceptors of electrons in a donor:acceptor pair such as $RS^{-}:Hg^{+2}$. Sulfur is acting here as a Lewis base. This definition is an extension of the usual acid concept which refers to a substance capable of releasing a proton. Examples of soft Lewis acids are Hg, Cd, Pb, Sb, Pt, i.e., many metal ions in the B-groups of the periodic table. They have a high acceptor strength and bind especially firmly with sulfur donors. Examples of hard Lewis acids are Be, Al, Mg, Cr, lanthanides, and actinides, i.e., many metal ions in the A-groups of the periodic table. They have a weak acceptor strength and bind oxygen donors. Intermediate cases are Cu, Ni, Fe, Zn, and the proton. Note that extreme hard and extreme soft metal ions are the major sources of known toxicity, and therefore other elements in these extreme classes are also potential sources of toxicity. Tl, In, and Te are examples of soft Lewis acids in this category, and Be and Ti are examples of hard Lewis acids of toxicological concern. At physiological pH, the SH group of some amino acids, such as cysteine in proteins, plays a primary role in the binding of the former metals in vivo, while oxygen centers from glutamate or OH centers from tyrosine bind the latter group (Williams, this volume).

A second chemical mechanism of importance concerns the role of stereospecificity in protecting biologically synthesized metal-binding ligands. The structural stability of such stereospecific molecular complexes for essential metals (i.e., Fe in heme) may hence protect biological systems from displacement reactions by toxic metals by forming nonexchangeable traps for essential metals. More reversible systems would be expected to have a greater susceptibility to toxic metal substitutions. Pb displacement of Zn from the enzyme δ-aminolevulinic acid (ALA) dehydratase is an example of such a substitution reaction. Certain Ca-, Mn-, and Mg-containing systems are also known to exchange readily with other metal ions with respect to displacement reactions. In contrast, Cu-containing enzymes show little capacity for exchange (e.g., superoxide dismutase).

Another chemical regulatory mechanism for preventing the substitution of toxic metals for essential metals in bioorganic complexes involves the separation of potentially sensitive reactions within organelles possessing selective unidirectional metal transport systems. This biochemical mechanism acts to maintain the integrity of normal essential metal-ligand interactions by reducing the accessibility of undesirable metal species while providing required metal ions in the correct redox state to the synthetic system of concern. An example of such a system is the ferrochelatase reaction which is localized in the mitochondrial inner membrane and results in heme formation after Fe^{3+} is reduced to Fe^{2+} during or subsequent to transport across the inner membrane.

In this review, we have not examined anion chemistry in detail. Thus, we do not describe the chemistry of such entities as SeO_4^{2-}, CrO_4^{2-}, MoO_4^{2-}, VO_4^{3-}, and AsO_4^{3-}, which would be related to the chemistry of the typical nonmetals SO_4^{2-}, PO_4^{3-}, etc. This is very different chemistry from the above since the binding sites would be positively charged but no new principles are involved.

Research Needs

1. Further research is needed to determine intracellular concentrations and binding constants of free metal ions and binding ligands in vitro and in vivo in order to assess more accurately the precise role of specific binding ligands in regulating metal ion bioavailability.

2. Studies are also needed to assess the in vivo localization or compartmentation of specific metal binding systems within cells in relation to the distribution and bioavailability of specific metal species.

3. The in vivo redox state and pH of cellular and subcellular systems should also be evaluated since these factors will markedly influence metal-ligand interactions and hence bioavailability.

4. Further research is needed to determine if intracellular or intraorganelle microdomains exist with marked differences in pH, redox, or binding ligand content such that the availability of metal ions within these regions is markedly different from other general compartments of the cell. This concept would be of particular importance for toxic metal species if such domains were also located in highly sensitive metabolic sites. Specific Ca-ATPase localization in synapse membranes may be an example of such a domain.

2. CAN WE UNDERSTAND AND PREDICT THE BIOCHEMISTRY OF METALS AT THE MOLECULAR, ORGANELLE, AND CELLULAR LEVELS, AND TO WHAT EXTENT DOES THIS UNDERSTANDING PROCEED FROM THE ABOVE SECTION?

The in vivo partitioning and transport (see next section) of metals in biological systems at the intracellular level is essentially a function of ligand composition and transport which varies with species, organ, cell, and organelle compartment. Binding constants of different metals for a given biological ligand also vary widely. If we restrict ourselves to ligands which are capable of complexing most of a given metal at physiological concentrations, then we limit our consideration to metal-ligand complexes with apparent stability constants of log $K_1 > 4$ under physiological conditions. Bioligands can be divided into low or high molecular weight compounds. The most prominent low molecular weight compounds are: reduced glutathione, cysteine, coenzyme A, dihydrolipoic acid, histidine, other amino acids, citrate, pyrophosphate, and nucleoside phosphates (ADP, ATP, GTP, etc.). Of the above ligands, only glutathione and the nucleoside phosphates occur intracellularly at millimolar concentrations. Hence, they are the most likely candidates of the low molecular weight ligands to bind both soft and hard Lewis acids, respectively. For some metals, formation of complexes with Cl^-, SCN^1, and HCO_3^- must also be considered.

Many metals of toxicologic concern are generally bound more firmly to macromolecular ligands. The most prominent high molecular weight ligands in this category are: metallothionein (MT), transferrin, ceruloplasmin, serum albumin, ferritin, intestinal calcium-binding protein, metal-binding enzymes, and nucleic acids. These high affinity metal-binding ligands provide for efficient control of metal homeostasis in biological systems and largely determine the bioavailability of many metal species at all levels of organization. The binding affinity of specific metals for ligands also determines which molecular targets are most sensitive to inhibition by these metals. Examples of this would include the vicinal thiols involved in pyruvate metabolism and the metal binding site of δ-ALA dehydratase.

The basic capacity of these metal-binding ligands to act in this manner is reasonably predicted from their general chemical characteristics (Kägi and Hapke, this volume). In addition, the rates of metal release from these sites are also controlled by the specific characteristics of the binding center. This latter factor might not be easily predicted from basic chemistry. An example would be the energetics of protein folding and location of the binding site(s) within the molecule.

Research Needs

1. There is a need for more accurate determination of the concentrations of ligands and the metal-binding constants of macromolecular biochemical ligands in vivo. Further research is also needed to determine more accurately the dissociation rate constants for the binding of metals to macromolecular ligands in vivo since this factor may also play a role in determining the extent to which a molecular ligand regulates metal ion bioavailability.

2. Factors which are responsible for observed discrepancies between in vitro and in vivo binding of metals to molecular ligands also need to be delineated in order to determine which parameters are important to the formation of multi-dentate sites.

3. The impact of protein synthesis, induction, and turnover on the ability of molecules such as metallothionein to bind both toxic and essential metals needs to be investigated over prolonged periods of metal exposure. This may be related to redistribution of bound metals to other cellular compartments with time.

3. CAN WE UNDERSTAND OR PREDICT THE BIOLOGICAL DISTRIBUTION OR EFFECTS OF METALS? DO THE ABOVE SECTIONS PROVIDE IMPORTANT INFORMATION?

In making the transition from metal biochemistry to physiology, it seems reasonable that metal-binding to macromolecules would a priori play a major role. These molecules should be expected to play a role in the physiological accumulation and transport of the various metals for which they possess high affinity. A number of the macromolecules identified in section 2 above are known to play a role in the intracellular and/or extracellular metabolism or transport of various metals. For many metal species of toxicologic concern, MT appears to be of primary importance with respect to explaining their accumulation in major target organs. The presence of higher levels of this protein in fetal liver may also explain the marked accumulation of Cu and Zn in fetal liver. In addition, Menkes disease which is a severe inborn error of copper metabolism may be connected with an abnormal distribution of Cu-MT. For toxic metals other than those which induce MT, we have relatively few data concerning their intracellular-binding, although high affinity cytosolic Pb-binding components from brain and kidney have been recently reported and proteins from Pb-intranuclear inclusion bodies of kidney have been isolated and partially characterized (5,7,8).

A simple model description of how these proteins participate in maintaining homeostasis through control of the intracellular concentration of a metal M is represented in Fig. 1. We assume that metal enters the cell as a free metal or a simple complex of M. Outward movement is considered to be possible by three modes as ML, and two cases of M_nL, where n is small (a metal cluster) or large (a metal precipitate) and where L is a protein. The rate of inward movement depends on $[M]_{out}$. The rate of outward movement depends on $[ML]_{in}$ or $[M_nL]_{in}$ which can be related to $[M]_{in}$ through binding constants. If L is produced in sufficient quantity, perhaps by feedback, then the buffering of $[M]_{in}$ (homeostasis) improves as *n* increases, see Fig. 1. The best buffer is clearly a precipitate, e.g., ferritin. A good buffer is a cluster complex, e.g., metallothionein. (The precipitate can also be any such material as a solid metal-phosphate, as is often found in vesicles.) Note that homeostasis may require ejection using a vesicle which is acidic and contains proteases, in which case the protein will be rapidly turned over and will not appear outside the cell. Figure 1 shows the dependence of $[M]_{in}$ on $[M]_{out}$ for the three cases of monomer, cluster, and precipitate outward pump. Note that the homeostatic

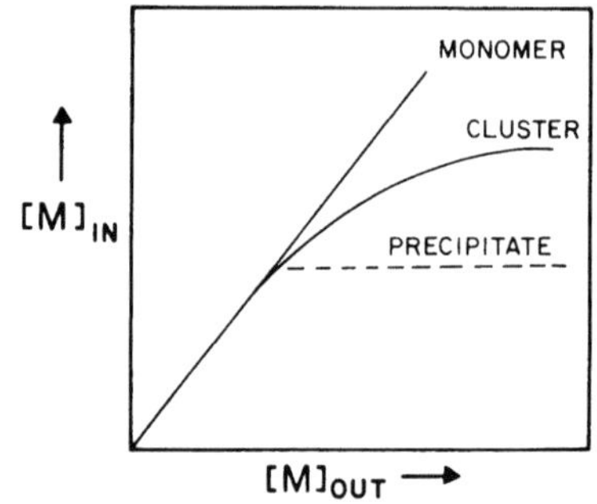

FIG. 1 – Possible relationships between the external free concentration of a metal $[M]_{out}$ and the internal free concentration $[M]_{in}$ are plotted. All cases refer to the movement inward of a monomeric metal complex of M. The plot labeled MONOMER refers to the relationship when the outward pump carries a monomer ML, which is a simple metal-protein complex of protein L. The plat labeled CLUSTER refers to the relationship when the outward pump carries a cluster, M_nL, where n is the number of metal atoms in a cluster board to a protein (e.g., metallothionein). The dashed line shows the relationship when the outward pump carries a PRECIPITATE of M inside a protein L M_nL, where n is now very large, e.g., ferritin.

mechanism requires ejection through a vesicle, which may be acidic, and partly destroys the ML unit, since free L must not become a component of the external fluids.

The relationships between the binding of various toxic metals to the proteins and other ligands as described above and actual mechanisms of toxic action at the intracellular level are obscure, and there is currently little specific information concerning precise chemical mechanisms of metal-induced toxicity in vivo. In this respect, it is important to note that while information from sections 1 and 2 above provide general guidance, there are few biochemical data which transcend the portal between the fundamental chemistry of metal-protein interactions and toxicological responses at the physiological level.

In terms of the entry of metals into cells, it seems clear that with the exception of uncharged chemical species ($HgCl_2$, $CdCl_2$) of certain metals which appear capable of traversing the lipid component of membranes, most membrane transports must occur via metal interactions with membrane proteins. Of the ligands discussed in the preceding section, MT, transferrin, calcium-binding proteins, metalloenzymes, amino acids, nucleoside phosphates, sugars, pyrophosphates, and peptides may be important in transporting metal species into cells. In bacterial systems, highly specific membrane-bound transport proteins for a number of toxic ions (^{+2}Hg, ^{+2}Cd, $^{-3}AsO_4$) have been characterized (Silver, this volume). In addition to these mechanisms, metal-binding to external membrane proteins or glycoproteins with subsequent endocytosis of the membrane also appears to be a possible mechanism for metal ion transport into cells, but specific data are currently lacking.

At present, data from sections 1 and 2 above provide explanations for the chemistry and specificity of metal-binding in biological systems but do not give much useful predictive information concerning the physiological transport of metals into cells.

Research Needs

1. Research is needed using isolated mammalian cells and cell culture systems to determine the nature of apparent cell-type specific differences in the membrane transport of toxic metal species.

2. Further research is needed to evaluate inborn errors of metal metabolism such as inherited copper toxicosis in dogs, Wilsons and Menkes diseases, and acquired diseases and nutritional disorders as opportunities for elucidating the normal function of proteins such as metallothionein in mammals.

3. The potential role(s) of high affinity metal-binding ligands in the organ or cellular distribution of metals in vivo need to be evaluated since transport ligands may play a critical role in determining both the sites and rates of uptake as well as mechanisms of action for toxic metals in specific organs or cell types.

4. More research is needed to identify and characterize currently unknown high-affinity metal-binding proteins for metals whose anthropogenic mobilization is expanding or expected to expand in the future.

5. Studies are also needed to determine whether amino acid transport systems provide a mechanism for the entrance of metals into target cells.

6. Detailed investigations concerning the potential role of high-affinity metal-binding proteins present at low concentrations in vivo, which act as biological receptors for transporting small amounts of toxic metals to highly sensitive sites, need to be undertaken.

7. There is a need to expand application of techniques such as nuclear magnetic resonance (NMR) to explore the ways in which metals influence intracellular processes resulting in toxicity.

8. The normal biological function(s) of high affinity metal-binding ligands such as metallothionein need to be determined. The impact of toxic metal interactions with these ligands in relation to their normal biological function(s) also needs to be elucidated.

9. There is a need for more research on homeostatic mechanisms for metals at the whole organism level.

4. WHAT PREDICTIONS FOR MAN CAN WE MAKE FROM THE PHYSIOLOGICAL EFFECTS OF METALS IN EXPERIMENTAL SYSTEMS?

For the purpose of this discussion, the toxicity of a given metal to an organism is operationally defined as any adverse effect occurring in any organ (6). Toxicity and threshold are also operationally defined and are not meant in a fundamental biological sense. This implies a statistical definition for populations at risk. A toxic effect occurs in this context subsequent to saturation of in vivo adaptation mechanisms (elimination, repair of damage, and storage) for a given metal. Mechanisms of adaptation could involve, for example, increased protein synthesis or neuronal sprouting. However, the threshold for toxicity is defined on the basis of a measurable deleterious effect. The concept of adaptation implies that a threshold for toxicity must exist and be exceeded prior to initiation of cell or organ injury. When considering adaptation to a toxic chemical we must also be aware that a single leak from a system could be amplified (e.g., gene damage) and that there is no fail-safe adaptation. DNA damage would be an example of the latter type of damage, which after one single event may lead to the development of cancer. The following diagram indicates the problem:

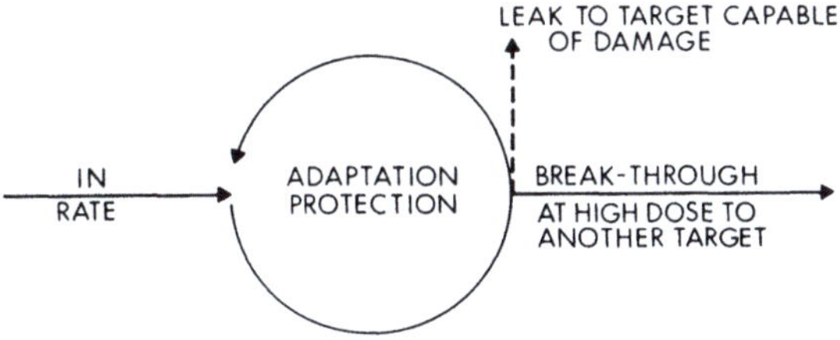

Adaptation is thus only a partial protection in that it does not protect against a high dose, and also cannot completely protect against a rare single event at a lower dose which is cooperatively amplified. There is also a biological cost of adaptation which may be particularly high for cell types such as neurons which by nature of their specialization have fewer available reserves for adaptation.

Problems in using the information derived in sections 1, 2, and 3 arise from variations in conditions of the receptor organism. In the first place, there are important species differences which make it difficult to extrapolate from animal experimental systems to humans. For example, rats have been shown to have completely different chemo-kinetics for inorganic arsenic when compared to most other mammals, including humans. Furthermore, other factors in animal experimental systems, such as circadian variation, patterns of test animal exposure, and stress, may also influence the result. Few studies have, however, examined these factors. In humans, factors such as age, sex, diet, race, geographical location, intestinal flora, and exposure to other toxic agents all clearly influence biological responses to metals. Nevertheless, data obtained from experimental systems (Sections 1,2,3) in animals have been crucial for the understanding of the mechanisms behind toxicity in humans from various toxic metals such as cadmium, mercury, lead, and manganese.

Based on sections 1, 2, and 3, biochemical and physiological information suggests that a number of metal(loid) species in addition to As, Cd, Hg, and Pb may pose significant risks. We cannot presently exclude any metal species from potential risk, based on the above discussions, but we would suggest relatively high research priorities for such metals as Be and Tl.

Research Needs

1. There is a clear need to delineate both external interactive factors (i.e., dietary status) and internal interactive factors (i.e., capacity to produce metal-binding proteins) which determine the biological availability and hence toxicity of metal species.

2. Further research is also needed to assess interactions between toxic metal species. Studies concerning interactions between toxic metal species and essential metals under conditions of excess or deficient intake also need to be undertaken.

3. Efforts should be made to evaluate different types of toxic effects in a number of organ systems since not all organ systems are equally sensitive to toxicity from metals such as those which have received little study but whose anthropogenic mobilization is increasing.

4. There is also a need to assess the reversibility of metal-induced toxicity in target organs and the usefulness of various treatment methodologies in this area.

5. Further research is also needed to determine the extent to which genetic variation in large groups may be used to detect sensitive or resistant populations in order to understand factors influencing mechanisms of action.

6. The mechanisms of homeostasis governing the toxicity of metal species at the organ, cell, organelle, and molecular levels of biological organization need to be elucidated in order to delineate which factors determining homeostasis are most susceptible to metal-induced perturbations.

5. WHAT INFORMATION ON MECHANISMS FROM EXPERIMENTAL SYSTEMS IS USEFUL FOR DEVELOPING TESTS TO ASSESS POTENTIAL HUMAN HEALTH EFFECTS?

Historically, for elements such as As, Cd, Hg, and Pb which have been used industrially for many years, incidents of human and domestic animal intoxication from these agents or their compounds have occurred prior to experimental studies concerning the mechanisms of toxicity (Elinder, this volume). The knowledge of human toxicity from these metals has, in fact, been the driving force behind the experimental research in this area. For other toxic metals, whose human use or exposure has been limited to date, we have relatively little experimental data or mechanistic understanding concerning their potential for adverse effects in humans. A number of these metals, however, are currently being mobilized secondarily into the environment from new or expanded anthropogenic activities and evolving technologies which specifically attempt to take advantage of their chemical properties. The conclusion to be reached from this evaluation is that for these metals there is currently little information from experimental systems of use in developing tests to assess potential human health effects. Data on chemical species of these metals from the types of tests recommended below should greatly facilitate judicious decisions concerning potential human health effects prior to extensive industrial use or entrance of these agents into the global biogeochemical cycle.

A review of currently available experimental test systems for rapid testing indicates that only in the area of mutagenesis-carcinogenesis are sensitive, validated tests available (Schoot Uiterkamp, this volume). It is particularly disturbing that no tests are currently available for nephrotoxicity or neurotoxicity since there are prominent effects of many metals on these organ systems in vivo. Some utilization of neuropharmacological test systems may be possible, such as the radioreceptor binding assays which have been validated for antipsychotic and analgesic drugs (4).

Research Needs

1. There is a clear need to expand our toxicological data base on relatively unstudied metals utilized in expanding energy or other technologies prior to their widespread dissemination or industrial use.

2. There is a need for assessment of the mutagenic potential and effects of these metals on cellular genetic excision/repair capabilities and their capacity to act as promoters of the carcinogenic response.

3. These agents should also be evaluated in experimental systems with known sensitivity to metal-induced toxicity such as the nervous system, kidney, cardiovascular, reproductive, reticuloendothelial systems, and the skin.

6. ARE THERE EARLY BIOLOGICAL INDICATORS FOR METAL EFFECTS AND CAN THESE BE USED IN HUMAN STUDIES?

In recent years, early biological indicators of metal toxicity within cells of target tissues have become increasingly important. These indicators of toxic metal interaction with subcellular biochemical systems are usually either metabolites of the system affected (porphyrins) or represent some specific function of a cellular system which is impaired (i.e., reabsorption of low molecular weight proteins by the kidney). Ideally these indicators should have the following characteristics in order to be of maximal use: (a) Biological Sensitivity – The parameter should exhibit a change long before overt clinical symptoms of metal toxicity appear. (b) Biological Specificity – Under ideal circumstances the indicator(s) should be relatively specific for a given metal so that in multi-element exposure circumstances, the biologically active element(s) may be discerned. On a practical basis, any significant change in these parameters among populations exposed to metal would indicate the need for further evaluation. (c) Relationship to Toxicity – In order to be of maximum value, these parameters should also be related to ongoing intracellular toxicity by both morphologic and biochemical evaluation at the target organ, cell, and organelle level.

While two examples of biochemical indicators for metal-induced target organ toxicity assessment are discussed below, it should also be noted that physiological tests such as nerve conduction velocity and pulmonary function tests may also serve as early indicators of toxicity.

Biochemical Indicators

Due to the fact that different porphyrins are produced during the synthesis of heme for either hemoglobin or intracellular enzymes and that several enzymes in the heme pathway are extremely sensitive to metal-induced inhibition, measurement of these metabolic products in either blood or urine is one major type of biological indicator test currently in use. Measurement of δ-ALA and zinc erythrocyte protoporphyrin in blood or δ-ALA and coproporphyrin in urine are widely used tests for lead toxicity in vivo. In experimental animal systems, prolonged oral exposure to arsenate has been shown to produce increased excretion of uroporphyrin and with lesser amounts of coproporphyrin but no excretion of δ-ALA (10). Methylmercury exposure in rats has been found to produce a porphyrinuria pattern dominated by coproporphyrin with lesser amounts of uroporphyrin in the urine but not elevated excretion of δ-ALA (9). The relationship between the above porphyrin patterns and ultrastructural morphometric/biochemical indices of cellular toxicity in target tissues has been reviewed elsewhere (1). A multifactorial study using combined exposures of rats to Pb, Cd, and As has demonstrated a differential pattern of porphyrinurias for each of the elemental combinations, suggesting that further evaluation of porphyrins for untested metal-species and multi-element exposures may be of value (3).

The other general category of biological indicators for assessing early metal-induced damage to the kidney involves the measurement of increased excretion of low molecular-weight proteins in the urine secondary to decreased proximal tubule cell reabsorptive capacity (2).

Biological Monitoring of Metal Toxicity Using Wild Animals or Livestock
Another possible approach to biological monitoring involves assessment of wild or domestic animal populations for metal accumulation or signs of toxicity. There is current evidence indicating metal accumulation or early signs of toxicity among animal populations living or grazing in geographical areas impacted by toxic metals, such as lead. Application of this approach to toxic metals, whose anthropogenic mobilization and entrance into global or local biogeochemical cycles is increasing, may lead to the early detection of potential health effects prior to production of harmful effects in man.

Research Needs
1. There is a need for the development of new and more sensitive early biological indicators of toxicity for metals.

2. Further research is needed to relate more precisely metal-induced target organ damage at the biochemical level to early biological indicators of toxicity in blood or urine.

3. The potential value of monitoring wild or domestic animal populations for signs of metal accumulation or toxicity needs to be evaluated further, since these populations may give biological responses to elevated levels of toxic metals prior to the onset of symptoms in man.

SUGGESTED READING

(1) Fowler, B.A. 1980. Ultrastructural morphometric/biochemical assessment of cellular toxicity. **In** Proceedings of the Symposium on the Scientific Basis of Toxicity Assessment, ed. H.P. Witschi, pp. 211–218. Amsterdam: Elsevier Publishing Company.

(2) Friberg, L.; Piscator, M.; Nordberg, G.F.; and Kjellström, T. 1974. Cadmium in the Environment, 2nd ed. Cleveland, OH: CRC Press.

(3) Mahaffey, K.R.; Capar, S.G.; Gladen, B.C.; and Fowler, B.A. 1981. Concurrent exposure to lead, cadmium and arsenic: Effects on toxicity and tissue metal concentrations in the rat. J. Lab. Clin. Med. **98**: 463–481.

(4) Molinoff, P.B.; Wolfe, B.; and Weiland, G. 1981. Quantitative analysis of drug-receptor interaction. Life Sci. **29**: 427–443.

(5) Moore, J.F.; Goyer, R.A.; and Wilson, M. 1973. Lead-induced inclusion bodies: solubility, amino acid content and relationship to residual acidic nuclear proteins. Lab. Invest. **29**: 488–494.

(6) Nordberg, G.F., ed. 1976. Effects and Dose-Response Relationships of Toxic Metals. Oxford, New York: Elsevier Scientific Publishing Company Amsterdam.

(7) Oskarsson, A.; Squibb, K.S.; and Fowler, B.A. 1982. Intracellular binding of lead in the kidney: the partial isolation and characterization of postmitochondrial lead binding components. Biochem. Biophys. Res. Commun. **104**: 290–298.

(8) Shelton, K.R., and Egle, P.M. 1982. The proteins of lead-induced intranuclear inclusion bodies. J. Biol. Chem. **257**: 11802–11807.

(9) Woods, J.S., and Fowler, B.A. 1977. Renal porphyrinuria during chronic methyl mercury exposure. J. Lab. Clin. Med. **90**: 266–272.

(10) Woods, J.S., and Fowler, B.A. 1978. Altered regulation of mammalian hepatic heme biosynthesis and urinary porphyrin excretion during prolonged exposure to sodium arsenate. Toxicol. Appl. Pharmacol. **43**: 361–371.

Standing, left to right:
Burton Bennett, Gerhard Winneke, Reinier Zielhuis, Kóji Nogawa, Tord Kjellström.

Seated, left to right:
Don Hornig, Jane Lin-Fu, Herb Needleman, Tom Clarkson.

Changing Metal Cycles and Human Health, ed. J.O. Nriagu, pp. 407–423.
Dahlem Konferenzen 1984. Berlin, Heidelberg, New York, Tokyo: Springer-Verlag.

PERSPECTIVES AND PROSPECTIVES ON HEALTH EFFECTS OF METALS

Group Report

T.E. Kjellström, Rapporteur
B.G. Bennett, T.W. Clarkson, D.F. Hornig, J.S. Lin-Fu, H.L. Needleman, K. Nogawa, G. Winneke, R.L. Zielhuis

METALS AND HEALTH – A GLOBAL CONCERN

Metals and other elements are fundamental to the basic survival of the human being and to the functioning of traditional and modern societies. Many of the metals and specific compounds of metals are essential for physiological and biochemical functions (e.g., Cu, Zn, Fe, Co, Mn). Others (e.g., Pb, Cd, Hg, As) have a proven toxicity to humans at exposure levels which have occurred in the environment. One can say that most metals are "essential" to the industrialized society in terms of their use in products which have become a part of everyday life. On the other hand, specific uses of a specific metal can often be eliminated by replacing the product or the metal by some alternative. The anthropogenic sources of metals in the environment can thus be influenced, and one ultimate aim of the evaluation of the health effects of metal exposure is to provide a basis for decisions about actions against anthropogenic metal emissions.

As discussed in the other workshop reports, the sources of metal exposures are often not related to the use of the metal itself (e.g., coal-fired power plants or geochemical power plants) or are natural geochemical phenomena. In any case, the evaluation of toxicity should be based on the total exposure to the metal itself or to specific metal compounds, taking into consideration known factors that interact with the mechanism of toxicity. There is no evidence so far that high exposures to specific toxic metal compounds due to "natural" sources are any less damaging to health than high exposures from anthropogenic sources, but this is an area of considerable interest as certain geographic areas have much higher "natural" exposures than other areas. The same can be said for low exposures of "essential" elements.

Metal exposures from many exposure routes (inhalation, ingestion, skin transfer, transplacental, parenteral) do occur, and health effects in humans have been reported for almost all metals. The reports of serious effects in humans are often the first indicators that the excess or lack of the metal constitutes a health hazard, as has also been pointed out in the report by Fowler et al. (this volume). Workers with occupational exposure have often had the highest exposures to toxic metals (e.g., Ni, Cr, Mn, Tl, Be, Pb, Cd, As), and they have served as "indicators" of effects to be seen later in other exposed groups.

Even though this workshop was not concerned with the evaluation of occupational exposures and the associated health risks, it is clear that the sources of occupational exposure are also often the major sources of general environment exposure at the local or regional scale. In some cases (e.g., Cd, As) high exposure to toxic metals can be transported by river water to populations at considerable distances from the source. The families of exposed workers may also be exposed via clothing or contaminated skin (e.g., to Be, Pb). The metal-containing products themselves (e.g., Pb-paint) can become a major health hazard, and when the products are disposed of, or leftover raw materials are disposed of, new emission sources develop (e.g., dumps, garbage incineration, sewage sludge).

Various incidents of severe metal poisoning in both workers and inhabitants of polluted environments have been documented. Severe deficiencies also occur and are usually related to dietary extremes. The association between soft water and heart disease is as yet an unresolved issue, though. For many of these metal-associated health effects, recent developments in modern science (e.g., biochemistry, physiology, analytical chemistry, toxicology, psychology, and biostatistics) have helped elucidate the metabolism of the metals and the mechanism of the health effects. The time it has taken from the discovery of the element to the discovery of its health effects and further to the elucidation of these mechanisms has decreased (1).

The development of biochemical indicators of early effects of the metals have shown that for two (Pb, Cd) of the four toxic metal compounds of major concern in the general environment (Pb, Cd, methyl-Hg, inorganic As), the health problem extends much further than the observed cases of clinical poisoning. If a health effect is defined as any measurable change in the biochemistry, physiology, or psychology of an individual which can be logically linked to the mechanism of the more severe overt ill health, one can measure the effects of lead in terms of changes in porphyrin metabolism, behavior, or intelligence, and one can measure cadmium effects in terms of specific changes in urinary protein excretion. Based on screening studies, estimates of the number of children in the USA with health effects of lead (high FEP) may be as high as several hundred thousand (12). In Japan the number of elderly people with health effects of cadmium (high urinary beta-2-microglobulin) may be as high as several thousand (9).

For methyl-Hg and inorganic As, no useful biochemical or other indicators have been found, and symptoms or signs have to be used to measure the health effects. Still, for methyl-Hg the health problem is great. Several thousand cases have been diagnosed as affected in Japan alone (Asami, this volume), and the poisoning incident in Iraq was of similar magnitude (Clarkson, this volume).

In spite of the magnitude of these health problems, important gaps exist in the knowledge about the metabolism and toxicity of the "big four" toxic metals (Pb, Cd, Hg, As). For other metals the lack of knowledge is even greater. The research to date has been reviewed in reports on individual metals (e.g., World Health Organization Environmental Health Criteria Documents) and in, e.g., the Handbook on the Toxicology of Metals (5). The latter report also discusses various aspects of toxicology which are common to all or several of the metals.

The other group reports from this workshop have identified metals of concern from the viewpoint of geochemical cycling, exposure routes, and the physical and chemical characteristics of the metals. In this group we concluded that key areas of concern in the evaluation of health effects are the accuracy of the measurement of exposure and effect, the background levels in different geographical areas and the transfer of exposures to new areas (particularly in developing countries), the selection of health effects to study, the selection of epidemiological methodology, the interpretation of the observations using models, and the criteria for action when the health risks have been identified.

The group would like to stress the need to avoid repeating the mistakes of the past over and over again. For specific population groups which have experienced high metal exposures serious health problems have occurred. It is a challenge to concerned scientists and administrators to make sure that this does not happen again. The following sections will discuss how we can improve our tools for this urgent job.

METHODS OF ANALYSIS AND SPECIATION

Among environmental chemicals of concern, metals have the peculiar characteristic that they remain undestroyed regardless of how they are treated, e.g., even at the high temperature in a garbage incinerator. If the effect of the metal is caused by the metal atom itself, then simple measurement of the total metal exposure may be sufficient to evaluate the health risk. For certain metals (e.g., Cr, As) the effects are strongly dependent on the valency state of the metal, and for some (e.g., As, Hg) the whole chemical compound determines the toxicity, not just the metal itself.

From the point of view of the chemistry of metals at different valency states and of metals in different compounds, there is reason to believe that these different "species" of each metal may in fact have very different toxicity. Depending on the natural stability of several toxic chemical species of a metal, one or the other may be seen as more prominent in observations of exposed populations.

It is clear that more research needs to be carried out on the speciation of metals in the environment and in biological systems. This may be of particular significance for the evaluation of carcinogenicity. Some Cr, Ni, and As compounds have a higher carcinogenic potential than others (1). Analytical methods for speciation have to be developed. It should be pointed out that the metal species may undergo changes in the environment (e.g., Hg to methyl-Hg) and in the human body (e.g., phenyl-Hg to Hg). When exposure is via ingestion, the acid environment of the stomach and the enzymes and microbial flora in the intestines may induce important changes in the metal species which determine, e.g., the absorption rate or the toxicity of the metal compound. Experimental gastric fluid systems may provide useful information in this context.

Still, the analysis of total metal content or concentration can provide sufficient information for the evaluation of the health effects and the underlying mechanism (e.g., Pb, Cd). Also, when it has been established that a particular metal species dominates (e.g., methyl-Hg in fish), then it may suffice in screening programs to measure the total metal content.

One of the most commonly used methods to measure total metal content, atomic absorption spectrophotometry, has undergone major developments during recent years. The sensitivity and accuracy have improved, and at the same time most laboratories have become increasingly aware of the need to avoid contamination and losses during sampling, handling, and final analysis. Intralaboratory and interlaboratory quality control procedures are a necessary part of any modern analytical activity. This applies both to the metal analysis and analysis of the effects of metals. The lack of such quality control in the past makes it difficult to utilize some of the data reported for health effect evaluations. In fact, even today analytical results with metal concentrations that are obviously in error pass the scrutinizing eyes of journal referees (15). The interlaboratory comparisons that have been published (e.g., (7, 10)) demonstrate the amazing differences in results between laboratories analyzing aliquots of the same sample.

All published data on metal concentrations should include an explanation of how the quality of the analysis has been assured. Rather elaborate quality assurance procedures for Pb and Cd in blood have been developed (15), but for everyday work the more common "round robin" systems may be sufficient. Such systems are currently operating, e.g., for Pb and Cd in blood (6) and for Hg in fish (4).

New methods that may be particularly useful as comparison methods in quality assurance programs are Isotope Dilution Mass Spectrometry (IDMS), Proton Induced X-ray Emission (PIXE), Nuclear Magnetic Resonance (NMR), and Laser-atomization with mass spectrometry. Some of these are micromethods which enable analysis in small parts of tissues. PIXE can be used, for instance, to measure the mercury or lead content in different layers of a strand of hair and to identify layers of varying lead concentrations in a tooth.

Non-invasive methods to measure metals in vivo in humans have increased our possibilities to measure metal doses and metal metabolism. In vivo neutron activation has been used to measure the cadmium content of kidneys and liver in highly exposed people. The sensitivity and accuracy need further development, but the method is promising, particularly for cadmium and lead.

The development of methods for total metal analysis and for metal speciation needs a multidisciplinary approach with close cooperation between chemists and toxicologists. In some cases simple physical separation methods may allow the identification of metal species. When new methods are needed, animal experiments may be the logical starting point, and in vitro systems such as the gastric juice test of species changes would give useful information.

More work should be devoted to evaluating the physicochemical features of different metal compounds. Solubility is one such feature which has been discussed in terms of possible differences in toxicity between cadmium compounds. The amount of metal intake and the volume of the fluid "pool" into which it will be dissolved should be considered. The intake is often only a few micrograms and, e.g., the gastric juice volume is several liters. At that concentration many "insoluble" compounds are in fact soluble. The same applies to inhaled particles of metal compounds. The rate of dissolution rather than just the equilibrium state should also be considered.

BACKGROUND EXPOSURE LEVELS

It has been mentioned in some of the papers at this workshop (Landrigan and Needleman, both this volume) that the "background" levels may have increased over the years due to local or widespread environmental pollution. This issue is discussed in the report by Andreae et al. (this volume), but we felt it was important to discuss it here also from the point of view of measuring differences in the occurrence of health effects in populations with different exposure levels.

The first question is whether the differences measured are true differences or are caused by lack of comparability of analytical results. If all the analysis is not carried out in the same laboratory, it is essential to apply strict quality assurance criteria to each laboratory and to check the comparability of measurement. This has been done for Pb and Cd in blood and Cd in kidney cortex in one recent study (15), and it shows that there are indeed some significant differences between similar population groups in different countries. A study of cadmium intake and cadmium in a number of human tissues in Sweden, Japan, and the USA (8) also employed interlaboratory comparisons and showed conclusively that Japan has about three times as high "background" cadmium exposure as the other two countries.

Health effect studies of cadmium in Japan use control groups within Japan to evaluate the effects, but it may be argued that the high "background" intake underestimates the differences between exposed and control groups. Similar queries have been raised concerning lead effect studies in, e.g., the USA, where there is a general exposure to lead much higher than that which occurs in remote areas of the world (Needleman, this volume). To get valid baseline data of the average exposures in different population groups according to age, sex, race, etc., and to find the frequency distributions of different exposure levels, it would be useful if the NHANES screening approach (12) were used in countries other than the USA. However, such screenings are not a substitute for more intense studies of high-exposure areas or high-risk groups.

Historical variations in metal exposures are just as important to measure as geographical variations. Evidence of an increasing exposure with time may indicate that the current exposures are not necessarily "no effect" exposures. Changes in exposure with time will also have an effect on the interpretation of observed accumulation curves in body tissues with long biological half-times for the metal. Some materials (e.g., grains) may be relatively easy to use for historical studies (11), but for exposure routes other than food, stored materials are seldom available.

Prospective studies can be used to study the contribution of particular sources to daily metal uptake. In Italy lead in petrol was labelled with particular stable isotopes during a period, and the changes of these isotopes in the blood lead after the labelling was terminated showed that a considerable proportion of blood lead was related to petrol lead (3). The interpretation of the findings is dependent on the metabolic model for lead that is used (see section on **METABOLIC MODELS**, below).

Changes of background metal exposure with time are not only dependent on the concentration of the metal in the environment, particularly in air and foodstuffs, but also on the changes in diet that occur over time. Such changes occur in all societies, and as the metal concentration in different foodstuffs varies considerably, the metal intake via food may change considerably. National diet surveys such as those carried out in Japan are useful sources of information. Rice is the main source of cadmium in the Japanese diet. As the average rice consumption per capita is decreasing, the cadmium intake may in fact decrease with time, even if the average cadmium level in rice remains the same.

Another example is methyl-Hg, the main source of which is fish. In New Zealand the shark has a particularly high methyl-Hg concentration of 1–5 ug/g (13). In 1972 the export of shark, mainly to Australia, was stopped because of the high methyl-Hg concentrations. The shark caught in New Zealand was instead sold inside the country and the local shark consumption increased dramatically. This is likely to have caused a considerable increase in the methyl-Hg intake by heavy fish consumers.

Background exposure levels and their historical development can be measured either by environmental monitoring or biological monitoring (see Thornton and Abrahams, this volume). There is a need to find suitable materials for the measurement of historical trends. The use of tooth analysis for assessing lead exposure in children and hair analysis for assessing methyl-Hg exposure in pregnant women are excellent examples of such materials.

The measurement of trends in health may be even more difficult as the early indicators of metal effects that are of particular interest have not been studied in the past. Trends of severe effects such as lung cancer may in some countries be measured retrospectively by using cancer registers.

Also, the study of geographical differences in "background" health effect prevalence is difficult, because in many cases there are racial, dietary, climatic, and other confounding factors which can obscure the possible effects of the metal. Studies of human migrants are one approach which to date have not been used for evaluations of health effects of metals but can yield interesting results.

EXPOSURE SOURCES IN DEVELOPING COUNTRIES

The information about metal exposures in developing countries is very limited, even though for some of them one can anticipate "hot-spots" of metal exposure due to mining, metal refining, cement production, chlor-alkali production, or other industry. In the major cities the air pollution is becoming considerable, and lead from petrol fumes or the indirect effects of acidic rain may increase the metal exposures considerably.

When industry is established in these countries as a part of the global "technology transfer" the same degree of emission controls may not be applied as in industrialized countries. In some cases obsolete and polluting plants may be shipped to developing countries to take advantage of a more attractive business climate. In some countries "Free Industrialization Zones" have been developed in which emission regulations do not apply. The impact of such policies on local and regional metal exposure levels can be imagined, but seldom are studies carried out. Recent reports from Zambia and Peru show that not only air, food, and drinking water can be important sources of intake, but also dust which settles through incomplete roofs in dry climates.

Chemical food poisoning via metal-containing pesticides are other imminent dangers in developing countries. The most famous and most devastating incident happened in Iraq in 1971 (Clarkson, this volume), but other similar incidents with other mercury pesticides have occurred, and without a constant vigilance by authorities in these countries more poisonings are likely to occur in the future.

From the geochemical transfer point of view each "hot-spot" with high exposure or each poisoning incident may appear local, but all the "local disasters" add up to a large combined exposed population. Furthermore, there are some sources of metal exposure which may have more widespread effects. For example, the high-Cd phosphates from West Africa are now unattractive in Western Europe because of their high Cd content. They are instead sold increasingly to other developing countries. Coal with high lead content is experiencing the same market pressure, and the coal-fired power plants in developing countries become "dirtier" than similar plants in industrialized countries.

From the epidemiological point of view, the populations of these countries may be at a particular risk for the following reasons:

1. High population density;
2. High proportion of local food consumption produced in the local area;
3. Large proportions of the population particularly susceptible due to age (many children and pregnant women), nutrition status, or health status (interaction with preexisting endemic diseases).

The Cd and methyl-Hg poisoning cases in Japan had all the features of this epidemiological situation.

Similar situations should be looked for in developing countries before serious health effects occur. Metal-producing countries and those with rice production and inshore or river fisheries may be of particular interest.

Studies in these countries can elucidate racial differences and interactions between metal exposures and other environmental, dietary, or health factors. These studies can be made prospectively and may therefore increase our understanding of the historical trends in industrialized countries. The studies will have to take a "total health" approach as it is likely that health hazards other than the metal in question may be of greater importance for the affected population or may obscure the metal-induced effect.

Within the terms of reference of this workshop we may have a particular responsibility as scientists in the environmental health field to make sure that risk areas are identified and that our knowledge of the hazards is made available in the developing countries concerned, to decision-makers, local scientists, and local exposed populations. There is also a need to make information on metal exposures and health effects in these countries available for the scientific community so that unnecessary repetition of metal poisoning incidents can be avoided. Recently a joint meeting of the Subcommittee of the Toxicology of Metals of the Permanent Committee and International Association of Occupational Health and the Venezuelan Society of Toxicology was held in Caracas (Clarkson, personal communication). Such joint exchanges of information may be one useful way of preventing major outbreaks of metal poisoning in developing countries. Other ways would be collaborative studies on specific problems with the training of local staff as an important element.

Informal or formal networks of exchange of information (including, for instance, provision of library material from industrialized countries) would also be useful. International agencies should encourage such cooperative activities.

TYPE OF EFFECT TO STUDY

In the first section of this report an attempt was made to define what we mean by health effect. Effects which are not known to be linked to more serious clinical effects sometimes are disregarded or are used as indicators of exposure. With increasing knowledge about the mechanism of toxicity, it may transpire that the previously disregarded effect is indeed linked to health. It is difficult to clearly define these concepts, and the preventive approach may be in the best interest of health. Regardless of how serious an early change is considered to be, the value of having early biochemical or physiological indicators of health effects can hardly be disputed. The first "adverse" effect has been termed the "critical effect" in metal toxicology (14). If this effect can be prevented, the more serious effects are also likely to be prevented.

There is a continuum of "health effects" from subtle changes to severe clinical disease and death. We have to come away from the threshold concept of health effect, meaning that either you are sick or you are well. Measurable (or non-measurable) slight changes may be even more important from the public health point of view.

It was pointed out that for the "big four" metals (Pb, Cd, Hg, As), we should not only study the generally accepted effects, regardless of whether these are the "critical" effects or more severe clinical effects. We should make sure that we do not miss important nonspecific effects without overt clinical signs. In particular, reproductive effects were mentioned. A workshop on such effects of metals was held in 1982 in Rochester, New York (Clarkson, personal communication). It should be pointed out that the effects on the reproductive system of both males and females are of similar concern. The emphasis on studying women in metal exposures may underestimate the reproductive effects.

Other effects of particular interest are renal effects and CNS effects. The developing CNS may be more sensitive than other organs due to its complexity. We need to utilize the methods developed by psychologists for measuring such effects. Otherwise, there is a possibility that behavioral aberrations, intelligence deficits, and psychological deficits may be overlooked.

Studies of CNS effects in children should choose sensitive populations and look at appropriate performances, e.g., classroom performance and measures of attention, cognition, or perception. These are likely to be more sensitive indicators of metal effects than standard neurological examinations.

In the different stages of life there are different effects and different risks. Very little is known about prenatal effects, transgenerational effects, and effects which show up as slowed childhood development or accelerated aging. It appears that for methyl-Hg, the effects at the lowest dose occur in the fetus of exposed pregnant women (Clarkson, this volume). Lead and cadmium can also produce prenatal effects, at least in animals. The methodology for the study of these types of effects needs further development and, just as for the metal analysis, it is necessary to carry out interlaboratory quality control studies on the early "subtle" effects.

EPIDEMIOLOGICAL METHODS

Because the administration of potentially toxic substances to human subjects cannot be ethically countenanced, observational studies of nonhuman exposure groups are required to identify and evaluate the effects of metals. Epidemiological techniques can add rigor to the quantitation of the strength and specificity of the observed relationships between exposure to the metal and the measured effects.

In the design and execution of epidemiological studies, certain methodological problems are encountered with regularity. If these issues are successfully addressed, inferences drawn from a given study will have increased precision and validity.

Among the more frequently encountered design problems are the following:

1. Difficulties in the correct classification of exposure.
2. Difficulties in identifying, measuring, and evaluating bias. Subjects who are excluded from a study may differ in some systematic fashion from those who are participants.
3. Problems in selecting the appropriate and sensitive measures of outcome.

4. Difficulties in measuring and controlling the covariates which may confound the effect of the agent under study.
5. Difficulties in selecting or developing a veridical causal model. If the causal model does not reflect the state of biological events, investigations may overcontrol or undercontrol for potential confounders.
6. Difficulties in evaluating the results of statistical analysis. Among these are the interpretation of the results of multiple tests of outcome. If many tests are employed, some of these may differ between groups on the basis of chance alone. On the other hand, the acceptance or rejection of differences predicated upon whether they differ by more or less than the arbitrary level of $P < 0.05$ may ignore effects of considerable importance. Studies which replicate each other, each showing differences at the $p = 0.1$ level, for example, may contain information of considerable merit.

Epidemiological studies, like other investigations in the natural sciences, measure only covariation; causality cannot be proven empirically. The assumption of a causal relationship can be made with some confidence if the following criteria for judging causal associations are met (2):

1. Time sequence of variables. The independent variable must precede the dependent variable in time.
2. Consistency of associations upon replication.
3. Strength of association. The stronger the association, the more likely it is to be causal.
4. Specificity of association.
5. Coherent explanation. Is there a dose-response relationship?
6. Presence of a plausible biological mechanism of causation.

The evaluations of the health effects of metal exposures can hardly be done without epidemiological studies. These provide the "final test" after animal experiments and theoretical evaluations have been carried out. Findings in epidemiological studies mean that effects do occur, that we got there "too late." On the other hand, if the early indicators of effects discussed in the previous section are used, maybe the situation can be remedied before any serious health effects occur.

METABOLIC MODELS

The metabolic modeling of accumulation of metal compounds in different tissues is very useful. It provides a framework for interpreting the data collected on metal concentrations in tissues and metal exposures or doses. One type of model depicts the absorption of the metal into the body, the transport between body compartments, and the excretion of the metal. Mathematical functions may be used to describe the transport in order that the metabolic kinetics of the metal can be calculated.

The metabolic model can also be extended to a dose-response model. This combines the kinetic calculations and information on the critical concentration distribution in a population in order to calculate the expected proportion of an exposed population with tissue levels above the critical concentration. The occurrence of health effects can then be interpreted within this framework. The model is likely to be most useful as an interpretation model rather than a predictive model, but careful estimates of possible future health risks under certain exposure conditions may provide a much better basis for decisions than pure guesswork. The model approach has mostly been used for cadmium and methylmercury. This method should be developed for other metal compounds. The model should take into account differences between sexes, age groups, etc.

The models need basic anatomical, physiological, and pharmacodynamic data (e.g., organ weights, urine volumes by age, urinary creatinine concentrations by age or sex, life span of blood cells, etc.) for meaningful calculations with estimates of errors. Such data are often lacking or incomplete. Further basic descriptive studies on these factors should be encouraged. There is a particular lack of data on physiology and biochemistry which determine metabolism of metals and/or susceptibility for children in different age groups, in spite of the fact that in children great changes take place. It is the same problem for metals and drugs when it comes to assessing the dose given and the tissue levels.

Age, sex, racial difference in metal absorption rates, half-time, etc., also need to be studied in order to construct good metabolic models. The lack of data on, e.g., the metabolism of lead in children is a major problem. The changing energy intake/kg b.w. in different age groups is another factor to consider. Maximum energy intake/kg b.w. occurs at age 5. This is important in epidemiological studies when groups of children are compared, for example, if the group exposed to methylmercury is on the average 8 years old and the average age for the control group is 5 years, erroneous conclusions may be drawn. Similar problems occur with age and respiratory rate or BMR.

There are other aspects of metabolism of particular concern. One is the problem of evaluating the contribution of different lead exposure routes to blood lead due to the long half-time for lead in bone. It has also been suggested that the lead mobility from bone varies with age and increases greatly at, e.g., pregnancy. These factors will affect the assessment of the health risks due to lead. The biliary excretion is another big unknown quantity for most metals. The existence of enterohepatic systems may be useful for treatment of poisoning.

Metabolic models help in identifying systematically the critical gaps in the data. They also have the advantage of making possible calculations of total uptake from several intake sources. This is essential when different hygienic standards are compared with each other or when an overall standard for all exposure routes is developed.

CRITERIA FOR ACTION

If we are surrounded by an imperfect data base, when and how do we decide that action against the exposure is needed or that further research is futile? One single study is seldom accepted as sufficient "proof" that a metal indeed causes the measured effect. Confirmatory studies are often needed, and such studies should be encouraged, because they form a part of the "tapestry" of evidence from animal and human studies we have to consider when evaluating the health effects of a metal.

As previously mentioned, epidemiologic investigations can never *prove* causal relationships. Indeed, causality is a construct not accessible to empirical proof. It is inherent to the very nature of epidemiologic work that all variables cannot be brought under complete control. This structural fact has allowed many controversies to simmer and occasionally fulminate. These controversies are particularly intractable in the field of environmental health, whenever abatement of some hazard imposes a severe cost in some sector of society.

Given these constraints, how should a prudent and rational society act when faced with evidence of risk from a pollutant, but possessing a body of evidence which does not achieve the rigor of mathematical proof? When the data base is of good quality, rigor, and objectivity, and meets the well-known criteria for causality codified by Bradford-Hill (see section on **EPIDEMIOLOGICAL METHODS**), a government, regulatory agency, or society that is unwilling to acknowledge this relationship and carry out the appropriate response invokes the risk of being judged unequal to its appointed role.

The final decision about action will always be based on judgement. Scientists can only provide as good a basis for the decision as possible, and we do not have a monopoly on interpretation of the data.

A phenomenon of great concern is the tendency for special interest groups to produce data or conclusions in a "quasi-scientific" way as to confuse the health effect evaluations. When unfounded assertions about the causal relationship between a certain metal and an effect, or the lack of such a relationship, are made often enough, people not familiar with all aspects of this metal's toxicity may be misled. This is one reason why research on known effects of metals has to be maintained and epidemiological studies have to be repeated in different population groups or in different countries.

Finally, the evaluations of health effects of metals within regulatory agencies in different countries are often made in an incoherent way because the responsibilities for clean water, air, food, etc., are divided between a number of separate agencies. A more coherent and coordinated approach is needed in many countries.

RECOMMENDATIONS

1. There is an urgent need for the development of methods for analysis of different metal species in biological systems. Cooperative work between chemists, physicists, and toxicologists would ensure that the methods are applied to metal exposure situations of concern to human health.

2. There is a need for collaborative studies of the background exposure levels and background effect levels in different countries. Studies of migrants between high and low "background" areas would also give useful data on the occurrence of health effects in areas not considered "polluted."

3. There is an urgent need to prevent serious metal poisoning outbreaks in developing countries. Information networks between scientists in industrialized and developing countries as well as collaborative studies on specific problems should be encouraged by scientific societies, international agencies, and research funding agencies.

4. There is a need to develop early biochemical or physiological indicators of effects in order to assist in preventing the occurrence of the more severe clinical effects.

5. There is a need for collaborative international epidemiological studies to carry out quality assurance of health effect measurements and to provide a broader dose range for dose-response analysis. Such studies will also make it possible to evaluate the effects of metals in different racial groups and different dietary conditions.

6. There is a need to carry out the tedious studies to collect the missing data on basic anatomical, biochemical, and physiological variables for children in different age groups. In addition, more descriptive studies of the basic metabolism and kinetics of metal compounds in children need to be carried out.

7. Regulatory agencies for quality control of one environmental compartment (e.g., ambient air, food, drinking water) should (a) closely cooperate with agencies for other compartments, (b) take into account total load, and (c) take into account different health risks from different (groups of) compounds of the same metal.

REFERENCES

(1) Belman, S. 1981. Proceedings of a workshop conference on the role of metals in carcinogenesis. Envir. H. Per. **40**: 1–252.

(2) Bradford-Hill, A. 1977. A Short Text Book of Medical Statistics. London: Hodder and Stoughton.

(3) Fachetti, S., and Griff, F. 1982. Isotopic Lead Experiment: Status Report. Brussels: CEC.

(4) Freshwater Institute. 1983. Mercury Quality Assurance Program. Winnipeg, Canada: Dept. of Fisheries and Oceans.

(5) Friberg, L.; Nordberg, G.F.; and Vouk, V.B., eds. 1979. Handbook on the Toxicology of Metals. Amsterdam: Elsevier.

(6) Gompertz, D., and Smith, N.J. 1982. The United Kingdom External Analytical Assessment Scheme (UKEQAS) for blood, lead and cadmium analysis. CEC-IUPAC International Workshop on Biological Indicators of Cadmium Exposure, Luxembourg.

(7) Herber, R.F.M.; Tonks, D.B.; and Stoeppler, M. 1980. Evaluation of the first round-robin exercise of cadmium in whole body by IUPAC, Division of Clinical Chemistry, Commission on Toxicology. Report by the Subcommittee of Environmental and Occupational Toxicology of Cadmium. Amsterdam, Montreal, and Julich: IUPAC.

(8) Kjellström, T. 1979. Exposure and accumulation of cadmium in people from Japan, USA and Sweden. Envir. H. Per. **28**: 169–197.

(9) Kjellström, T. 1984. Renal effects. **In** Cadmium and Health, eds. L. Friberg et al. Boca Raton: CRC Press, in press.

(10) Lauwerys, R.; Buchet, J.P.; Roels, H.; Berlin, A.; and Smeets, J. 1975. Intercomparison program of lead, mercury and cadmium in blood, urine and aqueous solutions. Clin. Chem. **21**: 551–557.

(11) Luepkee, N.P. 1979. Monitoring Environmental Materials and Specimen Banking, pp. 59. The Hague: Martinus Nijhoff Publishers.

(12) Mahaffey, K.R.; Annest, J.L.; Roberts, J.; and Murphy, R.S. 1982. National estimates of blood lead levels: United States, 1976–1980. N. Eng. J. Med. **307**: 573–579.

(13) Mitchell, J.W.; Kjellström, T.E.; and Reeves, R.L. 1982. Mercury in takeaway fish in New Zealand. NZ Med. J. **95**: 112–114.

(14) Nordberg, G.F., ed. 1976. Effects and Dose-response Relationships of Toxic Metals. Amsterdam: Elsevier.

(15) Vahter, M. 1982. Assessment of Human Exposure to Lead and Cadmium Through Biological Monitoring. Report prepared for the United Nations Environmental Program and the World Health Organization. National Swedish Institute of Environmental Medicine, Stockholm.

LIST OF PARTICIPANTS

ABEL, J.
Institut für Toxikologie
der Universität Düsseldorf
4000 Düsseldorf, F.R. Germany

Field of research: Biochemical toxicology, metallothionein

ANDREAE, M.O.
Dept. of Oceanography
Florida State University
Tallahassee, FL 32306, USA

Field of research: Biogeochemistry of trace metals in the marine environment, air/sea exchange chemistry

ASAMI, T.
Faculty of Agriculture
Ibaraki University
Ami-machi, Ibaraki, Japan

Field of research: Soil science

BALDI, F.
Istituto di Biologia Ambientale
Università di Siena
53100 Siena, Italy

Field of research: The contamination of biota and sediments by trace metals

BENNETT, B.G.
Monitoring and Assessment
Research Centre
London SW10 OQX, England

Field of research: Environmental pollutant behavior and human exposure assessments

BERNHARD, M.
Centro Studi Ambiente Marino
ENEA
19100 La Spezia, Italy

Field of research: Marine ecology, radioisotope and trace metal pollution

BERTINE, K.K.
Dept. of Geological Sciences
San Diego State University
San Diego, CA 92182, USA

Field of research: Biogeochemistry of trace metals in waters and sediments

BRINCKMAN, F.E.
Chemical and Biodegradation
Processes Group
Center for Materials Science
National Bureau of Standards
Washington, DC 20234, USA

Field of research: Ultratrace molecular characterization or speciation of key biogenic organometallic transport agents involved in rate-determining transformations and fate of toxic metals in the environment

BUAT-MÉNARD, P.E.
Centre des Faibles Radioactivités
Laboratoire mixte CNRS-CEA
91190 Gif-sur-Yvette, France

Field of research: Atmospheric and marine geochemistry of metals and organic carbon

CLARKSON, T.W.
Division of Toxicology
University of Rochester
School of Medicine
Rochester, NY 14642, USA

Field of research: Toxicology of mercury

DUCE, R.A.
Center for Atmospheric Chemistry
Studies, Graduate School of
Oceanography
University of Rhode Island
Kingston, RI 02881, USA

Field of research: Global atmospheric cycles of trace elements

ELINDER, C.-G.
Dept. of Environmental Hygiene
Karolinska Institute, and
The National Institute
of Environmental Medicine
104 01 Stockholm 60, Sweden

Field of research: Metabolism and toxicity of metals, epidemiology

FILIP, Z.
Institut für Wasser-, Boden- und
Lufthygiene des
Bundesgesundheitsamtes
Außenstelle Frankfurt
6000 Frankfurt 70, F.R. Germany

Field of research: Environmental microbiology

FÖRSTNER, U.
Technische Universität
Hamburg-Harburg
2100 Hamburg 90, F.R. Germany

Field of research: Environmental engineering/geochemistry, processes of metal mobilization, sediment/water interactions, technology of dredged material handling

FOWLER, B.A.
Environmental Toxicology Laboratory
National Institute of Environmental
Health Sciences
Research Triangle Park, NC 27709, USA

Field of research: Cellular mechanisms of trace metal toxicity with particular emphasis on lead, cadmium, and arsenic; characterization of lead-binding proteins from mammals and a unique cadmium-binding protein from marine molluscs

GOLDBERG, E.D.
Geological Research Division A-020
Scripps Institution of Oceanography
La Jolla, CA 92093, USA

Field of research: Marine chemistry

HAPKE, H.-J.
Tierärztliche Hochschule Hannover
3000 Hannover 1, F.R. Germany

Field of research: Toxicology of heavy metals and other environmental pollutants, in particular for farm animals; accumulation in food chains

HECHT, H.P.
Institut für Chemie und Physik
Bundesanstalt für Fleischforschung
8650 Kulmbach, F.R. Germany

Field of research: Carry-over of heavy metals, secondary contamination of foodstuffs by metals and metal content of meat and meat products

HEINRICHS, H.
Geochemisches Institut der
Universität Göttingen
3400 Göttingen, F.R. Germany

Field of research: Environmental and terrestrial geochemistry

HORNIG, D.F.
Interdisciplinary Programs in Health
Harvard School of Public Health
Boston, MA 02115, USA

Field of research: Environmental health and public policy

JAWORSKI, J.F.
National Research Council of Canada
Ottawa K1A OR6, Canada

Field of research: Environmental cycling and toxicology of metallic and semimetallic elements, selection of criteria for control measures

JERNELÖV, A.B.
I.V.L. — Swedish Water and Air
Pollution Research Institute
100 31 Stockholm, Sweden

Field of research: Metal transformation in biotic systems

KÄGI, J.H.R.
Biochemisches Institut der
Universität Zürich
8057 Zürich, Switzerland

Field of research: Protein chemistry, bioinorganic chemistry, structure and function of metallothionein

KJELLSTRÖM, T.E.
Dept. of Community Health
University of Auckland
Auckland, New Zealand

Field of research: Epidemiological studies of the human toxicity of cadmium, mercury, and lead

KLEIMINGER, J.
Städtischer Schlachthof Braunschweig
3300 Braunschweig, F.R. Germany

Field of research: Monitoring of environmental contaminations with heavy metals in different regions by using mammals and fish as bioindicators — specific working field, statistics

KLOKE, A.
Biologische Bundesanstalt für
Land- und Forstwirtschaft
1000 Berlin 33, F.R. Germany

Field of research: The contamination of plant and soil by heavy metals

KOJIMA, Y.
Dept. of Environmental Medicine
Graduate School of Environmental Science
Hokkaido University
Sapporo 060, Japan

Field of research: Chemical structure of metallothionein; metabolism of cadmium, zinc, and copper in the human body; inborn errors of metal metabolism

LEGOVIĆ, T.
Center for Marine Research
Institute "R. Bosković"
41001 Zagreb, Yugoslavia

Field of research: Mathematical ecology, modelling cycles of matter in marine environment

LIN-FU, J.S.
Dept. of Health and Human Services
Office for Maternal and Child Health
Rockville, MD 20857, USA

Field of research: Lead poisoning in children

McKENZIE, J.M.
Dept. of Nutrition
University of Otago
Dunedin, New Zealand

Field of research: Nutritional importance of zinc and copper in New Zealand adults, cadmium intake and health effects

NEEDLEMAN, H.L.
Depts. of Psychiatry and Pediatrics
University of Pittsburgh School of Medicine
Pittsburgh, PA 15213, USA

Field of research: Lead and neuropsychologic dysfunction

NOGAWA, K.
Kanazawa Medical University
Uchinada, Ishikawa, Japan

Field of research: Health effects of cadmium

NRIAGU, J.O.
National Water Research Institute
Environment Canada
Burlington, Ontario L7R 4A6, Canada

Field of research: Environmental cycling of pollutant metals

PACYNA, J.M.
The Norwegian Institute
for Air Research
2001 Lillestrøm, Norway

Field of research: Atmospheric chemistry in remote areas, emissions from anthropogenic sources

PAGE, A.L.
Dept. of Soil and Environmental Sciences
University of California
Riverside, CA 92521, USA

Field of research: Chemistry of trace metals in soils

SAUERBECK, D.R.
Institut für Pflanzenernährung und Bodenkunde
Bundesforschungsanstalt für Landwirtschaft
3300 Braunschweig, F.R. Germany

Field of research: Plant nutrition, soil chemistry, agricultural utilization of town waste materials

SCHOOT UITERKAMP, A.J.M.
Bureau of Environmental Projects
TNO
2600 AE Delft, The Netherlands

Field of research: Ecotoxicology of trace elements, technology of metal ion recycling

SILBERGELD, E.K.
Environmental Defense Fund
Washington, DC 20036, USA

Field of research: Lead neurotoxicity, neurobehavioral teratology

SILVER, S.
Biology Dept.
Washington University
St. Louis, MO 63130, USA

Field of research: Biological interactions with inorganic cations and anions; mechanisms of bacterial resistances to mercury, cadmium, arsenic, and other heavy metals; molecular genetics and biochemistry

SUMMER, K.H.
Institut für Toxikologie und Biochemie
Gesellschaft für Strahlen- und Umweltforschung mbH
8042 Neuherberg, F.R. Germany

Field of research: Biochemical toxicology, glutathione, metallothionein

THORNTON, I.
Applied Geochemistry Research Group
Imperial College
London SW7 2BP, England

Field of research: Applications of geochemistry and geochemical mapping to agriculture, water resources, soil and water pollution studies, and public health

TOBSCHALL, H.J.
Institut für Geowissenschaften der Universität Mainz
6500 Mainz, F.R. Germany

Field of research: Biogeochemistry of tin and platinum group elements, biogeochemistry of Sabkha environments

WASSERMAN, K.J.
Insolar Inc.
Port Jervis, NY 12771, USA

Field of research: Chemistry, drinking water pollution

WILLIAMS, R.J.P.
Inorganic Chemistry Laboratory
University of Oxford
Oxford OX1 3QR, England

Field of research: Inorganic biochemistry

WINNEKE, G.
Medizinisches Institut
für Umwelthygiene der
Universität Düsseldorf
4000 Düsseldorf, F.R. Germany

Field of research: Neurobehavioral effects of heavy metals (lead, cadmium) in animals, neuropsychological effects of lead in children

ZIELHUIS, R.L.
Coronel Laboratory for Occupational
and Environmental Health
University of Amsterdam
1054 BW Amsterdam,
The Netherlands

Field of research: Social medicine, occupational and environmental human toxicology

ZOLLER, W.H.
Chemistry Dept.
University of Maryland
College Park, MD 20742, USA

Field of research: Atmospheric chemistry in remote areas, emissions of anthropogenic sources

SUBJECT INDEX

AUTHOR INDEX

Dahlem Workshop Reports

Life Sciences Research Reports (LS)

LS 31 Microbial Adhesion and Aggregation.
Editor: K. G. Marshall (1984, in press)

LS 30 Leukemia.
Editor: I. L. Weissman (1984, in press)

LS 29 The Biology of Learning.
Editors: P. Marler, H. S. Terrace (1984, in press)

LS 28 Changing Metal Cycles and Human Health.
Editor: J. O. Nriagu (1984)

LS 27 Minorities: Community and Identity.
Editor: C. Fried (1983)

LS 26 The Origins of Depression: Current Concepts and Approaches.
Editor: J. Angst (1983)

LS 25 Population Biology of Infectious Diseases.
Editors: R. M. Anderson, R. M. May (1982)

LS 24 Repair and Regeneration of the Nervous System.
Editor: J. G. Nicholls (1982)

LS 23 Biological Mineralization and Demineralization.
Editor: G. H. Nancollas (1982)

LS 22 Evolution and Development.
Editor: J. T. Bonner (1982)

LS 21 Animal Mind - Human Mind.
Editor: D. R. Griffin (1982)

LS 20 Neuronal-Glial Cell Interrelationships.
Editor: T. A. Sears (1982)

Physical, Chemical and Earth Sciences Research Reports (PC)

PC 5 Patterns of Change in Earth Evolution.
Editors: H. D. Holland, A. F. Trendall (1984, in press)

PC 4 Atmospheric Chemistry.
Editor: E. D. Goldberg (1982)

PC 3 Mineral Deposits and the Evolution of the Biosphere.
Editors: H. D. Holland, M. Schidlowski (1982)

Springer-Verlag Berlin Heidelberg New York Tokyo

Dahlem Workshop Reports

Life Sciences Research Reports (LS)

LS 1 The Molecular Basis of Circadian Rhythms.
Editors: J.W. Hastings, H.-G. Schweiger

LS 2 Appetite and Food Intake.
Editor: T. Silverstone

LS 3 Hormone and Antihormone Action at the Target Cell.
Editors: J.H. Clark et al.

LS 4 Organization and Expression of Chromosomes.
Editors: V.G. Allfrey et al.

LS 5 Recognition of Complex Acoustic Signals.
Editor: T.H. Bullock

LS 6 Function and Formation of Neural Systems.
Editor: G.S. Stent

LS 7 Neoplastic Transformation:
Mechanisms and Consequences.
Editor: H. Koprowski

LS 8 The Bases of Addiction.
Editor: J. Fishman

LS 9 Morality as a Biological Phenomenon.
Editor: G.S. Stent

LS 10 Abnormal Fetal Growth: Biological Bases and Consequences.
Editor: F. Naftolin

LS 11 Transport of Macromolecules in Cellular Systems.
Editor: S.C. Silverstein

LS 12 Light-Induced Charge Separation in Biology and Chemistry.
Editors: H. Gerischer, J.J. Katz

LS 13 Strategies of Microbial Life in Extreme Environments.
Editor: M. Shilo

LS 14 The Role of Intercellular Signals: Navigation, Encounter, Outcome.
Editor: J.G. Nicholls

LS 15 Biomedical Pattern Recognition and Image Processing.
Editors: K.S. Fu, T. Pavlidis

LS 16 The Molecular Basis of Microbial Pathogenicity.
Editors: H. Smith et al.

LS 17 Pain and Society.
Editors: H.W. Kosterlitz, L.Y. Terenius

LS 18 Evolution of Social Behavior:
Hypotheses and Empirical Tests.
Editor: H. Markl

LS 19 Signed and Spoken Language:
Biological Constraints on Linguistic Form.
Editors: U. Bellugi, M. Studdert-Kennedy

Physical and Chemical Sciences Research Reports (PC)

PC 1 The Nature of Seawater (out of print)

PC 2 Global Chemical Cycles and Their Alteration by Man.
Editor: W. Stumm

Distributor for LS 1–19 and PC 1+2:
Verlag Chemie, Pappelallee 3, 6940 Weinheim,
Federal Republic of Germany

MIX
Papier aus verantwortungsvollen Quellen
Paper from responsible sources
FSC® C105338

If you have any concerns about our products,
you can contact us on
ProductSafety@springernature.com

In case Publisher is established outside the EU,
the EU authorized representative is:
Springer Nature Customer Service Center GmbH
Europaplatz 3, 69115 Heidelberg, Germany

Printed by Libri Plureos GmbH
in Hamburg, Germany